ORTON

thematics

APPLIED MATHEMATICS
a course companion

Tony Bridgeman
Gordon R. Baldock

Longman Scientific and Technical
Longman House, Burnt Mill, Harlow,
Essex CM20 2JE, England
and Associated Companies thoughout the world

First published 1989

British Library Cataloguing in Publication Data

Bridgeman, T.
Applied mathematics.———
1. Mathematics——Examinations, questions etc.
I. Title   II. Baldock, G. R.
510'.76   QA43

ISBN 0-582-25108-7

Library of Congress Cataloging-in-Publication Data

Bridgeman, T.
Applied Mathematics.
Includes index.
1. Mechanics, Analytic–Problems, exercises, etc.
I. Baldock, G. R.   II. Title.   III. Series
QA809.B76 1987   531'.076   87-4201
ISBN 0-582-25108-7

Produced by Longman Singapore Publishers (Pte) Ltd.
Printed in Singapore

Contents

Preface and Acknowledgements

This book has been devised, not as a text book, but as a guide for the candidate, in A-level single-subject Mathematics or in BTEC, on the types of question that can be encountered in mechanics and in elementary probability. Emphasis is placed on detailed points which are looked for by examiners and on guarding against the errors which are most easily made.

Each chapter is largely self-contained, so that the book can serve as a course companion no matter in what order the topics are treated in the course.

Mechanics is often thought to be harder than Pure Mathematics. This is partly because of the wider apparent variety of problems that appear in examination questions, but much of the difficulty stems from the need for intuition to convert the wording of a question into a mechanical picture and thence to derive appropriate mathematical statements. Intuition works very differently in different individuals. Consequently any explanation of a problem has a variable impact, one person's revelation is another's mystification. In this book we endeavour to provide easily understood explanations of the approaches to problems, and alternative solutions are often discussed. We also discuss those *wrong* methods which we know to be frequently used by candidates, and explain why they are wrong even though they sometimes seem to lead to correct results. The reader is thus warned how *not* to answer questions, as well as receiving guidance on the best ways of answering. We believe that this book will help the reader to establish a sound platform of basic knowledge and examination technique on which to build.

We are grateful to colleagues at the University of Liverpool and to many teachers and examiners for useful discussions.

The authors and publishers are grateful to the following Examination Boards for permission to reproduce past examination questions:

Associated Examining Board (AEB); Joint Matriculation Board (JMB); University of London School Examinations Board (LON); Northern Ireland Board (NI); Welsh Joint Examination Committee (WJEC).

The worked solutions of the examination questions are entirely the responsibility of the authors, and may not necessarily constitute the only possible solutions. They have been neither provided nor approved by the examining boards. The University of London School Examinations Board accepts no responsibility whatsoever for the accuracy or method of working in the answers given.

We are deeply indebted to Barbara Bridgeman and Joyce Baldock for their energetic cooperation in the preparation of the manuscript.

Tony Bridgeman, Gordon R. Baldock

Liverpool 1988

Chapter 1 Importance of the topic areas

1.1 AIMS

The aims of this book are as follows:

1. To help you to perceive exactly what is wanted by the examiner in a question.
2. To show you the technique of working systematically through the solution of a problem, providing full commentary along the way on the reasoning behind the choice of methods.
3. To map out the commonly encountered pitfalls and errors, and to provide you with an armoury of precautions against them.

A full explanation of the theory of any topic in applied mathematics, together with illustrative worked examples and exercises for the reader, may be found in any standard textbook. Here, in contrast, the aim is to attack the examination questions themselves, and so the book is really a handbook on the answering of questions which includes detailed advice on how *not* to answer them.

1.2 SCOPE OF THE TEXT

Each of the several examining boards responsible for the General Certificate of Education in England, Wales and Northern Ireland offers an A-level in mathematics in which about half of the syllabus is mechanics. With few exceptions, the mechanics topics of all the boards are treated in this book. There is complete coverage of the syllabuses of AEB, University of London, and JMB. The mechanics topics in Level III of the courses leading to the BTEC National Diploma in Science are also covered. The topics in Probability, which are included in the syllabus of the University of London and of some other boards, are covered in Chapter 15.

This book is not intended to cover the mechanics in all the

A-level syllabuses entitled Further Mathematics. It is, however, suitable for those syllabuses which comprise topics in both mechanics and statistics, such as JMB Further Mathematics (Pure and Applied Mathematics).

The table shows which of the topic areas defined by chapters in this book are treated in the various syllabuses. The abbreviations at the heads of the columns in the table are used throughout the text.

Table 1

Ch.	Course topic	GCE A-level Mathematics											BTEC
		AEB	LON	JMB			OXF	O&C	CAM	SMP	WJEC	NI	III
				MeI	AMI	AMII							
3	Rectilinear kinematics	✔	✔	✔	✔		✔	✔	✔	✔	✔	✔	✔
4	Rectilinear motion	✔	✔	✔	✔	✔	✔	✔	✔	✔	✔	✔	
5	Simple harmonic motion	✔		✔			✔	✔	✔	✔	✔	✔	✔
6	Two particles	✔	✔	✔	✔	✔	✔	✔	✔	✔	✔	✔	
7	Elastic strings	✔	✔	✔	✔	✔	✔	✔	✔	✔	✔	✔	✔
8	Relative motion		✔				✔	✔		✔	✔	✔	
9	Vector dynamics	✔	✔	✔	✔		✔	✔	✔	✔	✔	✔	
10	Projectiles	✔	✔	✔	✔		✔	✔	✔	✔	✔	✔	✔
11	Circular motion	✔	✔	✔	✔		✔	✔	✔	✔	✔	✔	✔
12	Coplanar forces	✔	✔	✔	✔	✔	✔	✔			✔		✔
13	Centres of mass	✔	✔	✔	✔	✔	✔	✔		✔	✔	✔	
14	Equilibrium	✔	✔	✔		✔	✔		✔	✔	✔	✔	
15	Systems of bodies	✔		✔		✔	✔						
16	Probability		✔		✔							✔	

Key

AEB	Associated Examining Board		Mathematics Paper 3 – Applied
LON	University of London		Mathematics B Paper 3; Applied Maths 2
JMB	Joint Matriculation Board	MeI	Mathematics (Pure Maths/Mechanics) II
		AMI	Mathematics (Pure & Applied) II
		AMII	Further Maths (Pure & Applied) II
OXF	Oxford Delegacy of Local Examinations		Mathematics 9850 Paper 2
O & C	Oxford and Cambridge Examinations Board		Mathematics 9650
CAM	Cambridge Local Examinations Syndicate		Mathematics Syllabus C Paper 2
SMP	School Mathematics Project		Mathematics
WJEC	Welsh Joint Education Committee		Mathematics Paper A2
NI	Northern Ireland GCE Examination Board		Mathematics
III	BTEC National Diploma in Science		Level III Mathematics

1.3 ARRANGEMENT OF THE SUBJECT MATTER

Each chapter contains one or more sections which begins with a collection of **essential facts**. These **facts** are presented as a concise reminder of work already known. Proofs of general results are not usually given in these sections, because it is assumed that you already have access to full explanations of the theory, either from tuition or from a textbook. The **facts** are followed by a few **illustrations** to show what they mean and how they can be applied. Finally, a large part of each chapter is devoted to the **solutions** of examination questions. Nearly all of the questions have been taken from recent examinations, but some new questions (marked by an asterisk ★ at the end) have been devised, to cover topics on which there are few published examples.

Examiners often combine different syllabus topics within one question. Any such question will be found in the chapter relevant to its major topic area. All the questions which involve, even if only to a minor extent, any one of the detailed syllabus items can be traced by consulting the Index.

In the work on many of the **illustrations** and **solutions** we have provided comments, usually on the right-hand side of the page, while the main argument proceeds on the left-hand side. The comments are labelled [A], [B] etc. referring to the relevant points in the argument. We use the following notation:

- ✕ Crosses enclose a false statement or an inadequate method. ✕
- ✓ Ticks enclose the correction of a previous error. ✓
- ? Question marks enclose a statement or method which, though not wrong, is either not useful or liable to lead to cumbersome work and possibly to error; to summarise, **not recommended**. **?**
- □ This sign marks the end of a question.
- ■ This sign marks the completion of the solution of a problem or part of a problem.
- *'25m'* At the end of the solution of each question an estimate of the time you may expect to be required for the solution is given. For example, if this time is estimated at 25 minutes the solution will end thus: *25m* ■
- ★ Specially devised question.

'See 3.3F7', or just '(3.3F7)', means 'see Section 3.3, Fact F7'
'See *PM* 11.5' means 'See Section 11.5 in our book *Pure Mathematics* (Longman 1986).

Chapter 2 Examination techniques

The assessment of your performance may be carried out by a conventional system of three-hour examinations covering the whole syllabus, possibly together with a short multiple-choice paper, or by a series of tests throughout the duration of your course. The aim of this chapter is to provide general guidance on how to lay the foundations of a good performance and on how to show your abilities to the best advantage in the course of the assessment.

2.1 PREPARATION FOR THE EXAMINATION

In mathematics you can gain no advantage by just memorising facts and hoping that you will be asked to deliver them in the examination. The only route to success is by striving to understand and master each topic of the syllabus as you study it during the course; you will find then that there are very few facts that you need to remember. Do not omit to work at any topic on the syllabus, even if it seems difficult at first. Once you have grasped the basic ideas you may well find that the examination questions on an apparently difficult topic are easier than many of those on simpler topics.

Most problems in mechanics have some practical bearing. Be ready to think about a topic from the point of view of common sense, not just formulae. It will help you to understand the implications of the formulae and to avoid outrageous errors.

SOLVING PROBLEMS

In each topic you must start by doing exercises from a standard textbook on the straightforward applications of the theory. Fluency with these must be attained before progressing to longer exercises which test your ability to select and apply the correct methods to less familiar problems. Here the examination questions worked in this book will help. To derive maximum benefit from them you should try them *before* looking at the solutions, and then compare your answers with the solutions and in the light of the notes which accompany the solutions. This comparison is very important, because it is just at the point when you perceive your own errors or invalid arguments that you make significant progress in understanding the topic. So you should not only look at the solution to see if your answer is right. Look at your *methods*, and read the comments to see if you have fallen into any of the traps. Whenever you fail to complete a question without looking at the solution you should try the question again the next day to reinforce what you have learned. If you encounter difficulties when working on a question from another source, look up the topic in the Index, and you will be directed to all the places in the book where similar questions are discussed.

It is a good habit to work out the basic equations of a problem in general form in cases where you have been given numerical data. Algebraic symbols are much easier to write and manipulate than numbers, and their use also shows up the pattern of an argument. In this way you get accustomed to the familiar expressions which recur from one problem to another. More important still, it enables you to conduct dimensional checks of your work; these checks cease to be available as soon as the numbers go in. So it pays to keep numbers out of your work until they are needed to yield the answers.

In the Solutions given in this book we provide examples both of this treatment and of directly dealing with the numerical data, which is often quicker but less safe.

PRESENTING SOLUTIONS TO PROBLEMS

In your work on problems, always include some explanation of what you are doing. Although it is often true that a problem can be solved by jotting down a few symbols and doing a quick calculation, keeping the thread of the argument in your head, this is not the way to present your work. Next time you look at what you have written you may not understand it yourself. On the other hand, there is no need to write long explanations. Rather you should write just enough to make clear your applications of physical laws to the problem and to make your mathematical arguments coherent, so that any reader could follow your work. Ultimately you will have to present your work in this way in the examination, so it is wise to develop this habit from the start.

CHECKING

Try to check your arithmetic and algebra as you work through a problem by asking yourself at each stage whether the result makes sense; do not wait until the end to find your answer is wrong. Checks take different forms; here is a list of useful procedures.

(a) A short step may sometimes be checked by using a quite different method.
(b) Indefinite integrals can be re-differentiated.
(c) Numerical answers must be credible. No cyclist should go at 200 km h^{-1}, and no tension should be negative.
(d) In a problem in which the data are not numerical, check dimensions as you go. Any two algebraic expressions which are to be added must have the same dimension.

Examples of these can be seen in the chapters which follow. If you make a habit of checking you will develop a valuable defence against going wrong in the examination.

CALCULATORS

You should equip yourself with a pocket calculator which provides values of the elementary transcendental functions ln, sin etc. and has one memory store. The calculator is helpful in the final stages of problems which demand a numerical answer.

REVISION

Towards the end of your course you will start a programme of systematic revision in preparation for the imminent examination. In mathematics revision is effective only if it consists largely in the working of examples. The number of different techniques which are available to be tested in the examination is limited, and therefore it is feasible for you to extend your practice to cover them all. Such coverage, together with the depth of understanding which will be enhanced by assiduous practice, will furnish you with the best equipment for success in the examination.

It is useful to make a brief summary (perhaps on a postcard) of the main facts and methods of each topic, just enough to refer to when solving problems. You should also make sure that you are familiar with the formula booklet which is usually provided by the examination authority, to the extent that you know which formulae it contains and where to find them quickly. It is even better to make sure that you *know* the pure-mathematical formulae that commonly arise in mechanics questions: like the standard integrals of simple functions, the formulae for $\sin(\theta + \alpha)$, $\sin\alpha + \sin\beta$, the conversion $A\cos\theta + B\sin\theta = R\cos(\theta - \alpha)$ in its various forms (a *very* important one); in answering mechanics questions especially you cannot afford to digress from the main line of your argument to have to think about these things, or to look them up, or you will lose track of the ideas.

Make out a timetable for your revision and keep to it so that no topics are left to be crowded into the last few days.

2.2 TAKING THE EXAMINATION

RUBRIC

Before looking at the questions, read the rubric on the front of the paper carefully. You must note the number of questions which may be attempted, and the number to be attempted in each section of the paper, the total time allowed, and any instructions there may be about crossing out work or handing in sections in separate books.

PAPERS WITH A CHOICE OF QUESTIONS

You may encounter, for example, a paper of duration $2\frac{1}{2}$ hours consisting of 8 questions of equal value with the instruction to answer 6. Scan through the paper quickly and select those questions which you think you can complete fairly rapidly. These will not necessarily be the questions which are most simply expressed. Questions which look long are often easier than those which can be read quickly, because they generally give you more information, to help you through the solution. However, before deciding to try a question which has several parts you must read to the end of it. Do not be tempted by an easy first part; the final part of the question may be much more difficult and may carry a high proportion of the marks. If you meet with difficulty in the course of answering a question it is wise to leave it for a while and move to another question so as to maintain your rate of accumulation of marks in the first hour of the examination. You will then have time to return to the difficulty later. Ultimately you should aim to complete 6 questions, but if you reach a stage when you are sure you can go no further with the 6 you have tried, you should attempt another one (or two if necessary). At the end of the examination, refer to the rubric if you have attempted more than 6 questions. If the rubric says ANSWER 6 QUESTIONS, you will have to decide what to cross out. If the rubric says ONLY THE BEST 6 ANSWERS WILL BE TAKEN INTO ACCOUNT, do not cross any question out. You must, however, cross out any work which you have replaced by another answer, which you think is better, to the same part of the same question.

PAPER WITH ALL QUESTIONS MARKED

Another type of paper consists of questions with different mark values arranged in order of increasing value, candidates being permitted to attempt all questions. Proceed as described in the paragraph above, by working first on the questions which you think you can complete, deferring difficulties until later, but in this case be especially careful not to spend too much time on any question which carries only a few marks. If you can 'see through' a 4- or 5-mark mechanics question, try it; otherwise defer it. Very often you can earn more marks faster on long questions even if your answers are incomplete. Then go through all the questions you have not yet attempted and answer any portions of them that can be done quickly. Finally go back to the questions that seemed difficult and try to complete them. In this way you will be using the paper to your best advantage, that is, you will be letting the examiner know the full extent of your knowledge and abilities in the subject.

UNDERSTANDING THE INDIVIDUAL QUESTIONS

Read the question carefully. You must be quite sure what the examiner is demanding before you start work. Examiners try hard to word their questions so that what is wanted is very plain.

The different parts of the question are ordered in a natural way – the very ordering often constitutes a series of hints on how to proceed. Whenever the statement or proof of a well-known result is demanded it is relevant to at least part of the rest of the question.

If the word 'hence' appears you must do the next part of the question using the result just obtained. 'Hence or otherwise' is an indication that use of the immediately preceding result is very likely to be the simplest method; you are allowed to use any other method but it may take longer.

There are a number of innocent-looking words, over which the eye can easily slide, which occur in the statement of problems and are exceedingly important. For example, **constant** (the same at all times), **uniform** (the same in all places), **smooth**, **light** etc. Pay attention to such words, or you may be trying to solve a problem more complicated than the one that has been set.

Logic is as important in mechanics as it is in pure mathematics. You must known the meanings of 'if', 'only if', 'provided that', etc. For example:

'Show that the particle comes to rest provided that $u^2 < 4ag$' means that, because it is *provided*, '$u^2 < 4ag$' may be assumed to start with. You are not being asked to show that $u^2 < 4ag$ if the particle comes to rest, and you must not waste time trying to.

When using data from the question at any stage in your answer you must be careful to extract them correctly. Many mistakes are caused by referring back to a question and reading the wrong information. In cases where the data are presented in a long verbal statement it can be helpful to extract the important items and note them on a page of your answer book so that you can easily refer to them. This is often best done by means of a diagram.

DIAGRAMS

In mechanics diagrams are almost always a valuable aid to thinking about a problem, much more so than in most pure mathematical problems. While you are reading a question you should be sketching out the diagram with appropriate labelling. Very often you will find your first sketch not quite right – things are badly out of proportion, or there is not enough room to put something in clearly, or something is the wrong way round. A bad diagram will upset all the work you do on the problem, so you cross it out and start a new one. Above all it is essential to have the diagram where you can see it while you are working, so do not start a new long question on a right-hand page of your answer book, for you will surely find yourself having to turn over the page to refer to the diagram, and this will waste time as well as expose you to the danger of making mistakes.

Mark forces, velocities, etc. by arrows either in different colours (*not red*, which examiners use) or by differently formed arrows. In this book forces (→) are marked more heavily than velocities (→), and the arrows for acceleration (↠) are double-headed. There is a collection of the symbols used on diagrams on page 244).

SYMBOLS

You will usually want to define some symbols at the beginning of the work. Get into the habit of doing this. But if you are short of time in the examination, just make the meaning of your symbols very clear on your diagram; then the examiner will be able to follow your work and you will not lose marks for obscurity. But the answers must be presented fully – it's no good giving '$v = 6$' as an answer to 'Find the velocity. . .'. It is up to you to say that your v does refer to the particular velocity you've been asked for.

Do not use any of the symbols printed in the question in a sense different from the defined one, or you will confuse both yourself and the examiner. Be ready to avoid this hazard when applying formulae.

UNITS AND DIMENSIONS

In presenting the answers to a question you should append the appropriate units, if units are mentioned in the text of the question.

In many questions units are not mentioned at all. Normally in such cases every physical quantity is represented by a symbol, not a number, and so the units are just assumed to be consistent, which means that all the well-known equations of mechanics apply in the system. In such cases DO NOT introduce units – they will only complicate your work unnecessarily.

You may find questions set in which some of the physical quantities are numbers and some are symbols, but units are unmentioned. Just carry on assuming that the system of units is consistent, but again don't introduce units yourself, for the examiner is not expecting it. Very occasionally you find questions in which some quantities are stated in units and some not. Here you just have to give the examiner the benefit of the doubt – assume the unit system to be consistent and work out the problem, preferably and conveniently without mentioning units along the way, finally presenting the answers in the correct units.

A lot of trouble can be saved, and dimensions checked as well, if you use symbols rather than numbers as far as possible in the working.

You must be careful with special units like tonnes and kilometres per hour. Convert to SI, if SI is used in the question; otherwise see if you can manage without conversion so as to avoid unnecessary work. Again, symbolic procedures save much anxiety, leaving conversion to the end.

In this book we have been broadly correct in the use of units in our Solutions, sometimes at the cost of extra writing. Occasionally we have suppressed the units, for convenience, in the course of a calculation, and made them reappear at the end. In the examination you have little time to spare for unnecessary writing, and nobody would penalise 'let the acceleration be f' when it should strictly have been '$f \text{ m s}^{-2}$'. Just be sure that the correct units are given in your final answers.

A list of units and dimensions is provided on page 244.

DECIMAL APPROXIMATIONS

If a question says 'find, correct to 2 significant figures', then you must do just that. Be very careful **not to round off prematurely**, which means that you must not evaluate an intermediate result to 2sf and then use that approximate result to calculate the answer, because doing so is very likely to lead to an error in the answer. If you have to use the intermediate result you should retain two more figures in the approximation so as to guard against this transmission of round-off error. Better still, leave the result in the memory of your calculator so that the round-off error is in, say, the 8th significant figure, and so harmless. Better still again, use the intermediate result in unsimplified exact form, or in algebraic symbols. Always postpone the introduction of numbers, or at least of numerical approximations, until as near to the end of the work as you can.

If a question says 'find' something, and the answer is a number, and the question says nothing about the form of answer required, it means that you can present answers in forms like 2π or e^2 or $\ln 3$ or $\arctan 2$, and you would be wasting your time providing decimal answers. Only if the exact answer is very complicated is a decimal approximate answer preferable. When, in such a case, you do decide to give a decimal approximation, 3 significant figures is reasonable, but you must state the degree of approximation; '4·73 (to 3sf)' is adequate. In this book we sometimes abbreviate by writing '4·73. .', the two dots signifying 'correct up to the last figure given'. But when we present Solutions to Examination Questions we state the degree of approximation properly.

If the question says 'find exactly' or 'the exact value' or 'in exact form' then you **must not** give a decimal approximation at all. In particular if it says 'show that the value of x is π' then a numerical calculation culminating in ✗ '$x = 3{\cdot}14159 = \pi$' ✗ will lose a mark, even if you scrupulously add '(to 5dp)', because you have *shown* it to be true *only approximately*.

CROSSING OUT

Cross out only work which you do not wish the examiner to read. You will often need to abandon an argument to replace it by a correct one. There is no penalty for any crossing out occurring in the body of your work provided that you make it quite clear what has been crossed out and what is to be read. DO NOT cross out in order to perform arithmetical or algebraic cancellations in the body of your work. Mistakes are often made during cancellations, and the examiner then has no way of knowing whether the work was correct before the cancellation took place, so that you thereby lose not only the marks for the result which is wrong, but also the marks you might have gained for the step before the cancellation. Certainly you will sometimes need to simplify expressions, but if you must cancel by crossing out, do so in a subsidiary calculation on the right-hand side of the page, well separated from the main argument. Above all, NEVER cancel by crossing out when proving a printed answer.

CHECKING YOUR WORK

As far as possible you should check your arithmetic and algebra as you go and, where appropriate, keep an eye on the dimensions of your expressions for physical quantities. If a calculation is becoming complicated you have almost certainly made a mistake, and it is better to look for the mistake than to continue. If you cannot find the mistake, leave the question and return to it later. The error may then become more apparent, or you may think of a new and more productive approach. When you have finished work on a question it is important to check that you have answered, precisely as demanded, each part of it that you could do. Marks are often lost by candidates just forgetting to complete a part of a question.

Chapter 3 Kinematics of a particle moving on a line

3.1 GETTING STARTED

Dynamics is the study of motion and its relation to the forces which act on the moving bodies. The mathematical description of the *motion alone* of a physical system, leaving aside the forces, is an important part of dynamics, and is called **kinematics**. In studying the motion of a system we must first have a method of specifying where the particles of the system are at any time. For this, the basic mathematical technique is vector algebra and geometry. Next, we wish to describe how each particle of the system moves, that is, how its position changes with time. In this respect the fundamental need is to be able to describe *acceleration*, because this is the kinematical quantity which is related to force through Newton's Second Law of motion. This entails dealing with *rates of change* with respect to time, for which the essential mathematical technique is the *differential calculus*, which was devised by Newton for this very purpose.

Vectors are not needed for the study of the motion of a particle **on a fixed straight line**, which forms the subject matter of this chapter, because the position of a particle can be specified by a single coordinate, just a number, which may take positive and negative values.

Although calculus is not needed for the treatment of problems on constant acceleration, it is fundamental to the understanding of kinematics, and it is indispensable in problems with variable acceleration. When working through this chapter you should have available for reference the essential facts and techniques of differentiation and integration (see *PM* Ch 10–11).

3.2 DISPLACEMENT, VELOCITY AND ACCELERATION

ESSENTIAL FACTS

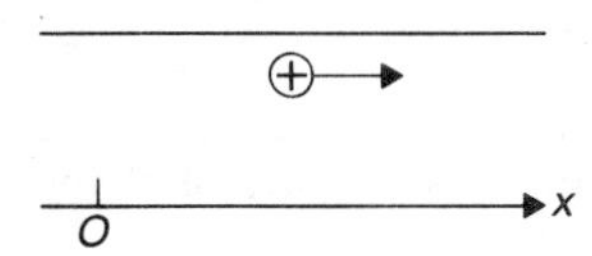

A particle P moves on a fixed straight line. A direction (sense) is chosen on the line and designated the **positive direction (sense)**. An arrow ⊕→ may be used to indicate the positive direction. It is often convenient to choose a fixed point O on the line as origin and to take the positive x-axis as the line from O drawn in the designated positive direction.

F1. Displacement along a line

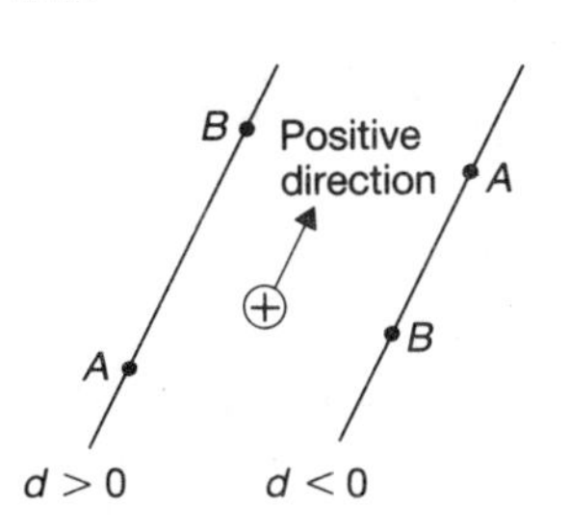

The displacement from a point A to a point B, written $\overrightarrow{AB}$, is a vector with magnitude the length AB and direction from A to B. When we are considering displacements which occur only on a given straight line on which a positive direction ⊕→ has been chosen, the displacement AB may be represented by a number d. The length $AB = |d|$, and
$d > 0$ means that AB is in the chosen positive direction on the line.
$d < 0$ means that AB is opposite to the positive direction on the line.

The **distance** from A to B means the length AB, and it is the **magnitude of the displacement** from A to B.

F2. Position of a point on a line

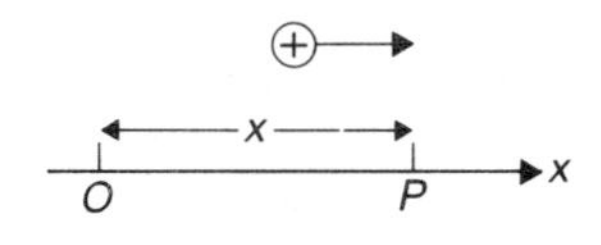

Let O be a fixed point on the line, and let the displacement of a point P from O be x. The number x defines the position of P. In the diagram the positive direction ⊕→ has been taken to the right; in this case $x > 0$ means that P is to the right of O and $x < 0$ means that P is to the left of O.

Taking O as the origin and the x-axis in the positive direction, P is the point with **coordinate** x. When an origin has been chosen, the term 'displacement of P' is taken to mean 'the displacement of P from O', which is the same as the coordinate x.

The **distance** of P from O is $|x|$, which is $\left\{\begin{array}{l} x \text{ when } x > 0 \\ -x \text{ when } x < 0 \end{array}\right\}$.

F3. Relative displacement

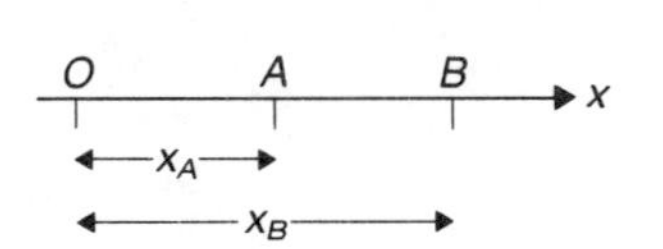

The following terms all mean the same thing.

Displacement of B relative to A
Displacement of B from A
Displacement from A to B
Displacement AB
Displacement undergone by P in moving from A to B

Let the coordinates of A and B with respect to an origin O be x_A and x_B. Then
the displacement of B relative to A is $x_B - x_A$.

F4. Velocity

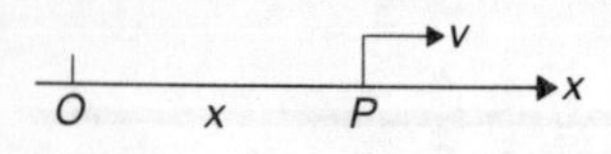

Velocity is the rate of change of displacement.

Let the coordinate of a moving point P with respect to a fixed origin O be x at time t. Then the

$$\text{velocity of } P \text{ is } v = \frac{dx}{dt} = \dot{x}.$$

Speed is the magnitude, $|v|$, of the velocity.

F5. Relative velocity

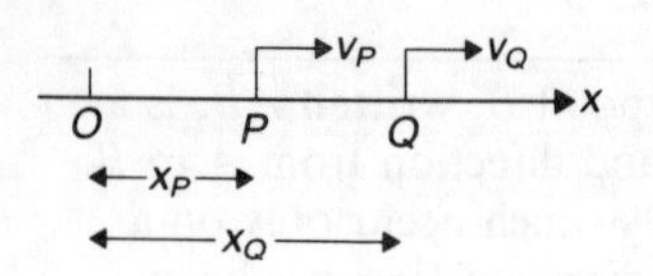

The velocity of a point Q relative to a point P is the rate of change of the displacement PQ.

Let the points P, Q have coordinates x_P, x_Q and velocities v_P, v_Q at time t. Then the velocity of Q relative to P is

$$\frac{d}{dt}(x_Q - x_P) = \dot{x}_Q - \dot{x}_P = v_Q - v_P.$$

The magnitude of the relative velocity, $|v_Q - v_P|$, is sometimes called the **relative speed**.

F6. Acceleration

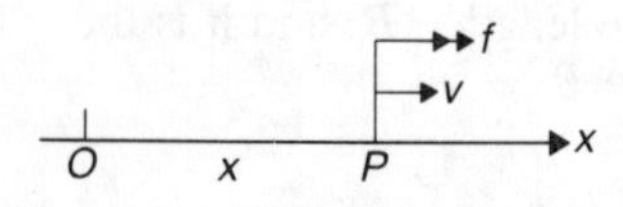

Acceleration is the rate of change of velocity.

Suppose that, at time t, the coordinate of a moving point P is x and that its velocity is v. Then the

$$\text{acceleration of } P \text{ is } f = \frac{dv}{dt} = \dot{v} = \frac{d^2x}{dt^2} = \ddot{x} = v\frac{dv}{dx} = \frac{d}{dx}\left(\frac{v^2}{2}\right).$$

F7. Relative acceleration

Relative acceleration is the rate of change of relative velocity.

Let the points P, Q have coordinates x_P, x_Q, velocities v_P, v_Q and accelerations f_P, f_Q at time t. Then the acceleration of Q relative to P is

$$\frac{d}{dt}(v_Q - v_P) = \dot{v}_Q - \dot{v}_P = f_Q - f_P = \ddot{x}_Q - \ddot{x}_P = \frac{d^2}{dt^2}(x_Q - x_P).$$

ILLUSTRATIONS

I1.

Points P and Q move on a straight line. Their displacements from a point O on the line at time t seconds ($t \geqslant 0$) are x m and y m respectively, where

$$P: \quad x = t^3 - 2t^2,$$

$$Q: \quad y = 2t^2 + 3t - 6.$$

(a) Determine the initial positions, velocities and accelerations of P and Q. □

Initially means when $t = 0$. Initially $x = 0$ and $y = -6$. Hence the initial position of P is at O, and the initial position of Q is Q_0, which has displacement -6 m from O.

$$\dot{x} = 3t^2 - 4t.$$

$$\dot{y} = 4t + 3.$$

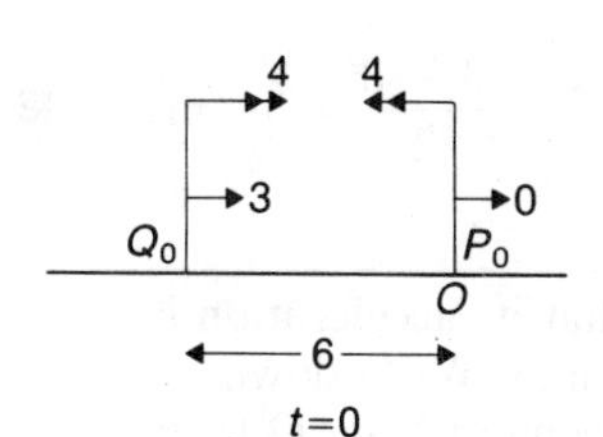

Hence the initial velocities of P and Q are $0\ \mathrm{m\,s^{-1}}$ and $3\ \mathrm{m\,s^{-1}}$ respectively.

$$\ddot{x} = 6t - 4.$$

$$\ddot{y} = 4.$$

Hence the initial accelerations of P and Q are $-4\ \mathrm{m\,s^{-2}}$ and $4\ \mathrm{m\,s^{-2}}$ respectively.■

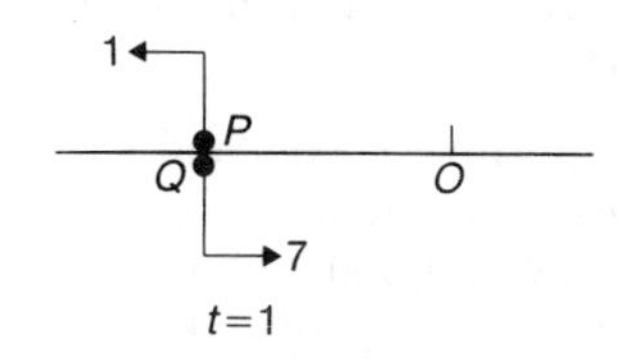

(b) State the displacement of Q relative to P at time t s and verify that P and Q meet when $t = 1$. What happens next? □

The displacement of Q relative to P is $(y - x)$ m, and

$$y - x = -t^3 + 4t^2 + 3t - 6.$$

This is zero when $t = 1$, so P and Q meet then.
When $t = 1$, $\dot{x} = -1$ and $\dot{y} = 7$.
So, as the diagram shows, Q then passes P and the displacement of Q relative to P becomes positive.■

(c) Find the velocity of Q relative to P at time t s, and determine when the displacement from P to Q is maximum. □

Let $v\ \mathrm{m\,s^{-1}}$ be the velocity of Q relative to P. Then

$$v = \dot{y} - \dot{x} = -3t^2 + 8t + 3.$$

The rate of increase of the displacement from P to Q is given by

$$\frac{\mathrm{d}}{\mathrm{d}t}(y - x) = -(3t^2 - 8t - 3) = -(3t + 1)(t - 3).$$

This is positive for $0 \leqslant t < 3$ and negative for $t > 3$. Hence $y - x$ rises to a maximum when $t = 3$. ■
The maximum value of this relative displacement is, from **(b)**,

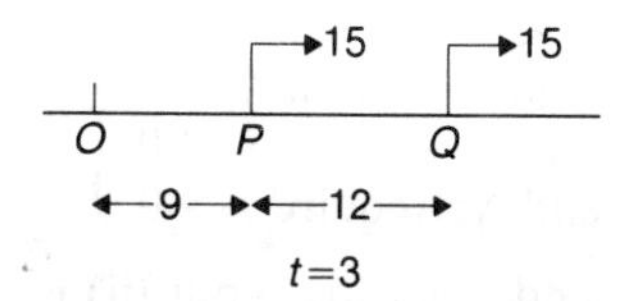

$$-27 + 4 \times 9 + 3 \times 3 - 6 = 12\ \mathrm{m}. ■$$

Note: After this, P catches up with and passes Q. The displacement eventually takes negative values of unlimited magnitude, and so the *distance PQ* will increase indefinitely.

---

12. The velocity of Q relative to P is $5\ \mathrm{m\,s^{-1}}$ and the velocity of Q relative to R is $8\ \mathrm{m\,s^{-1}}$. What is the velocity of P relative to R? Find the velocity of the midpoint of PQ relative to R. □

Using F5, $\quad v_Q - v_P = 5. \qquad v_Q - v_R = 8.$

$$\therefore\ v_P - v_R = (v_P - v_Q) + (v_Q - v_R) = -5 + 8 = 3.$$

Hence the velocity of P relative to R is $3\ \mathrm{m\,s^{-1}}$.

The midpoint M of PQ is the point with coordinate $x_M = \dfrac{x_P + x_Q}{2}$.

$$\therefore \text{ the velocity of } M \text{ is } v_M = \frac{v_P + v_Q}{2}.$$

$\therefore$ the velocity of M relative to R is

$$\frac{v_P + v_Q}{2} - v_R = \frac{v_P - v_R + v_Q - v_R}{2} = \frac{3+8}{2} = 5{\cdot}5 \text{ m s}^{-1}. \blacksquare$$

---

13. A particle P moves on a straight line so that its acceleration is always directed towards a point O on the line. Write down a differential equation relating x, the displacement from O to P, to the time t, in the case when

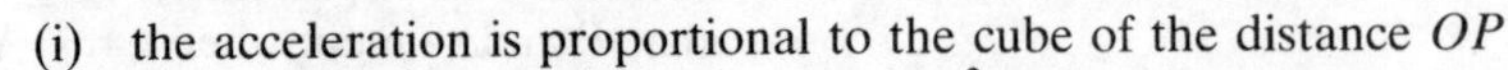

(i) the acceleration is proportional to the cube of the distance OP

(ii) the acceleration is proportional to OP^2. □

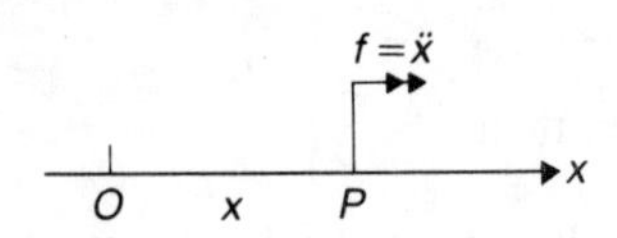

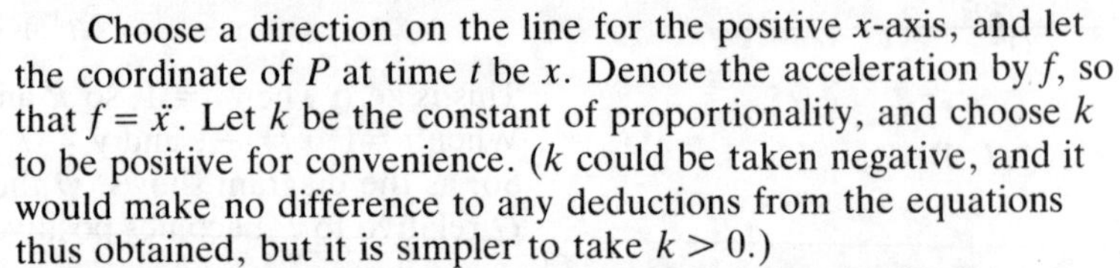

Choose a direction on the line for the positive x-axis, and let the coordinate of P at time t be x. Denote the acceleration by f, so that $f = \ddot{x}$. Let k be the constant of proportionality, and choose k to be positive for convenience. (k could be taken negative, and it would make no difference to any deductions from the equations thus obtained, but it is simpler to take $k > 0$.)

The direction of the acceleration, given as 'towards O', depends on whether x is positive or negative, and so we must carefully consider the two possibilities. First take $x > 0$. Then, since the acceleration is in the direction PO, f must be negative. We have chosen $k > 0$, and therefore the equations required are

$$\text{(i) } f = -kx^3 \qquad \text{(ii) } f = -kx^2,$$

that is, $$\text{(i) } \frac{d^2x}{dt^2} = -kx^3 \qquad \text{(ii) } \frac{d^2x}{dt^2} = -kx^2, \qquad \text{for } x > 0.$$

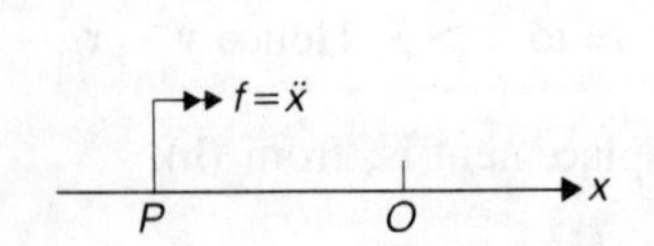

Now consider the case when x is negative. PO is now in the positive direction, and so f has to be positive. Our equations become

$$\text{(i) } f = -kx^3, \qquad \text{(ii) } f = kx^2, \quad \text{for } x < 0$$

(Remembering that $k > 0$, $x < 0$, and we require $f > 0$.)

We see that (i) is the same for x positive and x negative, but (ii) is not. It is generally true that, when the acceleration is given in a form like 'towards O', the parts of the motion for $x > 0$ and $x < 0$ have to be treated separately and then fitted together. The only exception to this rule is when the formula for $|f|$ in terms of $|x|$ corresponds to an *odd function*, as in the case with $|f| \propto |x|^3$.

In conclusion, the required differential equations are

$$\text{(i) } \frac{d^2x}{dt^2} = -kx^3 \qquad \text{(for all } x\text{);}$$

(ii) $\dfrac{d^2x}{dt^2} = -kx^2 \qquad (x > 0), \qquad \dfrac{d^2x}{dt^2} = kx^2 \qquad (x < 0),$

where k is a positive constant.■

I4.

The velocity of a particle P is v m s^{-1}. When the displacement of P from O is x m, $v^2 = 6x - x^2$. Find the acceleration of P and show that it is directed always towards a fixed point A, and find A. □

$$v^2 = 6x - x^2.$$

Differentiate with respect to x.

$$2v\frac{dv}{dx} = 6 - 2x.$$

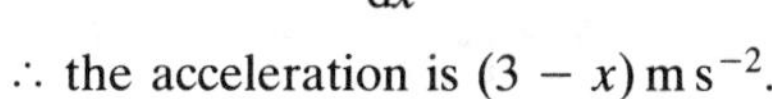

From F6, $\quad v\dfrac{dv}{dx} = f.$

∴ the acceleration is $(3 - x)$ m s^{-2}.

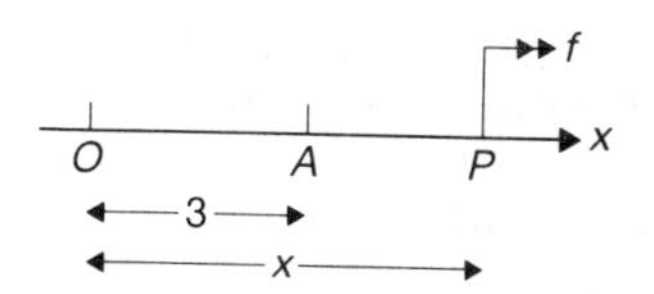

Now $f = -(x - 3)$. Let A be the point with displacement 3 m from O. If P is to the right of A, as in the diagram, $x - 3$ is positive and f is negative, and consequently the acceleration is to the left. Similarly, if P is to the left of A, $x - 3$ is negative and so f is positive and the acceleration is to the right. In either case the acceleration is directed towards A.■

3.3 ACCELERATION DEPENDENT ON THE TIME

ESSENTIAL FACTS

A particle P moves on the x-axis. At time t its coordinate is x, its velocity is v and its acceleration is f.
Initially, when $t = 0$, $x = x_0$ and $v = v_0$.

F1.

When f is given in terms of t, v is obtained by integrating, with respect to t, the equation $\dfrac{dv}{dt} = f$.

$$v = \int f\,dt + A, \qquad \text{where } A \text{ is a constant.}$$

A is found by setting $t = 0$, $v = v_0$ in this equation.
If the velocity is known at a particular time other than $t = 0$, A can also be found by the same method.

F2.

When v is given in terms of t, x is found by integrating $\dfrac{dx}{dt} = v$.

$$x = \int v\,dt + B, \qquad \text{where } B \text{ is a constant.}$$

B may be found if the value of x is known at any particular time.

F3. Average velocity in a time interval

The average (or **mean**) velocity of P in the time interval $t_1 \leqslant t \leqslant t_2$ is $\bar{v}$, where

$$\bar{v} = \frac{1}{t_2 - t_1}\int_{t_1}^{t_2} v\,dt.$$

F4.

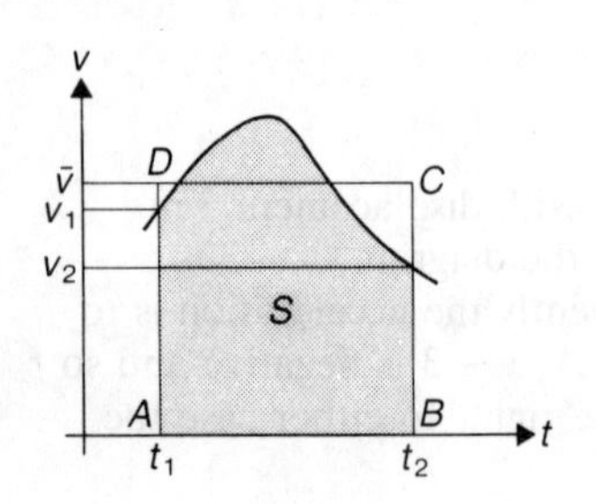

The displacement undergone by P in the time interval $t_1 \leqslant t \leqslant t_2$ is

$$S = \Big[x\Big]_{t=t_2}^{t=t_1} = \int_{t_1}^{t_2} v\,dt.$$

S is the area of the region shaded in the diagram. It is the area under the velocity–time graph between the ordinates $t = t_1$ and $t = t_2$.
Displacement = (Average velocity) × (Time interval).
[Shaded area = area of rectangle $ABCD$].

F5.

The average acceleration in the time interval $t_1 \leqslant t \leqslant t_2$ is

$$\frac{1}{t_2 - t_1}\int_{t_1}^{t_2} f\,dt = \frac{\Big[v\Big]_{t=t_1}^{t=t_2}}{t_2 - t_1}.$$

Increase in velocity in a time interval
= (Average acceleration) × (Time interval)

F6. Constant acceleration

When f is constant, the integrations in F1 and F2, followed by the insertion of the initial data, yield the formulae

$$v = v_0 + ft \qquad (1)$$
$$x = x_0 + v_0 t + \tfrac{1}{2}ft^2. \qquad (2)$$

F7.

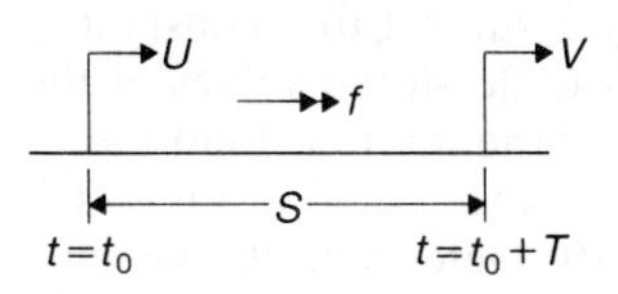

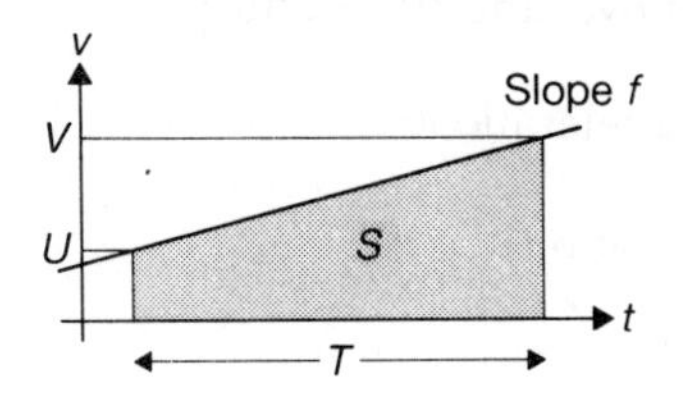

During a time interval T, a particle P is moving with constant acceleration f.
Let the velocity of P at the beginning of the interval be U.
Let the velocity at the end of the interval be V.
Let the displacement of P during the interval be S.

$$V = U + fT \tag{A}$$

$$S = \frac{(U + V)T}{2} \tag{B}$$

$$S = UT + \tfrac{1}{2}fT^2 \tag{C}$$

$$V^2 = U^2 + 2fS \tag{D}$$

(A) follows from F6 (1) (B) eliminating f from (A) and (C)
(C) follows from F6 (2) (D) eliminating T from (A) and (C).

ILLUSTRATIONS

I1.

The acceleration along Ox of a particle P at time t s is $f\,\text{m s}^{-2}$, where $f = 6t - 2$.
Find the velocity and the displacement of P from O at time t s in each of the following cases.

(i) Initially the displacement of P from O is 2 m and the velocity of P is $3\ \text{m s}^{-1}$.

(ii) When $t = 1$ the velocity of P is $6\ \text{ms}^{-1}$ and when $t = 2$ the displacement of P from O is 8 m. □

We integrate the acceleration to obtain the velocity $v\ \text{ms}^{-1}$.

$$v = \int f\,\mathrm{d}t = 3t^2 - 2t + A, \qquad \text{where } A \text{ is a constant.}$$

Integrate again to obtain the displacement x m.

$$x = \int v\,\mathrm{d}t = t^3 - t^2 + At + B, \qquad \text{where } B \text{ is a constant.}$$

(i) When $t = 0,\ v = 3$.
$\therefore 3 = 0 + A.$ $\qquad \therefore v = 3t^2 - 2t + 3.$

When $t = 0,\ x = 2$.

$\therefore 2 = 0 + B.$ $\qquad \therefore x = t^3 - t^2 + 3t + 2.$ ■

(ii) When $t = 1,\ v = 6$.
$\therefore 6 = 3 \times 1^2 - 2 \times 1 + A.$ $\qquad \therefore v = 3t^2 - 2t + 5.$

When $t = 2,\ x = 8$.

$\therefore 8 = 2^3 - 2^2 + 5 \times 2 + B.$ $\qquad \therefore B = -6.$

$\therefore$ the displacement is $(t^3 - t^2 + 5t - 6)$ m. ■

I2. A particle P moves from rest at a point O with constant acceleration $2\ \mathrm{m\,s^{-2}}$ for 6 seconds. It is then subject to a constant retardation of $0{\cdot}5\ \mathrm{m\,s^{-2}}$ for 4 seconds. Write down the values of the acceleration $f\ \mathrm{m\,s^{-2}}$ of P in the two parts of the motion. Find the displacement x m from O, and the velocity $v\ \mathrm{m\,s^{-1}}$ of P at time t s from the start of the motion, for $0 \leqslant t \leqslant 10$. Determine the values of the average velocity and acceleration over the 10 second interval.□

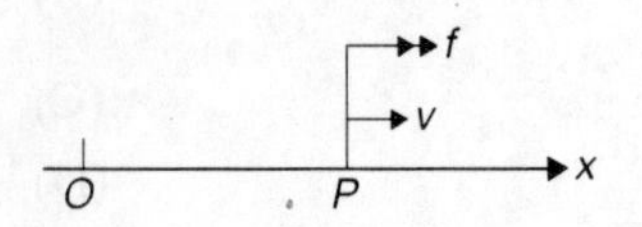

Take Ox in the direction of the initial acceleration.

Then for $0 \leqslant t \leqslant 6, \quad f = 2$

$6 < t \leqslant 10, \quad f = -\frac{1}{2}$ (retardation).

Method 1. Integration of the acceleration

$0 \leqslant t \leqslant 6$: $\quad f = 2.$

$\therefore v = 2t + A.$ When $t = 0, v = 0.$ $\therefore A = 0.$

Integrating again, $x = t^2 + B.$ When $t = 0, x = 0.$ $\therefore B = 0.$

$t = 6$: $\quad v = 2 \times 6 = 12, \quad x = 6^2 = 36.$

$6 < t \leqslant 10$: $\quad f = -\frac{1}{2}.$

$\therefore v = -\frac{1}{2}t + C.$ When $t = 6, v = 12.$

$\therefore 12 = -\frac{1}{2}(6) + C.$ $\therefore C = 15.$

$\therefore v = -\frac{1}{2}t + 15.$

Integrating again, $x = -\frac{1}{4}t^2 + 15t + D.$ When $t = 6, x = 36.$

$\therefore 36 = -\frac{1}{4}(6^2) + 15 \times 6 + D.$ $\therefore D = -45.$

$\therefore x = -\frac{1}{4}t^2 + 15t - 45.$

Hence the required results are

$0 \leqslant t \leqslant 6$:
Displacement $= t^2$ m, Velocity $= 2t\ \mathrm{m\,s^{-1}}$.

$6 < t \leqslant 10$:
Displacement $= (-\frac{1}{4}t^2 + 15t - 45)$ m, Velocity $= (-\frac{1}{2}t + 15)\ \mathrm{m\,s^{-1}}$. ■

When $t = 10$, $\quad v = -\frac{1}{2}(10) + 15 = 10$

$x = -\frac{1}{4}(10^2) + 15(10) - 45 = 80.$

From F3: Average velocity $= \dfrac{\text{Displacement}}{\text{Time interval}} = \dfrac{80}{10} = 8\ \mathrm{m\,s^{-1}}.$

From F5: Average acceleration $= \dfrac{\text{Increase in velocity}}{\text{Time interval}} = \dfrac{10 - 0}{10}$
$= 1\ \mathrm{m\,s^{-2}}.$ ■

Method 2. Using the formulae of F7

In using the formulae we must be careful to interpret the time interval 'T' of the formulae correctly in each part of the motion.

In the first part, 'T' means t.
In the second part 'T' means 'the time elapsed since $t = 6$', which is $t - 6$.

If space allows, such problems can be set out in tabular form.

Time	f	v	x
		Using F7 (A)	Using F7 (C)
$t = 0$		0	0
$0 \leqslant t \leqslant 6$	2	$0 + 2t$	$0 + (\frac{1}{2})2t^2$
$t = 6$		12	36
$6 < t \leqslant 10$	$-\frac{1}{2}$	$12 - \frac{1}{2}(t - 6)$	$36 + 12(t - 6) + \frac{1}{2}(-\frac{1}{2})(t - 6)^2$
$t = 10$		10	80

The results now follow as in method 1.

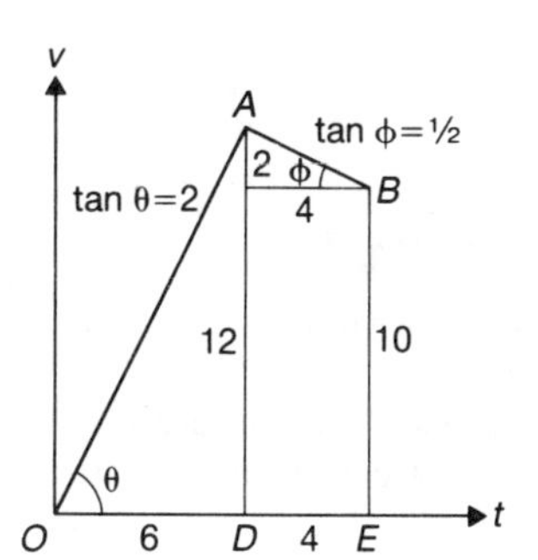

Note: In many cases we require only the final velocity and displacement. This can often be done by means of the velocity–time graph. Starting from the origin, we draw a line OA with gradient 2 for the acceleration, which persists for 6 seconds. Then AB, gradient $-\frac{1}{2}$, for 4 seconds. The velocity is then seen to rise to $6 \times 2 = 12\,\mathrm{m\,s^{-1}}$ at A, then fall by an amount $4 \times \frac{1}{2} = 2\,\mathrm{m\,s^{-1}}$ to B. So the height of B is $12 - 2 = 10$, as shown in the diagram, meaning that the final velocity is $10\,\mathrm{m\,s^{-1}}$.

The total displacement is the area under the graph for the 10 seconds.

Now ΔODA + trapezium $ADEB$

$$= \tfrac{1}{2}(6 \times 12) + \left(\frac{12 + 10}{2}\right)4 = 36 + 44 = 80.$$

Hence the total displacement is 80 m, as we found before.

3.4 ACCELERATION DEPENDENT ON THE DISPLACEMENT

ESSENTIAL FACTS

F1.

When the velocity v is known in terms of the displacement x, the relation between x and the time t can be obtained by an integration with respect to x.

$$\frac{dx}{dt} = v. \qquad \therefore \frac{dt}{dx} = \frac{1}{v}.$$

$$\therefore t = \int \frac{1}{v} \mathrm{d}x + A, \quad \text{where } A \text{ is a constant.}$$

A can be determined if the position of P is known at a particular time.

F2.

Provided the velocity does not become zero (that is, P does not come to rest), the time taken for P to undergo the displacement from $P_1(x_1)$ to $P_2(x_2)$ is

$$\int_{x_1}^{x_2} \frac{1}{v} \mathrm{d}x.$$

F3.

When the acceleration f is known in terms of the displacement x, the square of the velocity v can be obtained by an integration with respect to x. For, from 3.2F6,

$$\frac{\mathrm{d}}{\mathrm{d}x}\left(\tfrac{1}{2}v^2\right) = f.$$

$$\therefore \tfrac{1}{2}v^2 = \int f \mathrm{d}x + B, \quad \text{where } B \text{ is a constant.}$$

B can be determined if the speed $|v_0|$ is known when P is in any particular position x_0.
Note that this formula does not give the velocity v completely, only v^2, and hence the speed $|v|$ but not the direction of motion.

F4. Constant acceleration

When f is constant the integration in F3, followed by the insertion of the known speed $|v_0|$ at the position x_0, yields the formula

$$v^2 = v_0^2 + 2f(x - x_0),$$

which is equivalent to the formula 3.3F7 D, and can also be derived by eliminating t from 3.3F6(1) and (2).

ILLUSTRATION

I1.

A particle P moves so that when its displacement from O is x m its acceleration is $\frac{1}{8}x^3$ m s^{-2} away from O. Initially $x = 2$ and the velocity of P is 1 m s^{-1} towards O. Find the position of P after time t s. □

$$f = \frac{1}{8}x^3. \qquad \therefore \frac{\mathrm{d}}{\mathrm{d}x}\left(\frac{v^2}{2}\right) = \frac{1}{8}x^3.$$

Integrate with respect to x.

$$\frac{1}{2}v^2 = \frac{x^4}{32} + A, \quad \text{where } A \text{ is a constant.}$$

When $x = 2$, $v = -1$. $\qquad \therefore \frac{1}{2} = \frac{16}{32} + A. \qquad \therefore A = 0.$

$\therefore v^2 = \frac{x^4}{16}.$ Since $v = -1$ when $x = 2$, we select the negative root of this equation for v.

$$\therefore \frac{dx}{dt} = -\frac{x^2}{4}. \qquad \therefore \frac{dt}{dx} = -\frac{4}{x^2}.$$

Integrate with respect to x.

$$t = \frac{4}{x} + B, \quad \text{where } B \text{ is a constant.}$$

When $t = 0$, $x = 2$. $\qquad \therefore B = -2. \qquad \therefore t + 2 = \frac{4}{x}.$

$\therefore$ the displacement of P from O is $\frac{4}{t+2}$ m.■

3.5 ACCELERATION DEPENDENT ON THE VELOCITY

ESSENTIAL FACTS

A particle P moves on the x-axis. At time t its coordinate is x and its velocity is v. When the acceleration f of P is known in terms of v, both the v–t relation and the v–x relation can be obtained by integrations with respect to v.

F1.

$$\frac{dv}{dt} = f. \qquad \therefore \frac{dt}{dv} = \frac{1}{f}$$

$$\therefore t = \int \frac{1}{f} dv + A, \quad \text{where } A \text{ is a constant.}$$

A can be determined if the velocity of P is known at a particular time, and hence t can be expressed in terms of v.

F2.

Provided the acceleration does not become zero, the time taken for the velocity to change from v_1 to v_2 is $\int_{v_1}^{v_2} \frac{1}{f} dv$.

F3.

$$v\frac{dv}{dx} = f. \qquad \therefore \frac{dx}{dv} = \frac{v}{f}.$$

$$\therefore x = \int \frac{v}{f} dv + B, \quad \text{where } B \text{ is a constant.}$$

B can be determined if the velocity is known when P is at a particular point, and hence x can be expressed in terms of v.

F4. Provided the acceleration does not become zero, the displacement undergone by P as the velocity changes from v_1 to v_2 is

$$\int_{v_1}^{v_2} \frac{v}{f}\,\mathrm{d}v.$$

ILLUSTRATIONS

In these illustrations a particle P moves on the x-axis. At time t its coordinate is x, its velocity is v and its acceleration is f.

I1. $f = kv$, where k is a constant. Initially $v = u > 0$. Find v in terms of t. □

As v is required in terms of t, we write $f = \dot{v}$. Then

$$\frac{\mathrm{d}v}{\mathrm{d}t} = kv. \qquad \therefore k\frac{\mathrm{d}t}{\mathrm{d}v} = \frac{1}{v}.$$

Integrate with respect to v.

$kt = \ln(Av)$, where A is a constant. (This is more convenient than the form '$\ln v + C$', and ensures that logarithms of negative numbers do not occur – see *Pure Mathematics* Ch. 12.)

When $t = 0$, $v = u$. $\therefore \ln Au = 0$. $\therefore A = \frac{1}{u}$.

$$\therefore kt = \ln\frac{v}{u}.$$

$$\therefore v = u\,\mathrm{e}^{kt}. \blacksquare$$

I2. $f = kv^3$, where k is constant. When $x = a$, $v = u$. Find v in terms of x. □

As v is required in terms of x, we write $f = v\frac{\mathrm{d}v}{\mathrm{d}x}$.

Then $$v\frac{\mathrm{d}v}{\mathrm{d}x} = kv^3.$$

$$\therefore k\frac{\mathrm{d}x}{\mathrm{d}v} = \frac{1}{v^2}.$$ Integrate with respect to v.

$$kx = -\frac{1}{v} + A.$$ When $x = a$, $v = u$. $\therefore A = ka + \frac{1}{u}$.

$$\therefore kx = -\frac{1}{v} + ka + \frac{1}{u}. \qquad \therefore v = \frac{u}{ku(a-x)+1}. \blacksquare$$

13.

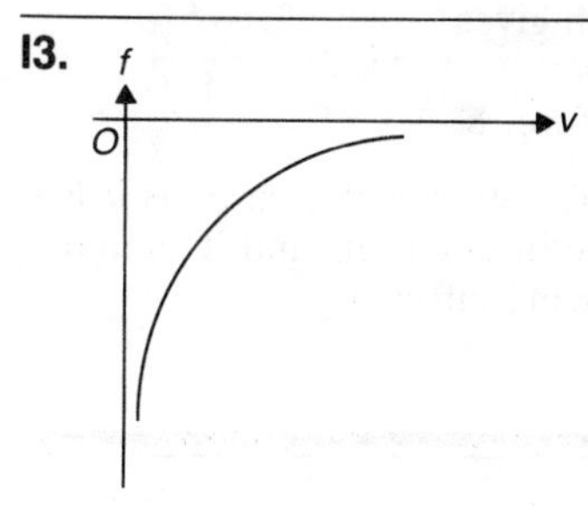

$f = -\dfrac{k}{v}$, where k is a positive constant. Initially P is at O and moving with velocity $u > 0$. Find the position of P at time t, where $0 \leqslant t < \dfrac{u^2}{2k}$. □

We first have to find v in terms of t or of x, and then integrate again to find x in terms of t. Here we show both routes to the result.

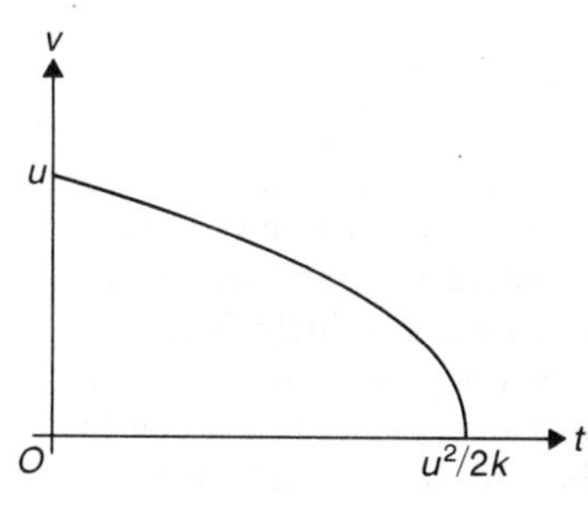

Using the method of F1: $\quad f = \dfrac{dv}{dt}$.

$$\therefore k\frac{dt}{dv} = -v. \qquad \text{Integrate with repect to } v.$$

$$kt = -\tfrac{1}{2}v^2 + A. \qquad \text{When } t = 0,\ v = u.$$

$$\therefore A = \tfrac{1}{2}u^2. \qquad \therefore v^2 = u^2 - 2kt.$$

Since $u > 0$, we require $v > 0$.

$$\therefore \frac{dx}{dt} = v = \sqrt{u^2 - 2kt}.$$

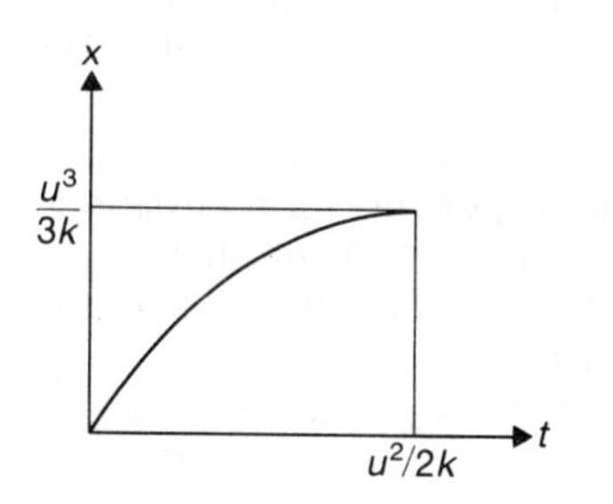

As v falls from the value u, the negative acceleration grows in magnitude, and so v falls faster. Now integrate with respect to t, choosing the constant so that $x = 0$ when $t = 0$.

$$x = -\frac{1}{2k}\left(\frac{2}{3}\right)(u^2 - 2kt)^{3/2} + \frac{1}{2k}\left(\frac{2}{3}\right)(u^2 - 0)^{3/2}.$$

Hence $\quad x = \dfrac{u^3 - (u^2 - 2kt)^{3/2}}{3k}$. ■

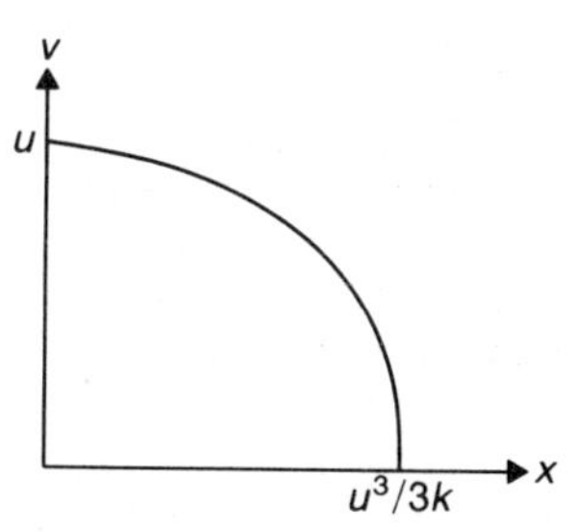

Alternatively using the method of F3: $\quad f = v\dfrac{dv}{dx}$.

$$\therefore k\frac{dx}{dv} = -v^2. \qquad \therefore kx = -\frac{1}{3}v^3 + \frac{1}{3}u^3.$$

$$\therefore \frac{dx}{dt} = v = (u^3 - 3kx)^{1/3}.$$

$$\therefore \frac{dt}{dx} = (u^3 - 3kx)^{-1/3}.$$

Integrate with respect to x, choosing the constant so that $t = 0$ when $x = 0$.

$$t = -\frac{1}{3k}\left(\frac{3}{2}\right)(u^3 - 3kx)^{2/3} + \frac{1}{3k}\left(\frac{3}{2}\right)(u^3 - 0)^{2/3}.$$

$$= \frac{u^2 - (u^3 - 3kx)^{2/3}}{2k}.$$

$$\therefore\ (u^3 - 3kx)^{2/3} = u^2 - 2kt, \qquad \text{which gives}$$

$$x = \frac{u^3 - (u^2 - 2kt)^{3/2}}{3k}, \qquad \text{as before. ■}$$

We see that the F1 route is slightly easier in this case, as it is in most cases, but the F3 route is sometimes easier, and it should be tried if a derivation using F1 is proving difficult.

3.6 EXAMINATION QUESTIONS AND SOLUTIONS

Q1. A particle A starts from the origin O with velocity $u\ \mathrm{m\,s^{-1}}$ and moves along the positive x-axis with constant acceleration $f\ \mathrm{m\,s^{-2}}$, where $u > 0$, $f > 0$. Ten seconds later, another particle B starts from O with velocity $u\ \mathrm{m\,s^{-1}}$ and moves along the positive x-axis with acceleration $2f\ \mathrm{m\,s^{-2}}$. Find the time that elapses between the start of A's motion and the instant when B has the same velocity as A, and show that A will then have travelled twice as far as B.

(JMB 1986)

Q2. A particle P moves in a straight line and experiences a retardation of $0{\cdot}01\ v^3\ \mathrm{m\,s^{-2}}$, where $v\ \mathrm{m\,s^{-1}}$ is the speed of P. Given that P passes through a point O with speed $10\ \mathrm{m\,s^{-1}}$, show that, when it is a distance 10 m from O, its speed is $5\ \mathrm{m\,s^{-1}}$.

Find the time taken for the speed of P to be reduced from $10\ \mathrm{m\,s^{-1}}$ to $5\ \mathrm{m\,s^{-1}}$.

(LON 1984)

Q3. A particle moves along a straight line ABC. The particle starts from rest at A and moves from A to B with constant acceleration $2f$. It then moves from B to C with acceleration f and reaches C with speed V. The times taken in the motions from A to B and from B and C are each equal to T. Find T in terms of V and f. Show that

$$AB = \frac{2}{5}BC.$$

(JMB 1984)

Q4. (i) The brakes of a train, which is travelling at $108\ \mathrm{km\,h^{-1}}$, are applied as the train passes point A. The brakes produce a constant retardation of magnitude $3f\ \mathrm{m\,s^{-2}}$ until the speed of the train is reduced to $36\ \mathrm{km\,h^{-1}}$. The train travels at this speed for a distance and is then uniformly accelerated at $f\ \mathrm{m\,s^{-2}}$ until it again reaches a speed of $108\ \mathrm{km\,h^{-1}}$ as it passes point B. The time taken by the

train in travelling from A to B, a distance of 4 km, is 4 minutes. Sketch the speed/time graph for this motion and hence calculate
(a) the value of f,
(b) the distance travelled at $36\ \text{km h}^{-1}$.
(ii) A particle moves in a straight line with variable acceleration $\dfrac{k}{1+v}\ \text{m s}^{-2}$ where k is a constant and $v\ \text{m s}^{-1}$ is the speed of the particle when it has travelled a distance x m. Find the distance moved by the particle as its speed increases from 0 to $u\ \text{m s}^{-1}$.
(LON 1983)

Q5.

A vehicle P moves along a straight track which passes through two points A and B which are at a distance of 80 m apart. A vehicle Q moves along a straight parallel track whose perpendicular distance from the first track is to be neglected. The vehicle P has an acceleration of $3\ \text{m s}^{-2}$ in the sense from A to B whilst Q has an acceleration of $1\ \text{m s}^{-2}$ in the sense from B to A. At time $t = 0$ s P passes through A moving towards B with a speed of $12\ \text{m s}^{-1}$. Find the distance of P from A six seconds later and its speed at this time.

Also at time $t = 0$ s Q passes B moving towards A with a speed of $8\ \text{m s}^{-1}$. Find an expression for the distance PQ at time t and determine when P and Q are at a distance of 32 m apart.

When P is moving at a speed of $30\ \text{m s}^{-1}$ the acceleration ceases, the brakes are applied and continue to be applied until the speed has dropped to $10\ \text{m s}^{-1}$. During this period the retardation produced when the speed is $v\ \text{m s}^{-1}$ is $\dfrac{v^2}{150}\ \text{m s}^{-2}$. Find the time taken for the speed to drop to $10\ \text{m s}^{-1}$ from $30\ \text{m s}^{-1}$. (AEB 1983)

Q6.

A particle moves on the positive x-axis. The particle is moving towards the origin O when it passes through the point A, where $x = 2a$, with speed $\sqrt{(k/a)}$, where k is constant. Given that the particle experiences an acceleration $k/(2x^2) + k/(4a^2)$ in a direction away from O, show that it comes instantaneously to rest at a point B, where $x = a$.

Immediately the particle reaches B the acceleration changes to $k/(2x^2) - k/(4a^2)$ in a direction away from O. Show that the particle next comes instantaneously to rest at A. (LON 1985)

Q7.

A particle P moves along a straight line such that, when its speed is $v\ \text{m s}^{-1}$, its retardation is $4v^{n+1}\ \text{m s}^{-2}$ where $n(> -1)$ is a constant. The speed of P at time $t = 0$ s is $u\ \text{m s}^{-1}$.
(a) Show that, for $n = 0$, $v = ue^{-4t}$.

(b) Find similarly an expression for v, in terms of u, n and t, when $n \neq 0$.

(c) When $n = 3$ obtain an expression for the speed with which P is moving when it has travelled a distance of s m from its initial position. (AEB 1984)

SOLUTIONS

S1.

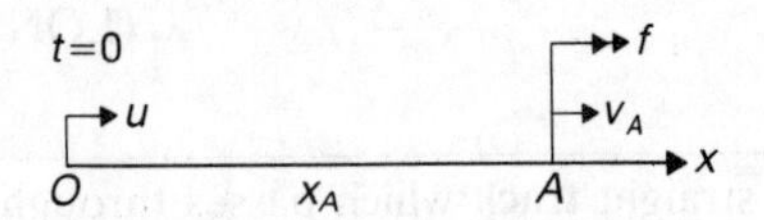

A

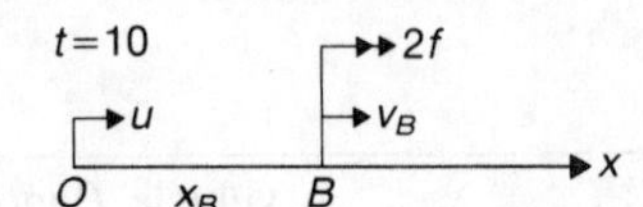

At a time t seconds after A starts, let:
Displacements of A and B be x_A m and x_B m,
Velocities of A and B, $v_A\,\mathrm{m\,s^{-1}}$ and $v_B\,\mathrm{m\,s^{-1}}$. B

Then $v_A = u + ft$

and $v_B = u + 2f(t - 10) \quad (t \geqslant 10)$. C

$\therefore v_A - v_B = ft - 2f(t - 10) \quad (t \geqslant 10)$
$= f(20 - t)$. D

$v_A = v_B$ when $t = 20$.

$\therefore$ B has the same velocity as A at an instant 20 seconds after the start of A's motion. E

At this time

$x_A = ut + \frac{1}{2}ft^2$ F

$= u(20) + \frac{1}{2}f(400) = 20(u + 10f)$.

$x_B = u(t - 10) + \frac{1}{2}(2f)(t - 10)^2$

$= u(10) + f(100) = 10(u + 10f) = \frac{1}{2}x_A$.

Hence A has travelled twice as far as B. *11m* ■

A We begin by carefully putting the data on to simple diagrams so that we can see what is happening without having to refer again to the text of the question.
B The meaning of the symbols should be clear from the diagram, but it is best to define them as well.
C Using 3.3F7 A.
D This is just as easy as equating the expressions for v_A and v_B, and enables you to check that the result is sensible. We see that $v_A - v_B$ decreases, due to the greater acceleration of B, as we should expect.
E Refer carefully to the question before presenting this result, to confirm that it was the time from A's (not B's) start that was demanded.
F Using 3.3F7 C.

S2.
Acceleration $f = -kv^3$, where $k = 0{\cdot}01$. A

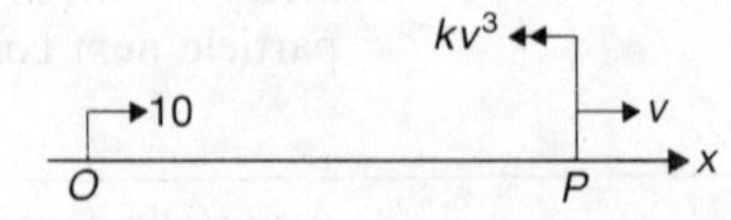

$\therefore v\dfrac{dv}{dx} = -kv^3$. B

$$\therefore k\frac{dx}{dv} = -\frac{1}{v^2}.$$

$$\therefore kx = \frac{1}{v} + A. \qquad \text{When } x = 0,\ v = 10.$$

$$\therefore \frac{x}{100} = \frac{1}{v} - \frac{1}{10}.$$

When $x = 10$, $\quad \dfrac{1}{v} = \dfrac{1}{10} + \dfrac{10}{100} = \dfrac{1}{5}.$ [C]

$\therefore$ When P is 10 m from O its speed is $5\ \mathrm{m\,s^{-1}}$. ■

$$\frac{dv}{dt} = -kv^3. \qquad \text{[D]}$$

$$\therefore k\frac{dt}{dv} = -\frac{1}{v^3}.$$

$$\therefore kt = \frac{1}{2v^2} - \frac{1}{200}, \qquad \text{taking } t = 0 \text{ when } v = 10.$$

When $v = 5$, $\quad \dfrac{t}{100} = \dfrac{1}{2 \times 25} - \dfrac{1}{200} = \dfrac{3}{200}.$

$\therefore$ the time taken for the speed to fall from $10\ \mathrm{m\,s^{-1}}$ to $5\ \mathrm{m\,s^{-1}}$ is 1·5 seconds. *25m* ■

[A] We take the x-axis in the direction of the velocity of P at O, and $t = 0$ when P is at O. In order to avoid repeatedly writing 0·01, which might lead to copying errors, call it k, and substitute for k when needed later.
[B] In the first part of the question both the data and the requirements concern *displacement* and velocity. We therefore use the method of 3.5F3.
[C] Alternatively use 3.5F4. Let the velocity when $x = 10$ be $v_1\ \mathrm{m\,s^{-1}}$. Then

$$k(10) = \int_{10}^{v_1}\left(-\frac{1}{v^2}\right)dv = \frac{1}{v_1} - \frac{1}{10}.$$

But ✕ $\quad \dfrac{1}{100}\Big[x\Big]_0^{10} = \displaystyle\int_{10}^{5}\left(-\frac{1}{v^2}\right)dv.$

$\therefore \dfrac{1}{10} = \dfrac{1}{10}$, hence the result ✕

would be unacceptable, because it assumes that which is to be proved.
[D] This part of the question is about velocity and *time*, so we use the method of 3.5F1.

S3.

→→ $2f$ →→ f

v_1 V

Time A ←—T—→ B ←—T—→ C

Let the velocity at B be v_1.
Then $v_1 = 2fT$. [A]
For the motion from B to C, $\qquad V = v_1 + fT.$

$$\therefore V = 2fT + fT = 3fT. \qquad \therefore T = \frac{v}{3f}. \ ■$$

$$AB = \frac{1}{2}(2f)T^2.$$

$$BC = v_1T + \tfrac{1}{2}fT^2.$$

$$\therefore AB = fT^2$$

and $BC = 2fT^2 + \frac{1}{2}fT^2 = \frac{5}{2}fT^2$.

$\therefore AB = \frac{2}{5}BC.$ *11m* ■

[A] The first part of the question is about velocity and time, so 3.3F7 A is useful. If you wish to quote any of the formulae for constant acceleration, do so at the side, not within the body of the work. Otherwise you may produce something like the following argument.
?Motion in AB:

$$V = U + fT;\ U = 0, f = 2f,\ T = T,\ V = ?$$

$$\therefore V = 0 + 2fT = 2fT.$$

Motion in BC:

$$V = U + fT;\ U = 2fT, f = f,\ T = T,\ V = V.$$

$$\therefore V = 2fT + fT. \qquad \textbf{?}$$

Luckily we have reached the right answer, even after statements like '$f = 2f$'! But if anything had gone wrong, as it well might when the meanings of symbols keep changing, the examiner would be unable to guess how much of it *was* correct, and many marks would be lost.

S4.

[A] (i) $108\ \mathrm{km\,h^{-1}} = 30\ \mathrm{m\,s^{-1}}$.
$36\ \mathrm{km\,h^{-1}} = 10\ \mathrm{m\,s^{-1}}$. [B]

From the diagram, the time taken during the retardation is $\frac{20}{3f}$ s, and the time taken to accelerate again to $30\ \mathrm{m\,s^{-1}}$ is $\frac{20}{f}$ s. [C]

Since the total distance travelled is 4000 m the area under the graph is 4000.

The area $JMHG$ is $240 \times 10 = 2400$.

$\therefore\ \triangle GDC + \triangle EHF = 4000 - 2400 = 1600.$

Now $EH = 3GD$, and so $\triangle EHF = 3\triangle GDC$.

$\therefore\ 4\triangle GDC = 1600. \qquad \therefore\ \triangle GDC = 400.$

$\therefore\ \frac{1}{2}(GD)20 = 400. \qquad \therefore\ GD = 40.$

$\therefore\ EH = 3 \times 40 = 120.$

$\therefore\ \frac{20}{f} = 120. \qquad \therefore\ f = \frac{1}{6}.$ ■

$DE = 240 - 40 - 120 = 80.$

$\therefore$ area $KLED = 10 \times 80 = 800.$

$\therefore$ the distance travelled at $36\ \mathrm{km\,h^{-1}}$ is 800 m. ■

(ii) $v\dfrac{dv}{dx} = \dfrac{k}{1+v}$. [D]

$\therefore\ k\dfrac{dx}{dv} = v + v^2.$

$\therefore$ the distance moved when v changes from 0 to u is $\dfrac{1}{k}\displaystyle\int_0^u (v + v^2)\,dv = \dfrac{1}{k}\left(\dfrac{u^2}{2} + \dfrac{u^3}{3}\right)$. [E] 25m ■

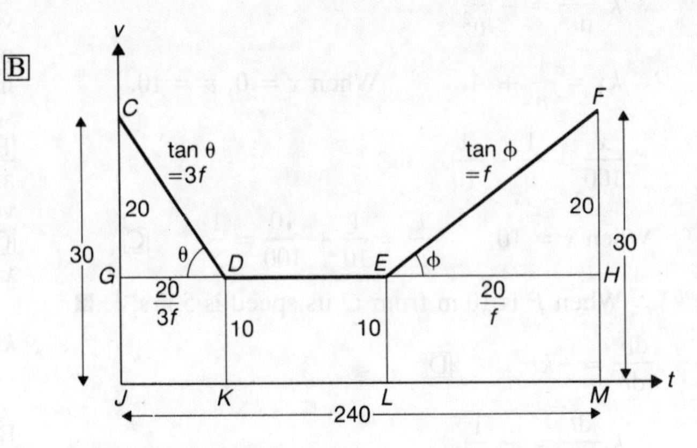

[A] Read the question through before starting it. Part (ii) is not connected with (i).

[B] The acceleration is quoted in $\mathrm{m\,s^{-2}}$, so it is advisable to convert the speeds to $\mathrm{m\,s^{-1}}$ immediately.

[C] We cannot draw the diagram accurately because the value of f is not known. Draw the line CD, falling from 30 to 10, and then draw another line FE, also falling from 30 to 10, with a less steep gradient. Label these lines with their resepective slopes corresponding to the retardation and acceleration. Then mark in the time intervals GD and EH.

[D] The question is about the relation between speed and distance, so we use the method of 3.5F3.

[E] Using 3.5F4.

S5.

[A]

Let the displacement of P from A at time t s be x m. Then $\ddot{x} = 3$.

$\therefore \dot{x} = 12 + 3t$ and $x = 12t + \frac{1}{2}(3)t^2$.

$\therefore$ when $t = 6$,
the distance AP is $72 + 54 = 126$ m,
and the speed of P is $12 + 18 = 30 \text{ m s}^{-1}$. ■
Let the displacement of Q from A at time t s be y m. [B]
Then $\ddot{y} = -1$.

$\therefore \dot{y} = -8 + (-1)t$ and $y = 80 - 8t - \frac{1}{2}t^2$.

The displacement PQ in the direction AB is

$$y - x = 80 - 8t - \tfrac{1}{2}t^2 - (12t + \tfrac{3}{2}t^2)$$
$$= 80 - 20t - 2t^2. \quad \text{[C]}$$

P and Q are 32 m apart when $y - x = 32$, that is, at time t_1 s, where

$32 = 80 - 20t_1 - 2t_1^2$.

This simplifies to $(t_1 - 2)(t_1 + 12) = 0$.

$\therefore$ the distance PQ is 32 m when $t = 2$ and when $t = -12$. [D]
✗ (complete answer) ✗ [E]
✓ $x - y = 32$ at time t_2 s, where

$-32 = 80 - 20t_2 - 2t_2^2$, that is

$t_2^2 + 10t_2 - 56 = 0$.

$(t_2 + 14)(t_2 - 4) = 0$. ✓

$\therefore$ the distance PQ is 32 m also when $t = 4$ and when $t = -14$. ■

$$\frac{dv}{dt} = -kv^2, \quad \text{where } k = \frac{1}{150}. \quad \text{[F]}$$

$$\therefore \frac{dt}{dv} = -\frac{1}{kv^2}.$$

$\therefore$ the time for v to change from 30 to 10 is

$$\int_{30}^{10} \left(-\frac{1}{kv^2}\right) dv = \left[\frac{150}{v}\right]_{30}^{10} \quad \text{[G]}$$

$$= 15 - 5 = 10 \text{ s.}$$ *25m* ■

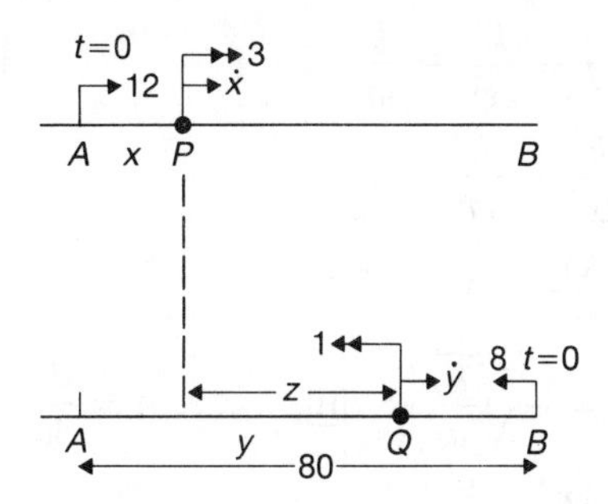

[A] The first task is to pick out the essential data from the first two paragraphs of the question and to enter them on a diagram so that we can see what is happening. It is possible to manage without the diagram, but dangerous.
[B] It is best to work in terms of AQ, not the distance BQ moved by Q, because it will then be easier to work out PQ without error.
[C] Alternatively let the displacement of Q relative to P be z m. Then

$\ddot{z} = (\text{accn of } Q) - (\text{accn of } P)$ (by 3.2F8)

$= -1 - 3 = -4$.

Initially $z = 80$ and $\dot{z} = -8 - 12 = -20$.
Integrating we obtain $\dot{z} = -4t - 20$,
and so $z = -2t^2 - 20t + 80$.

Consideration of the relative motion is often useful in problems like this, although in this example it does not make much difference to the work.
[D] We include $t = -12$ because the question does not state that t is not to be negative. Consequently you could lose a mark by omitting it, even though it can be seen to imply driving the vehicles in reverse.
[E] The *distance* PQ is $|y - x|$. We must remember to consider the case $x - y = 32$, which occurs after P and Q have passed one another.
[F] Introducing k to save undue copying of '150'. Here it is the velocity–time relation which is needed, so we use the method of 3.5F1.
[G] Using 3.5F2.

S6.

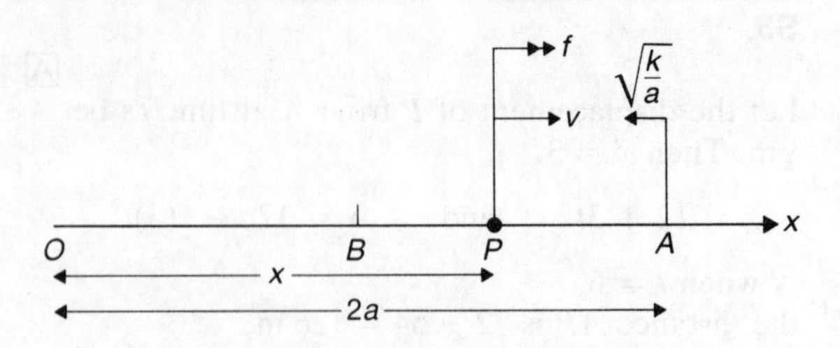

The acceleration $f = \frac{k}{2x^2} + \frac{k}{4a^2}$.

$$\therefore \frac{d}{dx}\left(\frac{1}{2}v^2\right) = \frac{k}{2x^2} + \frac{k}{4a^2}.$$

$$\therefore \frac{v^2}{2} = -\frac{k}{2x} + \frac{kx}{4a^2} + A. \quad \boxed{A}$$

When $x = 2a$, $v = -\sqrt{\frac{k}{a}}$. $\boxed{B}$ $\quad \therefore A = \frac{k}{4a}$.

$$\therefore \frac{v^2}{2} = -\frac{k}{2x} + \frac{kx}{4a^2} + \frac{k}{4a}.$$

When $x = a$, $\quad \frac{v^2}{2} = -\frac{k}{2a} + \frac{k}{4a} + \frac{k}{4a} = 0.$

✕ $\therefore$ the particle comes instantaneously to rest when $x = a$. ✕ $\boxed{C}$

When $v = 0$, $\quad 0 = -\frac{1}{2x} + \frac{x}{4a^2} + \frac{1}{4a}$,

that is, $\quad x^2 + ax - 2a^2 = 0.$

$$(x - a)(x + 2a) = 0. \quad \boxed{D}$$

✓ $\therefore$ the particle comes instantaneously to rest when $x = a$. ✓ ■

Subsequently $\frac{d}{dx}\left(\frac{1}{2}v^2\right) = \frac{k}{2x^2} - \frac{k}{4a^2}$.

$$\therefore \frac{v^2}{2} = -\frac{k}{2x} - \frac{kx}{4a^2} + \frac{3k}{4a} \quad (v = 0 \text{ when } x = a)$$

$$= \frac{k(-x^2 + 3ax - 2a^2)}{4a^2x} = \frac{-k(x-a)(x-2a)}{4a^2x}.$$

Therefore it next comes to rest when $x = 2a$, that is, at A. *25m* ■

$\boxed{A}$ Using 3.4F3

$\boxed{B}$ The positive direction is away from O, and the motion at A is towards O, so the negative root must be chosen. Because the results depend on v^2, not v, they would be unaltered by the choice of the positive root, but the argument would be wrong and you could thus lose marks.

$\boxed{C}$ We have shown that, if the particle ever reaches B, it will come to rest there. But we have not shown that it will get there at all, because the speed might become zero at some point between A and B. To decide this issue, we must find all the points where the expression for v^2 is zero.

$\boxed{D}$ v^2 can be zero only if $x = a$ or $x = -2a$. So there is *no* point between A and B at which the particle stops, and our conclusion is now valid.

S7.

$$f = -4v^{n+1}.$$

(a) $\frac{dv}{dt} = -4v. \qquad \therefore \frac{dt}{dv} = -\frac{1}{4v}.$

$\therefore t = -\frac{1}{4}\ln Av. \qquad$ (using the method of 3.5I1)

When $t = 0$, $v = u$. $\quad \ln Au = 0. \qquad \therefore Au = 1.$

$$\therefore t = -\frac{1}{4}\ln\frac{v}{u} = \frac{1}{4}\ln\frac{u}{v}.$$

$$\therefore \frac{u}{v} = e^{4t}. \quad \therefore v = ue^{-4t}. \text{ ■}$$

(b) $\frac{dv}{dt} = -4v^{n+1}$. $\quad \therefore \frac{dt}{dv} = -\frac{1}{4}v^{-n-1}$.

$$\therefore t = -\frac{1}{4}\left(\frac{1}{-n}\right)v^{-n} + A.$$

When $t = 0$, $v = u$. $\quad \therefore t = \frac{v^{-n}}{4n} - \frac{u^{-n}}{4n}$.

$$\therefore v^{-n} = u^{-n} + 4nt.$$

$$\therefore v = (u^{-n} + 4nt)^{-1/n} = u(1 + 4nu^{n}t)^{-1/n}. \blacksquare$$

(c) $v\frac{dv}{ds} = -4v^4$. $\quad \therefore \frac{ds}{dv} = -\frac{1}{4v^3}$.

$\therefore s = \frac{1}{8v^2} - \frac{1}{8u^2}$ $\quad$ because $v = u$ when $s = 0$.

$$\therefore v^{-2} = u^{-2} + 8s.$$

$$\therefore v = (u^{-2} + 8s)^{-1/2} = \frac{u}{\sqrt{1 + 8u^2s}}.$$

25m ■

Chapter 4

Rectilinear motion of a particle

4.1 GETTING STARTED

Particles act upon each other in various ways. Sometimes there is a mutual attraction, such as gravitation or a pull exerted by means of a string, and sometimes repulsion as in the case of like magnetic poles. These influences which bodies have on one another are called **forces**. They can be measured by many means, such as spring balances and by comparison with weights, or even by the accelerations to which they give rise. The study of the motions of bodies which exert forces on each other is simplified if we first think of a single particle which is subject to certain prescribed forces which we imagine are just present without explanation, or perhaps exerted by bodies, such as Earth, which are so massive that they can be assumed to be unmoved by the particle whose motion is being studied. In this chapter we treat the simplest case; that of a single particle which moves along a fixed straight line.

4.2 LAWS OF MOTION

ESSENTIAL FACTS

All the following Facts relate to a particle P of mass m moving along a fixed straight line subject to forces acting along the line.

F1. Momentum

The momentum of a particle of mass m moving with velocity v is mv.
Momentum has dimension MLT^{-1} (SI unit $kg\,m\,s^{-1}$).

F2. Newton's First Law of Motion

If no force acts on P, then the velocity of P is constant. Equivalently, P moves with constant momentum.

F3. Newton's Second Law of Motion

A particle of mass m subject to a single force F moves with acceleration f, where

$$F = mf.$$

This relation between the force and the acceleration is also called the **equation of motion** of P.
Equivalently, the force is equal to the rate of change of the momentum of the particle.
Force has dimension MLT^{-2}.
The SI unit of force is the **newton** (N) $\equiv$ (kg m s^{-2}).

F4. Composition of forces

A particle subject to two forces F_1 and F_2 moves with the acceleration f it would have under the action of a single force $F_1 + F_2$.

$$mf = F_1 + F_2.$$

Another way of stating this fact is to say:

The **resultant** of two forces acting on a particle is a force equal to the sum of the two forces.

F5.

The resultant force R of a set of n forces $F_1, F_2, \ldots, F_n$ acting on a particle is given by

$$R = F_1 + F_2 + \cdots + F_n.$$

F6. Equilibrium

Equilibrium is the state of permanent rest. By F4, if a particle is in equilibrium the resultant force acting on it is zero.

F7. Average force in a time interval

The average of the force F in the time interval $t_1 \leqslant t \leqslant t_2$ is

$$\frac{1}{t_2 - t_1} \int_{t_1}^{t_2} F \, dt.$$

F8. Impulse

The impulse of a force F in the time interval $t_1 \leqslant t \leqslant t_2$ is

$$I = \int_{t_1}^{t_2} F \, dt.$$

F9. Impulse and momentum The increase in momentum of a particle in a time interval is equal to the impulse of the resultant force acting on the particle during that interval. This follows from F3 by integration.

$$I = \int_{t_1}^{t_2} F\,dt = \int_{t_1}^{t_2} m\frac{dv}{dt}\,dt = \Big[mv\Big]_{t=t_1}^{t=t_2}.$$

Denote the velocities of P at t_1 and t_2 by v_1 and v_2 respectively. Then $I = mv_2 - mv_1$.
Impulse has dimension MLT^{-1} (SI unit $N\,s \equiv kg\,m\,s^{-1}$).

ILLUSTRATION

I1.

A particle P of mass 2 kg, moving along a straight line, is subject to two forces. One force has magnitude 8 N and acts towards a fixed point O of the line. The other force acts away from O and, at time t s, has magnitude $(4 + 6t)$ N. Initially P is at rest at a point A which is at a distance 5 m from O.

(a) Form the equation of motion of P. □

Choose Ox in the direction OA, and let x m be the coordinate of P. Then the forces which act on P are F_1 N and F_2 N, where

$$F_1 = -8 \quad \text{and} \quad F_2 = 4 + 6t.$$

$\ddot{x}$ $\dot{x}$ 8 $4+6t$ O B P A

By F4, $\quad 2\ddot{x} = F_1 + F_2 = -8 + 4 + 6t.$

Therefore the equation of motion is $\quad 2\ddot{x} = -4 + 6t. \quad (1)$ ■

(b) Find when the resultant force on P is zero, and when and where P comes to rest instantaneously. □

The resultant force is zero when $t = \frac{2}{3}$.

Integrate the equation of motion (1) with respect to t.

$$2\dot{x} = -4t + 3t^2 + 0 \qquad (2)$$

(because P is at rest, that is, $\dot{x} = 0$, when $t = 0$).

$$\therefore 2\dot{x} = t(3t - 4).$$

The particle comes to rest when $\dot{x} = 0$, that is, when $t = \frac{4}{3}$.

To find the displacement, integrate (2) with respect to t.

$$2x = -2t^2 + t^3 + 10 \qquad \text{(because } x = 5 \text{ when } t = 0\text{)}.$$

When $t = \frac{4}{3}$, $\quad x = -\left(\frac{4}{3}\right)^2 + \frac{1}{2}\left(\frac{4}{3}\right)^3 + 5 = \frac{119}{27}.$

$\therefore$ P comes to rest at B, when its displacement from O is $\frac{119}{27}$ m. ■

(c) Find the impulse of each of the forces in the interval $1 \leqslant t \leqslant 3$. □

The impulse of F_1 in the time interval is I_1 N s, where

$$I_1 = \int_1^3 (-8)\,dt = -16.$$

Note that in this case, because the **force is constant**,

$$(\text{impulse}) = (\text{force}) \times (\text{time interval}).$$

For F_2, similarly, we have

$$I_2 = \int_1^3 (4 + 6t)\,dt = \Big[4t + 3t^2\Big]_1^3 = 32.$$

Therefore the impulse of F_2 is 32 N s. ■

Here, as in **most cases** where impulse is used, particularly in collisions, the **force is not constant**, and so it is **not correct to use the relation (impulse) = (force) × (time interval).**

Check: From (2), the increase in momentum of P in the time interval $1 \leqslant t \leqslant 3$ is

$$\Big[2\dot{x}\Big]_{t=1}^{t=3} = \Big[-4t + 3t^2\Big]_1^3 = 16\ \text{kg m s}^{-1}.$$

This is equal to the sum of the impulses of the two forces:

$$-16 + 32 = 16, \text{ which agrees.}$$

4.3 ENERGY

ESSENTIAL FACTS

A particle P of mass m moves on the x-axis and is subject to forces acting along the x-axis. At time t the coordinate of P is x and the velocity of P is v.

F1. Kinetic energy

The kinetic energy of P is $\frac{1}{2}mv^2$.
Energy has dimension ML^2T^{-2}.
The SI unit of energy is the **joule** ($J = \text{kg m}^2\text{s}^{-2}$)

F2. Power

The power exerted by a force F acting on P is Fv.
Power has dimension ML^2T^{-3}.
The SI unit of power is the **watt** ($W \equiv \text{kg m}^2\text{s}^{-3} \equiv \text{J s}^{-1}$).

F3. Work

The work done in an interval $t_1 \leqslant t \leqslant t_2$ by a force which exerts power S is $\int_{t_1}^{t_2} S\,dt$.

Work has dimension ML^2T^{-2}. The SI unit of work is the **joule** (J).

F4. Rate of working

Let the work done by a force F from a fixed instant to time t be W. Then:

The power exerted by F is the rate of working of F: $$S = \frac{dW}{dt}.$$

F5. Work done in a displacement

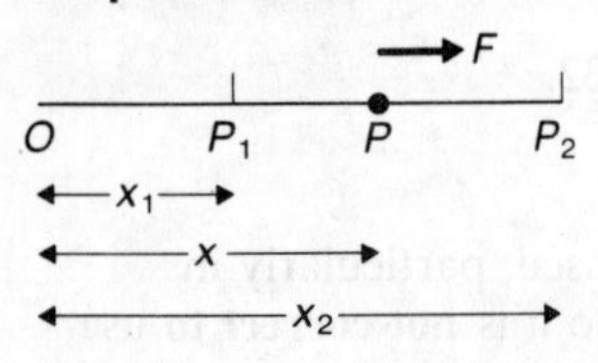

A particle P undergoes a displacement from $P_1(x_1)$ to $P_2(x_2)$. During the displacement a force, which takes the value F when the coordinate of P is x, acts on P. Then:

The work done by F during the displacement from P_1 to P_2 is equal to $\int_{x_1}^{x_2} F\,dx$.

F6. Principle of energy

The increase in the kinetic energy of P during an interval is equal to the total work done in that interval by all the forces acting on P.

F7. Potential energy

In the case when a force F acting on P **depends only on** x, the potential energy V of the force F is defined as **minus the work done by** F in a displacement to P from some fixed point. Consequently:

Potential energy $$V = -\int F\,dx,$$

in which the constant of integration may be determined by arranging for V to be zero at the chosen fixed point.

Equivalently, $$F = -\frac{dV}{dx}.$$

F8. Conservation of energy

In the case when V is the potential energy of the *resultant* force acting on a particle, V is called the **potential energy of the particle**. It then follows from the principle of energy (F6) that

$$\tfrac{1}{2}mv^2 + V = E \text{ (constant)}$$

Kinetic energy of P + potential energy of P = total energy of P.

This is the **principle of conservation of energy**.

The total energy E is constant throughout the motion, but its constant value depends on the constant of integration chosen in the potential energy V (see F7).

F9.

As a particle P of mass m moves from a point P_1 to a point P_2 its potential energy changes from V_1 to V_2 and its speed from v_1 to v_2. By F8,

$$\tfrac{1}{2}mv_2^2 + V_2 = \tfrac{1}{2}mv_1^2 + V_1.$$

Equivalently, $\tfrac{1}{2}mv_2^2 - \tfrac{1}{2}mv_1^2 = V_1 - V_2$,

that is, (Gain in K.E.) = (loss in P.E.)
in the displacement from P_1 to P_2.

F10. If the forces acting on P are $F_1, F_2, \ldots F_n$ with corresponding potential energies $V_1, V_2 \ldots V_n$, then the principle of conservation of energy may be written

$$\tfrac{1}{2}mv^2 + V_1 + V_2 + \cdots + V_n = E.$$

ILLUSTRATIONS

I1. A particle P of mass 3 kg moves on the x-axis so that at time t s its displacement x m from the origin O is given by $x = t^3 - t^2$. It is subject to two forces; F_1 N, where $F_1 = 6t$, and a force F_2 N.

(a) Find F_2 in terms of t. □

By 4.2F4, $3\ddot{x} = F_1 + F_2$.

Differentiate x twice with respect to t.

$\dot{x} = 3t^2 - 2t.$ $\ddot{x} = 6t - 2.$

$\therefore 3(6t - 2) = 6t + F_2.$

$\therefore F_2 = 12t - 6.$ ■

(b) Determine the power of the force F_1 at time t s and the work done by F_1 in the time interval $1 \leqslant t \leqslant 2$. □
Let the power of F_1 be S_1 W.

By F2, $S_1 = F_1\,\dot{x} = 6t(3t^2 - 2t).$

$\therefore$ the power of the force F_1 at time t s is $6(3t^3 - 2t^2)$ W.

Let the work done by the force F_1 in the interval be W_1 J.

By F3, $$W_1 = \int_1^2 S_1\,dt = 6\int_1^2 (3t^3 - 2t^2)\,dt = \left[\frac{9t^4}{2} - 4t^3\right]_1^2.$$

$\therefore$ the work done by F_1 is $40 - \dfrac{1}{2} = 39{\cdot}5$ J. ■

(c) Find the work done in the interval $1 \leqslant t \leqslant 2$ by the force F_2. □
Knowing F_2, we can use the method in (b) to find the work, W_2, done by F_2. However, having found W_1, we can avoid another integration by proceeding as follows.
Let the velocity of P be v_1 m s^{-1} when $t = 1$ and v_2 m s^{-1} when $t = 2$.
By the principle of energy F6,

$$\tfrac{1}{2}(3)\,v_2^2 - \tfrac{1}{2}(3)\,v_1^2 = W_1 + W_2.$$

$$\therefore \tfrac{3}{2}\Big[(3t^2 - 2t)^2\Big]_1^2 = 39{\cdot}5 + W_2.$$

$\therefore$ the work done by F_2 is $\tfrac{3}{2}(64 - 1) - 39{\cdot}5 = 55$ J. ■

I2. Obtain a potential energy for each of the following forces acting along the x-axis on a particle P with coordinate x m on the axis, where $x > 0$.

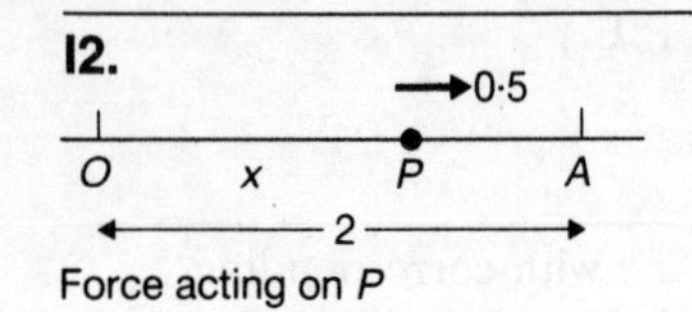

Force acting on P

(a) A force 0·5 N along Ox. □

$$F = \tfrac{1}{2}.$$

$$\therefore V = -\int (\tfrac{1}{2})\,dx = -\tfrac{1}{2}x + C, \qquad \text{where } C \text{ is a constant.}$$

Any value may be chosen for C. If it suits us to make $V = 0$ when $x = 0$, we take, in this problem, $C = 0$, and consequently $V = -\tfrac{1}{2}x$.

If, however, the point A, where $OA = 2$ m, happened to be an important reference point in our problem, we might find it convenient to choose A as the **zero of potential energy**, that is, the point at which $V = 0$. The potential energy would then appear as $V = 1 - \tfrac{1}{2}x$. ■

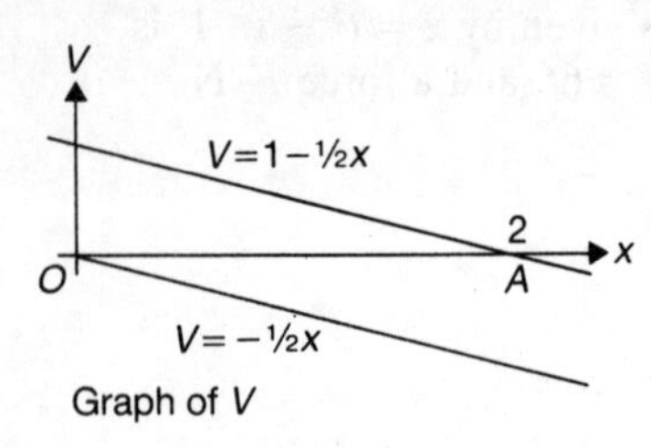

Graph of F

Graph of V

(b) A force $2x$ N towards O. □

$$F = -2x.$$

$$\therefore V = -\int (-2x)\,dx = x^2,$$

taking O as the zero of potential energy. ■

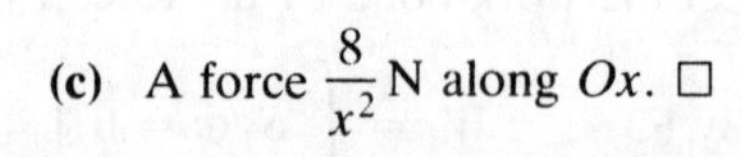

Force acting on P

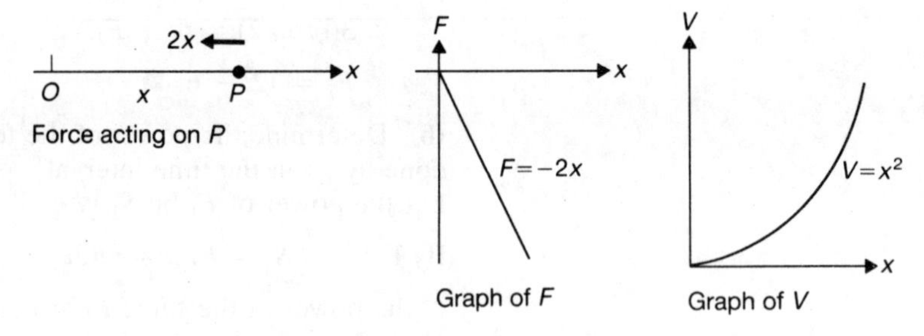

Graph of F Graph of V

(c) A force $\dfrac{8}{x^2}$ N along Ox. □

$$F = \frac{8}{x^2}.$$

$$\therefore V = -\int \frac{8}{x^2}\,dx = \frac{8}{x} + C. \ ■$$

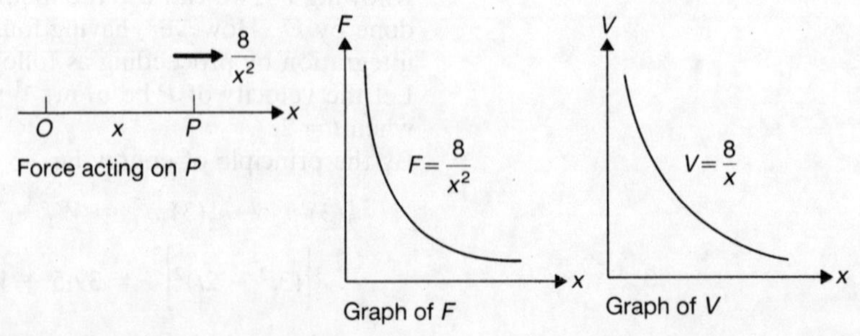

Force acting on P

Graph of F Graph of V

If we take $C = 0$, which is likely to be convenient in this case, then $V = \frac{8}{x}$, and there is no value of x for which $V = 0$.
As $x \to \infty$, $V \to 0$, and so it is customary to say 'the potential energy is zero at infinity'.

I3.

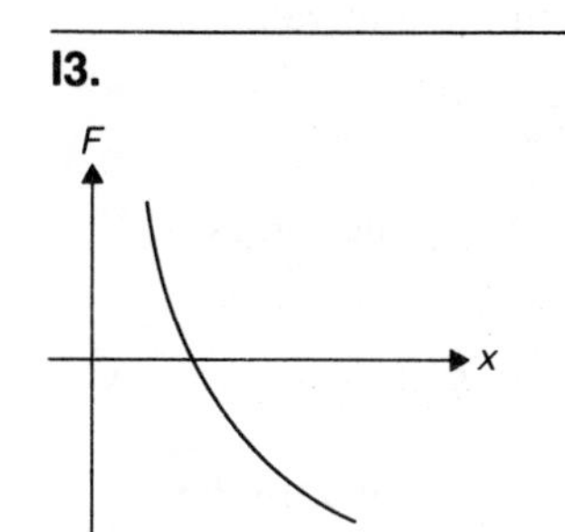

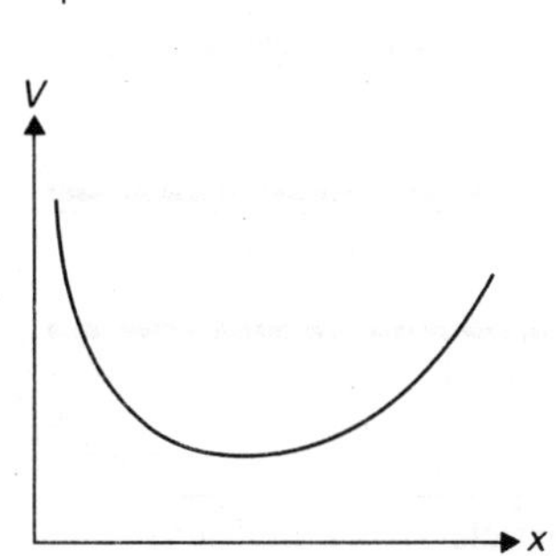

A particle P of mass 4 kg is subject to a force $\left(-2x + \frac{8}{x^2}\right)$ N along Ox. When the displacement of P from O is 2 m the velocity of P is 4 m s^{-1}. Obtain an equation of conservation of energy for P and find the speed of P when the displacement from O to P is 4 m. □
Following the method of I2b,c, we take the potential energy of P as $V = x^2 + \frac{8}{x}$.
Let the velocity of P be $v\text{ m s}^{-1}$. Then the energy equation is

$$(\tfrac{1}{2})4v^2 + \left(x^2 + \frac{8}{x}\right) = E, \qquad \text{where } E \text{ is constant.}$$

When $x = 2$, $v = 4$. $\quad \therefore \frac{1}{2}(4)4^2 + 4 + \frac{8}{2} = E. \quad \therefore E = 40.$
$\therefore$ the equation of conservation of energy is

$$2v^2 + x^2 + \frac{8}{x} = 40.$$

When $x = 4$, $\quad 2v^2 + 16 + 2 = 40. \quad \therefore 2v^2 = 22.$
Hence the speed when $x = 4$ is $\sqrt{11}\text{ m s}^{-1}$. ■
Note that in this example P never reaches a point at which $V = 0$. We could make any point we wish the zero of potential energy by adding a suitable constant to V, but, as we see, any required further results can be derived without doing so.

I4.

Deduce F5 from F2 and F3. □
Suppose $t = t_1$ when $x = x_1$ and $t = t_2$ when $x = x_2$. Then

$$\text{Work done} = \int_{t_1}^{t_2} S\,\mathrm{d}t = \int_{t_1}^{t_2} Fv\,\mathrm{d}t = \int_{t_1}^{t_2} F\frac{\mathrm{d}x}{\mathrm{d}t}\mathrm{d}t = \int_{x_1}^{x_2} F\,\mathrm{d}x. \blacksquare$$

I5.

Assuming F4 and F5, deduce F2. □

From F4, $\quad S = \frac{\mathrm{d}W}{\mathrm{d}t}$. $\qquad$ From F5, $\quad \frac{\mathrm{d}W}{\mathrm{d}x} = F$.

$$\therefore S = \frac{\mathrm{d}W}{\mathrm{d}t} = \frac{\mathrm{d}W}{\mathrm{d}x}\frac{\mathrm{d}x}{\mathrm{d}t} = Fv. \blacksquare$$

I6. Deduce the principle of energy (F6) from the laws of motion. □

Let $v = v_1$ when $t = t_1$ and $v = v_2$ when $t = t_2$.

The rate of change of kinetic energy is $\frac{d}{dt}(\frac{1}{2}mv^2) = mv\frac{dv}{dt}$.

Let the forces acting on P be $F_1, F_2, \ldots, F_n$, doing work $W_1, W_2, \ldots, W_n$ respectively, in the interval.

By 4.2F5, $\quad m\frac{dv}{dt} = F_1 + F_2 + \cdots + F_n.$

Multiply by v and use F2, denoting the powers of the forces by $S_1, S_2, \ldots, S_n$.

$$\therefore \frac{d}{dt}(\tfrac{1}{2}mv^2) = (F_1 + F_2 + \cdots + F_n)v = S_1 + S_2 + \cdots + S_n.$$

Integrate both sides with respect to t from t_1 to t_2, and use F3.

$$\tfrac{1}{2}mv_2^2 - \tfrac{1}{2}mv_1^2 = W_1 + W_2 + \cdots + W_n = \text{total work done.}$$

4.4 PHYSICAL FORCES

ESSENTIAL FACTS

F1. Newton's Third Law

The force exerted on a particle B by a particle A is equal and opposite to the force exerted on A by B, and acts in the line AB.

F2. Tension

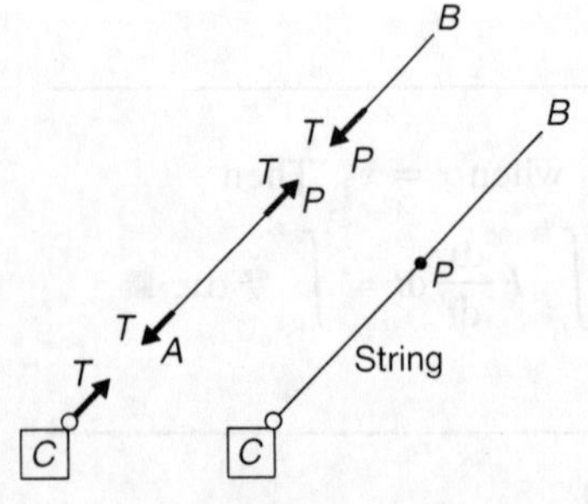

Let AB be a piece of string which is taut. At any point P of the string the segment PB of the string exerts on the segment AP a force T in the direction AB. By F1, the segment AP exerts on the segment PB a force T in the direction BA. The force T is called the tension in the string at the point P.

If the string is **light**, and if no other force acts on it at any point between A and B, the tension has the same value at all points of the string in AB.

If the string is fastened at A to a body C, then the string exerts a force on C which is equal to T in the direction AB.

The tension in a string is always positive or zero, and never negative.

F3. Tension and thrust

R R R P R R R
A B

A **light rod**, freely pivoted at its ends, is like a string, except that the tension can be positive or negative. Negative tension implies that the rod is subject to a **compressive force**, R, which is called the **thrust** in the rod; $R = -T$.

F4. Reaction

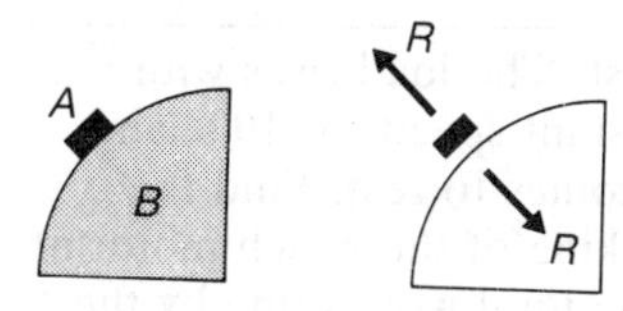

When a particle A is in contact with the surface of a rigid body B, the force, R, exerted by the body on the particle is called a reaction. The force exerted by A on B is, by F1, equal and opposite to this reaction.

While contact continues, the reaction exerted by B on A must be directed outwards from the surface.

The reaction of a **smooth** surface is always along the **normal** (perpendicular) to the surface.

F5. Resistance

A particle moving through a fluid or along a surface may be subject to a **resisting force** exerted by the fluid or surface. The resisting force always acts in a **direction opposite to the velocity** of the particle. The **magnitude of the resisting force** is called the **resistance**.

By F1, the particle exerts an equal and opposite force on the resisting material.

F6. Weight

The force of gravity acting on a particle of mass m near the surface of the earth is called the weight of the particle. It is equal to mg vertically downwards, where g is the **acceleration due to gravity**.

A particle in a vacuum, isolated from all other forces, moves with downward acceleration g.

The acceleration due to gravity is for all practical purposes constant in any one locality. It varies from about $9{\cdot}78\ \mathrm{m\,s^{-2}}$ at the Equator to $9{\cdot}83\ \mathrm{m\,s^{-2}}$ at the Poles.

F7. Gravitational energy

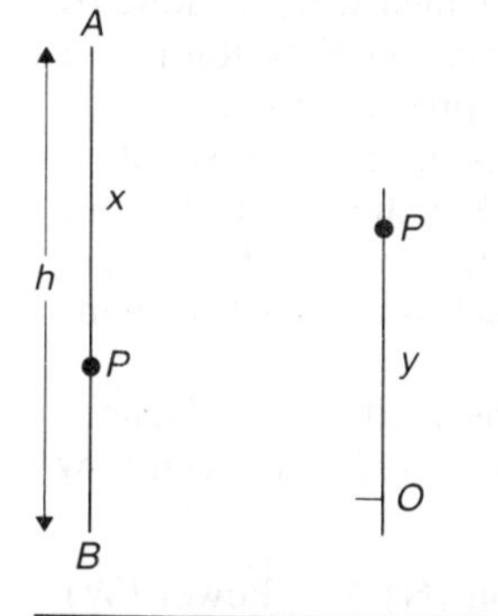

The work done by gravity on a particle P of mass m when P moves vertically downwards through a displacement h is

$$\int_0^h mg\,\mathrm{d}x = mgh.$$

When P is at a height y above a point O (chosen as the zero level) the **gravitational potential energy** of P is mgy.

F8. Motion on an incline

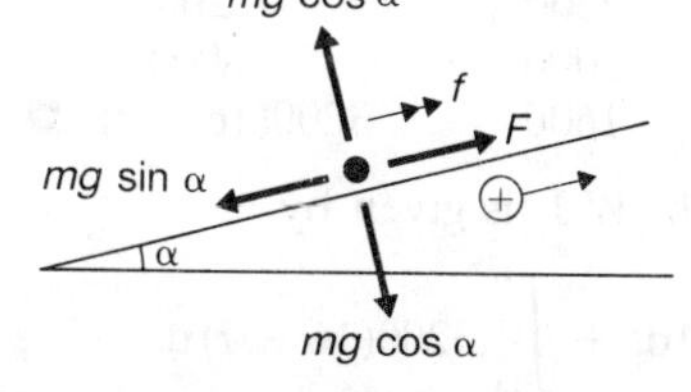

When a particle moves on a straight line inclined at an angle α to the horizontal, the forces causing its acceleration f up the line are the weight component $-mg\sin\alpha$ together with the sum, F, of the components along the line of all the other forces acting.

$$mf = F - mg\sin\alpha.$$

ILLUSTRATIONS

In the following Illustrations we take $g = 10 \text{ m s}^{-2}$.

I1. A winch raises a load of 200 kg from rest. The load rises with acceleration 1 m s^{-2} for 4 s, then at constant speed for 10 s, and finally with retardation 2 m s^{-2} until it comes to rest. Find the tension in the cable and the rate of working of the winch motor at all times during the ascent. Calculate the total work done by the winch. □

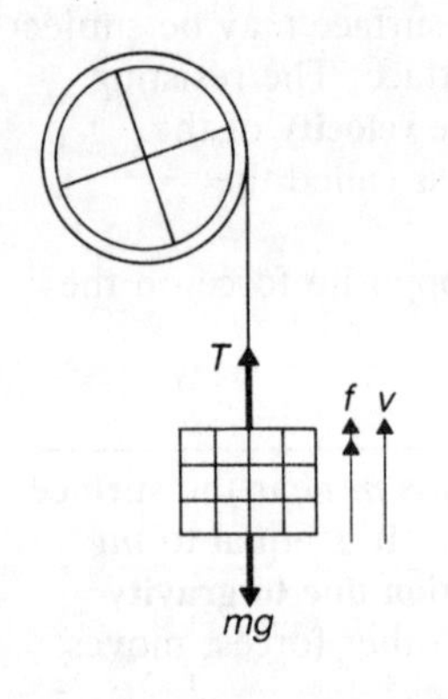

Let the tension in the cable be T N. Then, with the other symbols defined as shown in the diagram, the equation of motion of the load is

$$T - mg = mf.$$

We have $m = 200$, and f takes 3 values successively. Therefore, at time t s after the start of the motion:

$0 < t < 4$: $f = 1$; $T = m(g + f) = 200\,(10 + 1)$;
$\therefore$ tension $= 2200$ N,

$4 < t < 14$: $f = 0$, so $T = mg$; $\therefore$ tension $= 2000$ N,

$14 < t < t_0$ (time when it stops): $f = -2$; $T = 200(10 - 2)$;
$\therefore$ tension $= 1600$ N. ■

We can show the value of T on a graph. First determine t_0.
Velocity acquired during acceleration $= (1)4 = 4 \text{ m s}^{-1}$.
Time taken for speed to fall to zero at retardation 2 m s^{-2}

is $\dfrac{4}{2} = 2$ s. $\quad\therefore\ t_0 = 16.$

Let the rate of working of the winch motor be S W. 'The rate of working of the winch' means the rate at which work is done by the winch on the moving load. This is the usual convention in the statement of problems about engines which provide power to vehicles and machinery. S is not the same as the rate (Z W) at which the winch motor is using up energy. For the motor will not be 100% efficient. It will radiate energy in the form of heat. The **efficiency** of the motor is the ratio S/Z, which is always less than one.

The force which the winch exerts on the load is T N. Hence the power supplied by the winch is given by $S = Tv$. We can now express the result in tabular form.

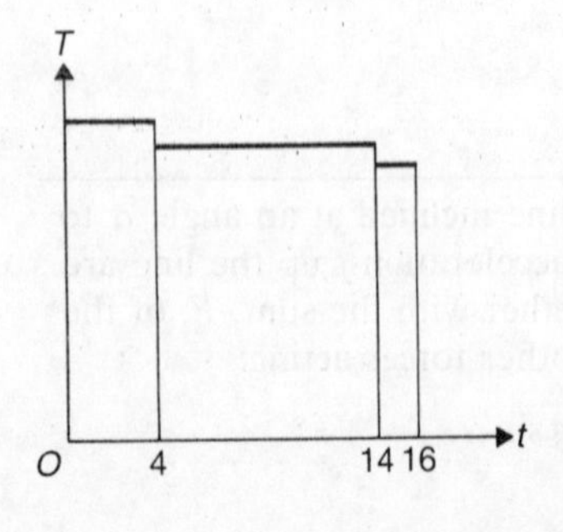

Time (s)	Velocity (m s^{-1})	Tension (N)	Power (W)
t	v	T	H
$0 < t < 4$	t	2200	$2200t$
$4 < t < 14$	4	2000	8000
$14 < t < 16$	$2(16 - t)$	1600	$3200(16 - t)$. ■

The total work done by the winch, W J, is given by

$$W = \int_0^{16} S\,dt = \int_0^4 2200t\,dt + \int_4^{14} 8000\,dt + \int_{14}^{16} 3200(16 - t)\,dt.$$

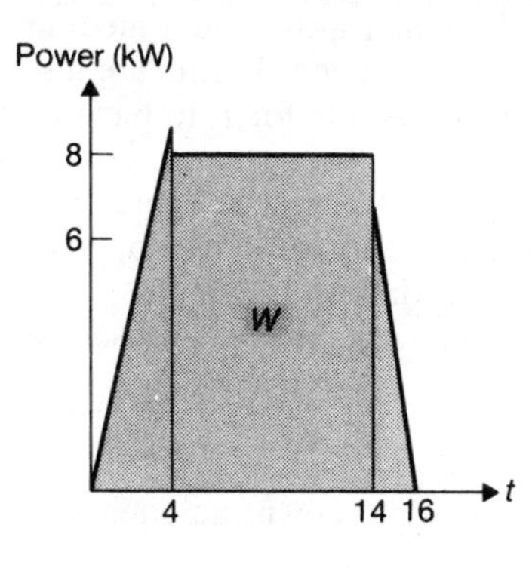

These integrals can be evaluated either analytically or by finding the area of the shaded region under the S–t graph. We obtain
Total work done $= \frac{1}{2}(4)(8{\cdot}8) + (10)(8) + \frac{1}{2}(2)(6{\cdot}4) = 104$ kJ. ■

Alternatively, using the principle of energy (4.3F6),
(work done by T) + (work done by mg) = increase in K.E.
$\therefore$ (work done by the winch) $- mgh = 0 - 0$ (using F7),
where h m is the height through which the load is raised.
Using 3.3F7, we find

$$h = \tfrac{1}{2}(1)4^2 + 4(10) + \tfrac{1}{2}(2)2^2 = 52.$$

Hence the work done is $mgh = 200 \times 10 \times 52 = 104\,000$ J.
$= 104$ kJ.

12.

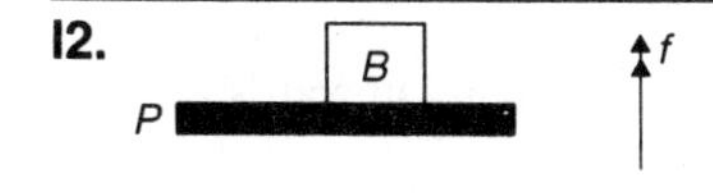

A machine causes a horizontal platform P of mass M to move vertically with upward acceleration f. A box B of mass m rests on the platform. Find

(i) the force exerted on the platform by the box,
(ii) the force which is being applied to the platform by the machine,
(iii) the condition that the box remains on the platform. □

(i) Let the force exerted by B on the platform be R downwards. Then, by F1, the force exerted *on B by P* is R upwards. The only other force acting on B is its weight, mg downwards. Therefore the equation of motion of the box is

$$R - mg = mf.$$

$$\therefore R = m(g + f).$$

Therefore the force exerted on the platform by the box is $m(g + f)$. ■

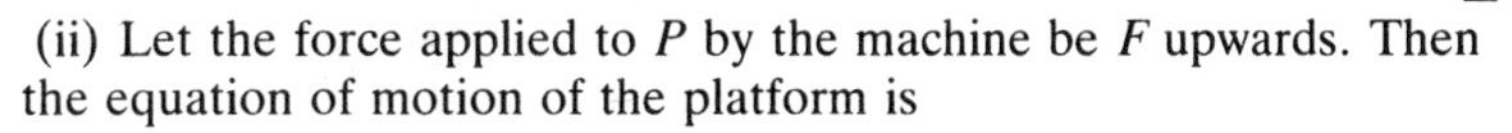

(ii) Let the force applied to P by the machine be F upwards. Then the equation of motion of the platform is

$$F - R - Mg = Mf.$$

$$\therefore F = R + M(g + f) = m(g + f) + M(g + f).$$

Therefore the force applied to the platform is $(M + m)(g + f)$. ■

Note that this result could be obtained also by treating the (box + platform) as a single particle of mass $M + m$.

(iii) Unless the platform is sticky, the force it can exert on the box cannot act downwards. $\therefore R \geqslant 0$.
Consequently, if $m(g + f) < 0$, the assumption that the box has the same acceleration f as the platform is not valid. Hence the condition for the box to remain on the platform is that $g + f \geqslant 0$, that is $f \geqslant -g$.
In words, the downward acceleration of the platform must not exceed g. ■

I3. A particle P of mass m moves vertically under gravity in a medium which offers a resistance proportional to the speed. Formulate an equation of motion for P. Show that it is possible for P to have a constant downward velocity. □

Suppose the resistance is k times the speed. Resistance being the magnitude of the resisting force, k is necessarily positive. Suppose P is at a depth x below a fixed point O at time t.

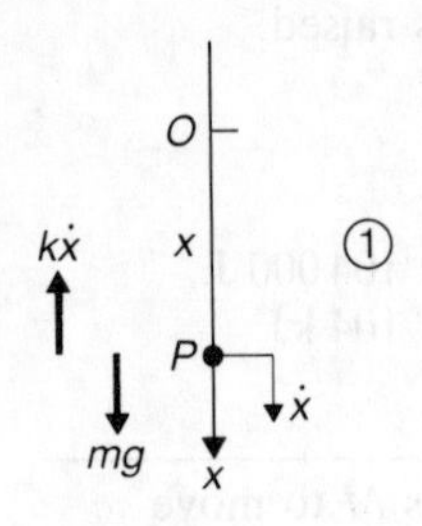

Downward motion: When P is descending, x is positive and so the speed is $\dot{x}$. Since k also is positive:

Resisting force is $k\dot{x}$ upwards.

This is shown in diagram 1. We can now write down the equation of motion.

$$mg - k\dot{x} = m\ddot{x}. \tag{1}$$

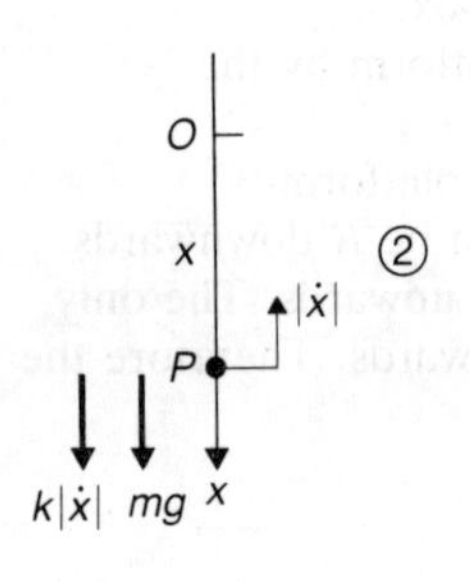

Upward motion: In diagram 2, P is shown moving upwards at speed $|\dot{x}|$. In this case, therefore:

Resisting force is $k|\dot{x}|$ downwards.

Since $\dot{x}$ is now negative, this is the same as a force $k\dot{x}$ upwards. So diagram 1 is correct also for the upward motion, and (1) remains valid. We conclude that the equation of motion is

$$mg - k\dot{x} = m\ddot{x},$$

whichever way the particle is going. ■

Suppose the particle has zero acceleration when $\dot{x} = V$.

Then $\quad mg - kV = 0.$

$$\therefore V = \frac{mg}{k}.$$

Hence P can have a constant downward velocity of magnitude $\dfrac{mg}{k}$. ■

This is called the **terminal speed** for the particle in the medium, because when the particle moves under gravity in the medium its speed ultimately tends to the value V.

I4. Consider the case of a particle as in I3, but subject to resistance proportional to the *square* of its speed. □

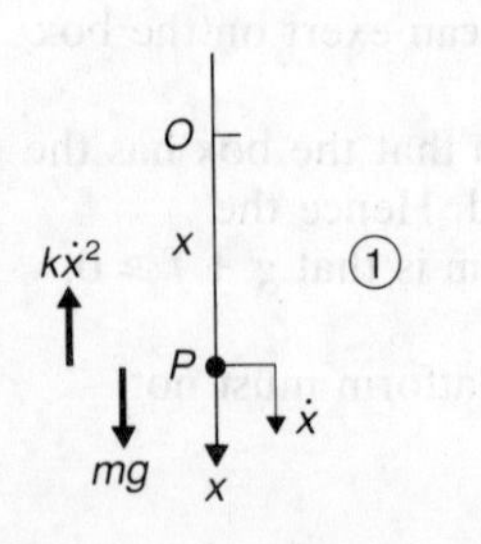

Downward motion: The resisting force is $k\dot{x}^2$ upwards (diagram 1). Hence the equation of motion is

$$mg - k\dot{x}^2 = m\ddot{x}, \qquad \text{for } \dot{x} > 0.$$

Upward motion: The resisting force is $k\dot{x}^2$ downwards (diagram 2). The modulus sign is not needed because $\dot{x}^2$ is the same as $|\dot{x}|^2$. Hence the equation of motion is

$$mg + k\dot{x}^2 = m\ddot{x}, \qquad \text{for } \dot{x} < 0. \ ■$$

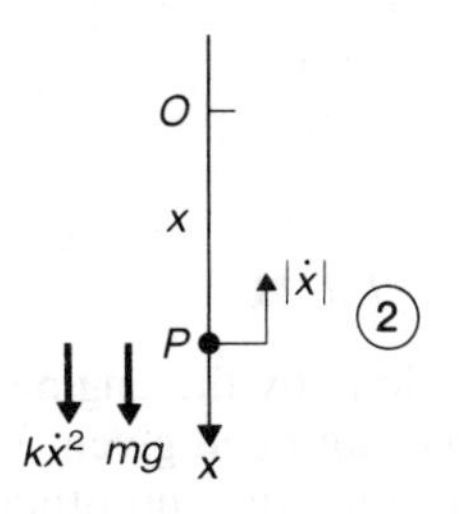

Note that the equations governing the downward and the upward motions differ from one another. This is so in all problems on resisted motion except when, as in I3, the formula giving the resistance in terms of the speed is one which corresponds to an *odd* function. For example, the equations would not differ if the resistance had the form $\alpha|\dot{x}| + \beta|\dot{x}|^3$.

The terminal speed in this Example is $V = \sqrt{\dfrac{mg}{k}}$. ■

I5.

A car of mass m ascends an incline at an angle α to the horizontal. At time t, the distance travelled from a point O on the road is x, the speed is v, the acceleration up the hill is f, the resistance to the motion is R, and the rate of working of the engine is P.

(**a**) Write down the equation of motion of the car. □

We interpret the 'rate of working of the engine' as the power contributed by the engine to the motion of the vehicle as a whole. Let the corresponding pull exerted on the car (which is called the **tractive force**) be T.

Then the power $P = Tv$.

The forces acting on the car in the line of motion, up the hill, are:

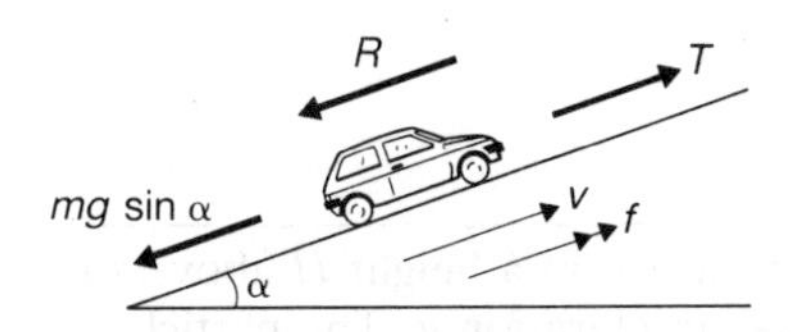

The tractive force	$T = \dfrac{P}{v}$	
The weight component	$-mg\sin\alpha$	(from F8)
The resisting force	$-R$	

There are also forces which act perpendicular to the road surface, but we know that they will cancel out. For the car moves only on the road, and so the resultant force on the car is along the road. Hence the equation of motion of the car is

$$mf = \frac{P}{v} - mg\sin\alpha - R. \blacksquare$$

(**b**) Given: R is a known constant and the car increases its speed from v_1 to v_2 while travelling a distance d up the road. Find:

(i) the total work done by the forces acting on the car,
(ii) the work done against gravity,
(iii) the work done against the resistance,
(iv) the work done by the engine. □

(i) By 4.3F6, the total work W done is the increase in kinetic energy:

$$W = \tfrac{1}{2}mv_2^2 - \tfrac{1}{2}mv_1^2\,. \blacksquare$$

(ii) Let the initial displacement of the car from O be x_0. Then the work done by gravity

$$W_1 = \int_{x_0}^{x_0+d} (-mg\sin\alpha)\,\mathrm{d}x = -mgd\sin\alpha.$$

$\therefore$ the work done *against* gravity is $W_1 = mgd\sin\alpha$. ■

(iii) The work done by the resistance is

$$W_2 = \int_{x_0}^{x_0+d} (-R)\,dx = -Rd.$$

$\therefore$ the work done against the resistance is $-W_2 = Rd$. ■

(iv) We cannot directly calculate the work W_3 done by the engine because the power P (which may be variable) has not been given. But we can obtain it from the results of (i)–(iii). For, since no other forces are acting, the total work done $W = W_1 + W_2 + W_3$.
$\therefore$ the work done by the engine is $W_3 = W - W_1 - W_2$, which is $\frac{1}{2}m(v_2^2 - v_1^2) + mgd\sin\alpha + Rd$. ■
Notice that this can also be written

Work done by engine
= K.E. gain + total work done *against* resistance and gravity.

4.5 EXAMINATION QUESTIONS AND SOLUTIONS

Q1. A particle of mass m is dropped from rest at a height H above soft ground and falls under gravitational acceleration g. The particle comes to rest at a depth D below the surface.
(i) What is the potential energy of the particle relative to the ground when it is at height H?
(ii) Find the kinetic energy of the particle when it hits the ground.
(iii) Assuming that the resistive force R exerted by the ground on the particle is constant, find D (the depth to which the particle penetrates). (NI 1984)

Q2. A cyclist moves against a resistance which is proportional to his speed. When the cyclist is working at the rate of 64 W, his maximum speed along a horizontal road is 4 m s^{-1}. Given that the total mass of the cyclist and his machine is 60 kg, find the maximum speed, in m s^{-1}, which the cyclist reaches when he travels down a slope of angle θ, where $\sin\theta = 1/30$, when he is working at the rate of 56 W.

Show also that, when he is moving along a horizontal road with speed 2 m s^{-1} and working at the rate of 46 W, his acceleration is 0·25 m s^{-2}.

[Take g as 10 m s^{-2}] (LON 1983)

Q3.

A lift which, when empty, has mass 600 kg is carrying a parcel of mass 50 kg. The lift is ascending with an upward acceleration of $0{\cdot}8\ \mathrm{m\,s^{-2}}$. Calculate the tension in the lift cable and the reaction of the floor of the lift on the parcel.

[*Take g as* $9{\cdot}8\ \mathrm{m\,s^{-2}}$, *and ignore friction.*]

During this motion of the lift, a piece of loose material falls from the roof of the lift and strikes the floor. How long does this take given that the roof is 3 m above the floor? Calculate also the equivalent time for this to happen when the lift has a downward acceleration of $0{\cdot}8\ \mathrm{m\,s^{-2}}$.

The lift cable should not bear a tension greater than 12 000 N. What is the maximum number of parcels of mass 50 kg which the lift can safely carry up when its acceleration is $0{\cdot}8\ \mathrm{m\,s^{-2}}$?

(AEB 1986)

Q4.

A body of mass m is released from rest and falls under gravity against air resistance. The body reaches a speed v after falling through a height h. Find the work done by the body against the air resistance. (LON 1982)

Q5.

A fast cruiser is propelled at a constant speed of $60\ \mathrm{km\,h^{-1}}$ when its engines deliver power of 30 000 kW. Calculate the resistance to the motion of the cruiser.

Given that the resistance to motion is proportional to the square of the speed of the cruiser, calculate the resistance when the cruiser is moving at a speed of $72\ \mathrm{km\,h^{-1}}$. (AEB 1986)

Q6.

A submarine of mass m, when moving in a horizontal line Ox with speed v, experiences a resistance kv^2, where k is a constant. The power P of the engines is constant and is just sufficient to maintain a speed u. Show that

$$mv^2\frac{\mathrm{d}v}{\mathrm{d}x} = \frac{P}{u^3}(u^3 - v^3).$$

Show also that the distance moved while the submarine accelerates from a speed $\frac{1}{4}u$ to a speed $\frac{3}{4}u$ is

$$\frac{mu^3}{3P}\ln\frac{63}{37}.$$

Find the time taken for the speed of the submarine to reduce from $\frac{3}{4}u$ to $\frac{1}{4}u$ when moving against the resistance with no power supplied by the engines. (JMB 1983)

Q7. The resistance to the motion of a train of total mass 300 tonnes is 18 000 N. The greatest tractive force which the engine can exert is 30 000 N and the greatest power available is 450 kW. The train starts from rest and moves along a level line with the greatest possible acceleration. Show that the engine first develops its maximum power when the speed is 54 km h^{-1}. Show also that the speed of the train cannot exceed 90 km h^{-1}.

Find the acceleration of the train in m s^{-2} when its speed is 72 km h^{-1}. (JMB 1985)

Q8. An electric train of mass M kg moves from rest along a straight level track. The tractive force of the motors, initially P N, decreases uniformly with time to R N over a period of T s and then remains constant at R N. The total resistance to motion is R N. Show that the acceleration a of the train at time t seconds after it starts to move is given, for $t \leqslant T$, by

$$Ma = P + (R - P)t/T - R.$$

Find the maximum speed achieved by the train, the distance it travels before reaching that speed and show that its average speed over that distance is $\frac{2}{3}$ of the maximum speed.

Find the power developed by the motors (i) at time $T/2$, (ii) at time $3T/2$. (WJEC 1983)

SOLUTIONS

S1.

Potential energy relative to the ground when the particle is at A is mgH. [A] ■

Kinetic energy at B (on the ground) is equal to the loss of potential energy in falling from A to B. [B]

$\therefore$ K.E. at ground $= mgH$. ■

In the displacement from B to C, where the particle stops,

Work done: by gravity $= mgD$,
by resistance $= -RD$.

Increase in K.E. $= -mgH$. [C]

$\therefore -mgH = mgD - RD$.

$\therefore D = \dfrac{mgH}{R-mg}$. [D] *11m* ■

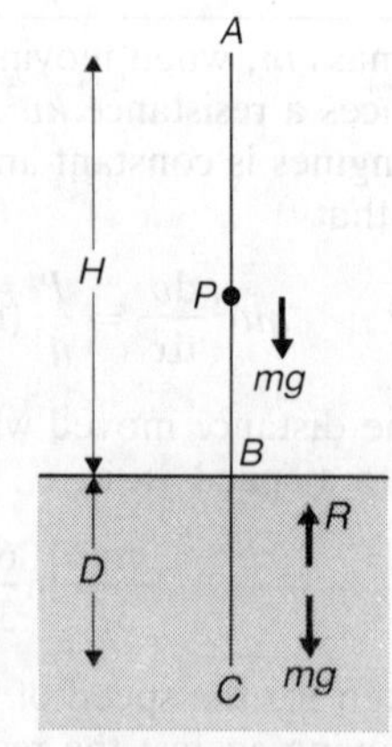

[A] ‘Relative to the ground’ means taking B as the zero of potential energy. The force (mg downwards) is constant, and so V decreases by an amount mgH as P falls from A to B (see 4.4F7).

[B] Using 4.3F9.

[C] The question strongly hints that an energy argument is suitable, and so we prepare to use the principle of energy (4.3F6). Where there is *resistance* it is best not to try to use energy *conservation* (4.3F8), because resistance always removes energy from the system.

[D] Check any answer which could be zero or negative. If D is to be the penetration depth it must be positive, and so the result makes sense only if $R > mg$. This is quite in accordance with reality, because the resistance must exceed the weight or the particle would never stop sinking, as in a bog which offers very low resistance.

S2.

Define the following symbols, taking the positive direction down the slope. [A]

Velocity of cyclist (C)	$v\ \text{m s}^{-1}$.
Acceleration of C	$f\ \text{m s}^{-2}$.
Resisting force	$-kv$ N.
Rate of working of cyclist	P W.
Mass moving	m kg

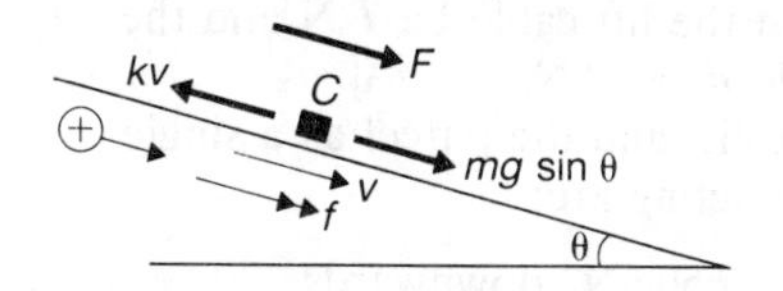

Then the tractive force acting $= \frac{P}{v}$ N.

The component of the weight $= mg\sin\theta$ N.

Therefore the equation of motion is

$$\frac{P}{v} + mg\sin\theta - kv = mf. \quad \text{[B]}$$

Now $\quad m = 60$ and $g = 10$.

Case I $\quad P = 64, f = 0, \theta = 0, v = 4.$ [C]

$$\frac{64}{4} + 0 - k(4) = 0. \qquad \therefore k = 4.$$

Case II $\quad f = 0, \sin\theta = \frac{1}{30}, P = 56.$

Let the maximum speed be $V\ \text{m s}^{-1}$.

$$\frac{56}{V} + 600\left(\frac{1}{30}\right) - 4V = 0.$$

$$\therefore V^2 - 5V - 14 = 0.$$

$$\therefore (V - 7)(V + 2) = 0.$$

Reject the solution $V = -2$ because it corresponds to motion *up* the slope. Therefore the maximum speed down the slope is $7\ \text{m s}^{-1}$. ■

Case III $\quad \theta = 0, v = 2, P = 46.$

$$\frac{46}{2} + 0 - 4 \times 2 = 60f. \quad \text{[D]}$$

$$\therefore f = \frac{1}{4}.$$

Therefore the acceleration is $0{\cdot}25\ \text{m s}^{-2}$. *25m* ■

[A] If you read only the first two sentences of the question you may think that to start with you should draw a diagram of the bicycle on a horizontal road so as to find k. But read the whole paragraph, and you will see that the first thing that you have to *find* concerns motion on a slope. Then in the last paragraph the *acceleration* is required. The best way of tackling this question is to aim first at an equation of motion for the (bicycle + man) which is general, and then to substitute the data in it.

[B] Now we are ready to consider all three of the adventures of the cyclist, using the same equation in each case.

[C] In cases of resisted motion you must be careful about the interpretation of the word 'maximum'. It does not mean that the speed rises to a maximum and then falls. What it means is that it is a speed V such that, if the initial speed is less than V, then the speed will go on increasing, *tending* to the limiting value V which in fact will not actually be achieved (see 4.4I3).

[D] Do not substitute $f = 0{\cdot}25$ and prove that $0 = 0$; this would be to risk loss of marks for a faulty argument.

S3.

Let the tension in the lift cable be T N and the reaction of the floor be R N. [A]
Consider first the lift and the parcel as a single body. The forces acting are:

Weight	$650g$ N	downwards
Tension	T N	upwards.

$\therefore$ the equation of motion is

$T - 650g = 650\ (0{\cdot}8)$. [B]

$\therefore T = 650\ (9{\cdot}8 + 0{\cdot}8) = 650 \times 10{\cdot}6$.
$\therefore$ the tension is 6890 N.

Now consider the parcel alone. The forces acting on it are:

Weight	$50g$ N	downwards

Reaction exerted by the floor, R N upwards.
$\therefore$ the equation of motion of the parcel is

$R - 50g = 50(0{\cdot}8)$.

$\therefore R = 50(9{\cdot}8 + 0{\cdot}8)$. (1)

Therefore the reaction of the floor on the parcel is $50 \times 10.6 = 530$ N. ■
Let the depth below the roof of the loose piece P at time t s after its release be x m. [C]
The acceleration of P is $9{\cdot}8\ \text{m s}^{-2}$ downwards.
The acceleration of the lift is $0{\cdot}8\ \text{m s}^{-2}$ upwards.
$\therefore$ the acceleration of P relative to the lift is $(9{\cdot}8 + 0{\cdot}8)\ \text{m s}^{-2}$ downwards. [D]

$\therefore \ddot{x} = 10{\cdot}6$.

Initially $x = 0$ and $\dot{x} = 0$, because P was (loosely) attached to the roof when $t = 0$.

$\therefore x = \frac{1}{2}(10{\cdot}6)\,t^2$.

Therefore, when $x = 3$, $t = \sqrt{\dfrac{2 \times 3}{10{\cdot}6}}$.

Therefore the time to reach the floor

is 0·75 s (to 2dp).

When the lift has a downward acceleration of $0{\cdot}8\ \text{m s}^{-2}$, $\quad \ddot{x} = 9{\cdot}8 - 0{\cdot}8 = 9$.

The time taken to reach the floor is then

$\sqrt{\dfrac{2 \times 3}{9}} = 0{\cdot}82$ s (to 2dp).

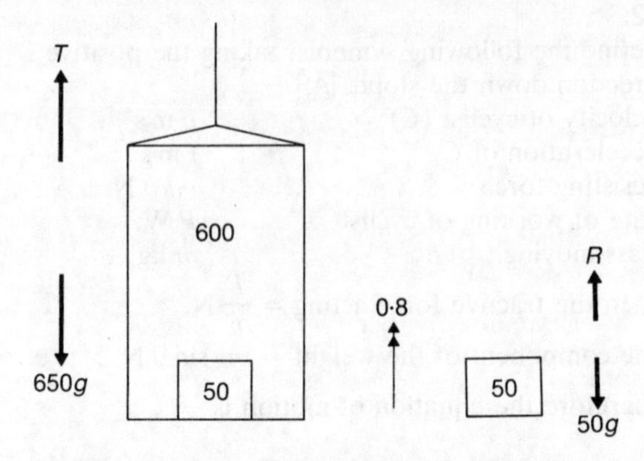

[A] Start by drawing a diagram showing the acceleration and the forces acting on the bodies in motion. We have chosen to draw first the (lift + parcel) system, being careful *not* to indicate the reaction of the floor, which does not act *on* the system. The reaction R should appear only in a diagram depicting either the lift alone or the parcel alone.

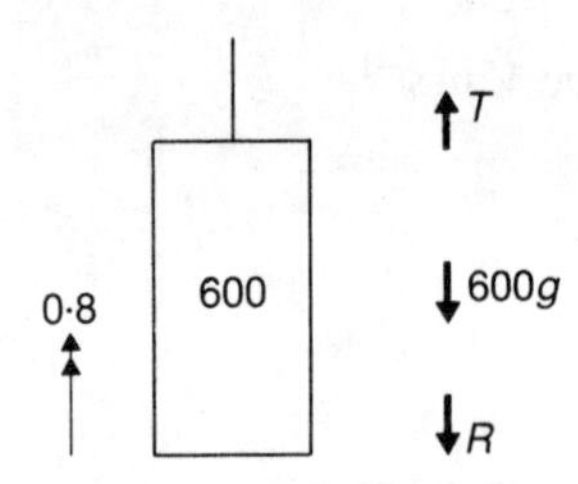

[B] Alternatively consider the lift alone. The reaction exerted *on* the lift *by* the parcel is R downwards. So $600(0{\cdot}8) = T - R - 600g$. R can be eliminated using (1) and then T can be found.

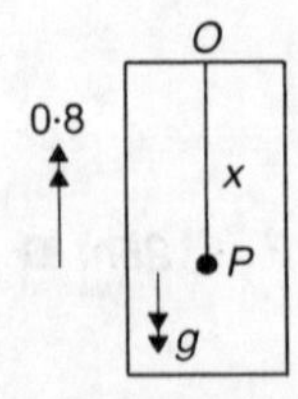

[C] Since the particle P of loose material is confined within the lift, the simplest way of studying its motion is to consider its displacement relative to the lift.
[D] Using 3.2F7.

When there are k parcels and the *upward* acceleration is $0{\cdot}8\ \mathrm{m\,s^{-2}}$, the equation of motion of the lift and parcels together is
$T - (600 + 50k)g = (600 + 50k)(0{\cdot}8)$.

$\therefore\ T = (600 + 50k)(9{\cdot}8 + 0{\cdot}8)$. Now $T \leqslant 12\,000$.

$\therefore\ 530k \leqslant 12\,000 - 6360 = 5640$.

$\therefore\ k \leqslant \dfrac{5640}{530} = 10{\cdot}64\ldots$

$\therefore$ the maximum number of parcels is 10. [E] *27m* ■

[E] Do not take the nearest integer to 10·64, which is 11; this might break the cable.

S4.

Let the work done by the body against the resistance be W.
$\therefore$ the work done by the resistance

$= -W$. [A]

The work done by gravity $= mgh$. [B]
Increase in kinetic energy is equal to the total work done on the body.

$\therefore\ \frac{1}{2}mv^2 - \frac{1}{2}m(0) = -W + mgh$. [C]

$\therefore$ the work done against the resistance is

$W = mgh - \frac{1}{2}mv^2$. *6m* ■

[A] Work done against resistance is always positive, because resistance always acts opposite to the velocity, but here we cannot check our result by inspecting its sign because no numerical magnitudes are given. So we need to be sure that we have set up the problem correctly from the start. For both forces, consider the work done *on* the body, not *by* the body, and then the energy equation will be right.
[B] Using 4.4F7.
[C] Using 4.3F6.

S5.

When the speed is v and the power delivered by the engines is P, the forward force is $\dfrac{P}{v}$. [A]

[A] Write this formula down before calculating the force. It is very easy to write down the wrong numerical fraction at this point. At least you will inform the examiner that you do know the correct relation.

The equation of motion of the cruiser when its speed is constant (zero acceleration) is

$$\frac{P}{v} - R = 0. \qquad \therefore R = \frac{P}{v}.$$

The power is 3×10^4 kW. [B]

[B] Or 3×10^7 W. But do not express it as 30 000 000 W, or you will risk an error.

The speed is $\dfrac{60}{3{\cdot}6} = \dfrac{100}{6}\,\mathrm{m\,s^{-1}}$. [C]

[C] The relation $1\,\mathrm{m\,s^{-1}} = 3{\cdot}6\,\mathrm{km\,h^{-1}}$ is so often needed that it is worth memorising. Get used to the idea that a metre per second is faster than a kilometre per hour.

$$\therefore \text{the resistance is } 3 \times 10^4 \left(\frac{6}{100}\right) = 1800\text{ kN}. \quad \text{[D]}$$

[D] kN, remembering that we had kW.

$$\frac{\text{Resistance at } 72\text{ km h}^{-1}}{\text{Resistance at } 60\text{ km h}^{-1}} = \frac{72^2}{60^2} = (1{\cdot}2)^2. \quad \text{[E]}$$

[E] Since it is only the *ratio* of the speeds that matters there is no need to convert the speeds to $\mathrm{m\,s^{-1}}$ at this point.

$$\therefore \text{Resistance at } 72\text{ km h}^{-1} = 1800(1{\cdot}2)^2 = 2592\text{ kN}.$$

13m ■

S6.

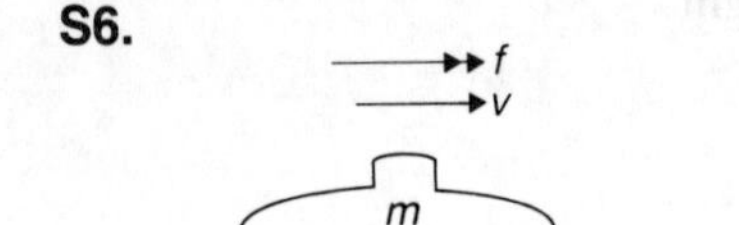

Let the acceleration of the submarine be f. Then

$$mf = \frac{P}{v} - kv^2. \qquad (1)$$

The power P is just sufficient to maintain a speed u. This means that $f = 0$ if $v = u$.

$$\therefore \frac{P}{u} - ku^2 = 0. \qquad \therefore k = \frac{P}{u^3}.$$

Substitute $f = v\dfrac{\mathrm{d}v}{\mathrm{d}x}$ and $k = \dfrac{P}{u^3}$ in (1).

$$mv\frac{\mathrm{d}v}{\mathrm{d}x} = \frac{P}{v} - \frac{Pv^2}{u^3} = \frac{P}{v}\left(\frac{u^3 - v^3}{u^3}\right),$$

which is equivalent to the required result. ■

$$\therefore \frac{\mathrm{d}x}{\mathrm{d}v} = \frac{mu^3}{P}\left(\frac{v^2}{u^3 - v^3}\right) = \frac{mu^3}{3P}\left(\frac{3v^2}{u^3 - v^3}\right).$$

$\therefore$ the distance moved while v increases from $\frac{u}{4}$ to $\frac{3u}{4}$

$$\text{is} \int_{u/4}^{3u/4} \frac{mu^3}{3P}\left(\frac{3v^2}{u^3 - v^3}\right) dv = \frac{mu^3}{3P}\Big[-\ln(u^3 - v^3)\Big]_{u/4}^{3u/4}$$

$$= \frac{mu^3}{3P}\ln\frac{u^3 - (u/4)^3}{u^3 - (3u/4)^3}$$

$$= \frac{mu^3}{3P}\ln\frac{4^3 - 1}{4^3 - 3^3}$$

$$= \frac{mu^3}{3P}\ln\frac{63}{37}. \qquad (2) \blacksquare$$

When the engines are not working the equation of motion is

$mf = -kv^2$. Substitute $f = \frac{dv}{dt}$ and $k = \frac{P}{u^3}$.

$$m\frac{dv}{dt} = -\frac{Pv^2}{u^3}.$$

$$\therefore \quad \frac{dt}{dv} = -\frac{mu^3}{Pv^2}.$$

$\therefore$ the time taken for the speed to change from $\frac{3u}{4}$ to $\frac{u}{4}$ is

$$\int_{3u/4}^{u/4}\left(-\frac{mu^3}{Pv^2}\right) dv = \frac{mu^3}{P}\left[\frac{1}{v}\right]_{3u/4}^{u/4}$$

$$= \frac{mu^3}{P}\left(\frac{4}{u} - \frac{4}{3u}\right)$$

$$= \frac{8mu^2}{3P}. \qquad \text{25m} \blacksquare$$

Dimensional check: The result (2), printed in the Question, tells us that $\frac{mu^3}{P}$ is a distance, having dimension L.

Now u is a speed, and so $\frac{mu^2}{P}$, which is $\frac{mu^3}{P}\left(\frac{1}{u}\right)$, has dimension $\frac{\text{L}}{\text{LT}^{-1}} = \text{T}$; a time, as required.

Note that in this question the changes in x and t while the speed is changing from one fixed value to another are demanded. We have not been asked to express x or t in terms of v. It is therefore best to use the definite–integral methods.

S7.

When the speed of the train is v m s^{-1} and the engine is exerting maximum tractive force the power being used is $30 \times 10^3 v$ W. [A]
The train starts with $v = 0$. As v increases, the power increases, reaching its maximum when $30 \times 10^3\, v = 450 \times 10^3$, i.e. $v = 15$.
This occurs at speed

$15 \times 3{\cdot}6 = 54$ km h^{-1}. [B][C] ■

After reaching this speed the engine continues to deliver maximum power P W. Write m kg for the mass and f m s^{-2} for the acceleration of the train. Then the equation of motion is

$$mf = \frac{P}{v} - R, \qquad (1)$$

where R N is the resistance. The speed increases only when $f > 0$, and so

$$v \leqslant \frac{P}{R} = \frac{450 \times 10^3}{18 \times 10^3}.$$

Hence the speed cannot exceed 25 m s^{-1}
= 90 km h^{-1}. ■

When $v = \dfrac{72}{3{\cdot}6}$, $\quad \dfrac{P}{v} = \dfrac{450 \times 10^3}{20} = 22{\cdot}5 \times 10^3.$

Then $f = \dfrac{1}{m}\left(\dfrac{P}{v} - R\right)$

$$= \frac{22{\cdot}5 \times 10^3 - 18 \times 10^3}{300 \times 10^3} = \frac{4{\cdot}5}{300}.$$

Hence the acceleration at 72 km h^{-1}
is 0·015 m s^{-2}. *20m* ■

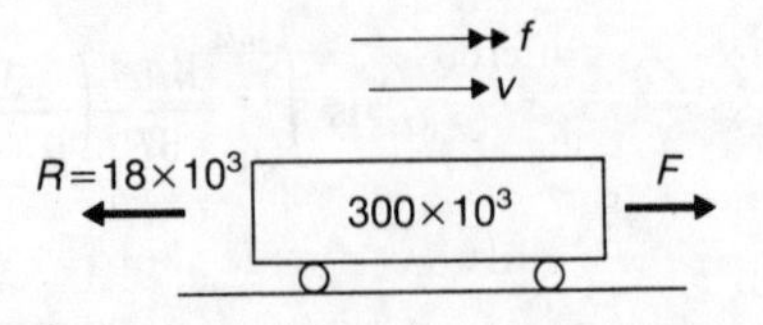

[A] The occurence of terms like 'greatest' and 'maximum' make this look harder than it is. Start by entering the mass and the resistance on a diagram. The equation of motion is

$$F - R = mf$$

which means that, to maximise the acceleration, the greatest possible tractive force F is needed. This is 30×10^3 N, and is fully available at first, while v is low, without exceeding the maximum power.
[B] Use of factor 3·6: see S5 [C].
[C] Above this speed the limitation on the power prevents the maximum tractive force from being applied. The acceleration is kept as high as possible by using the maximum power, but equation (1) shows that f will fall as v increases, because P is now constant.

S8.

Let the tractive force be F N.
Then the equation of motion of the train is

$Ma = F - R.$ (1)
For $t \leqslant T$, $F = P - kt$, (2)
where k is a constant. [A]
When $t = T$, $F = R$. (3)

$\therefore R = P - kT$, that is, $k = \dfrac{P - R}{T}$.

$\therefore Ma = P - \dfrac{P - R}{T}t - R.$ (4) [B]

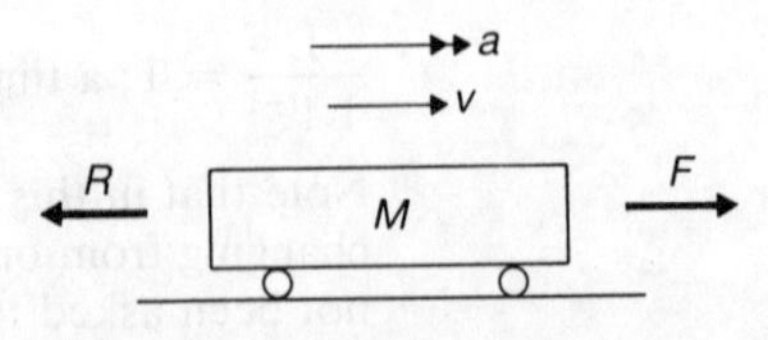

[A] 'Decreases uniformly' means that the rate of change of F is a negative constant. So $\dfrac{\mathrm{d}F}{\mathrm{d}t} = -k$, and $F = P$ when $t = 0$.

Let the speed of the train be $v \text{ m s}^{-1}$.
When $t = 0$, $a > 0$, because $P > R$, and so v increases from zero. As t increases, $\dfrac{P - R}{T}t$ increases, and so a decreases. When t reaches the value T, $a = 0$ and so v stops increasing. After that, we are told, $F = R$, and so a remains zero and v is then constant.
Therefore the maximum speed occurs when $t = T$.

Integrate (4) with respect to t.

$$Mv = Pt - \frac{1}{2}\left(\frac{P - R}{T}\right)t^2 - Rt + 0. \qquad (5) \text{ [C]}$$

$$\therefore v = \frac{P - R}{MT}\left(Tt - \frac{t^2}{2}\right). \qquad (6)$$

When $t = T$, $\quad Tt - \dfrac{t^2}{2} = \dfrac{T^2}{2}$.

$\therefore$ the maximum speed $v_m = \dfrac{(P - R)T}{2M} \text{ m s}^{-1}$. [D]

Integrate (6) with repect to t from 0 to T. [E]
Distance travelled in the interval $0 \leqslant t \leqslant T$

is $$\int_0^T v\,dt = \frac{P - R}{MT}\left[\frac{Tt^2}{2} - \frac{t^3}{6}\right]_0^T = \frac{(P - R)T^2}{3M} \text{ m.}$$

$$\text{Average speed} = \frac{\text{distance}}{\text{time interval}} = \frac{(P - R)T}{3M} \qquad \text{[F]}$$

$$= \frac{2v_m}{3} \text{ m s}^{-1}. \ ■$$

When $t = \dfrac{T}{2}$,

power $= Fv$

$$= \left(\frac{P + R}{2}\right)\left(\frac{P - R}{MT}\right)\left(\frac{T^2}{2} - \frac{T^2}{8}\right) \qquad \text{[G]}$$

$$= \frac{3(P^2 - R^2)T}{16M} \text{ W.}$$

Dimensional check: $\dfrac{(\mathrm{MLT^{-2}})^2\mathrm{T}}{\mathrm{M}} = \mathrm{ML^2T^{-3}}$.

(power)√

When $t = \dfrac{3T}{2}$, power $= Rv_m$ [H]

$$= \frac{R(P - R)T}{2M} \text{ W.} \qquad \textit{27m} \ ■$$

[B] This result has been printed in the question, so you must show clearly that you have derived it, by writing down statements (1), (2) and (3), all with correct signs. However, less than 4 minutes should be devoted to it.

[C] The train starts from rest, so $v = 0$ when $t = 0$. The constant of integration we require is therefore 0.

[D] v_m could alternatively have been obtained from (4) as the definite integral $\int_0^T a\,dt$, but the result could not be used, as (6) can, to find the distance.

[E] Here, as we do not need to know the displacement at all values of t, the definite integral $\int_0^T v\,dt = \dfrac{P - R}{MT}\int_0^T \left(Tt - \dfrac{t^2}{2}\right)dt$ suffices.

[F] This is the point at which, after carefully avoiding the mistake of assuming constant acceleration at any stage, you could fall into the error

✗ Average speed
$= \frac{1}{2}$ (initial speed + final speed) $= \frac{1}{2}v_m$. ✗

This formula is valid only when the v–t graph is a straight line.

[G] Since F decreases *uniformly* from P to R in time T, it must equal the mean of P and R after time $T/2$. Alternatively, at this time

$$F = P - \frac{P - R}{T}\left(\frac{T}{2}\right) = P - \frac{P}{2} + \frac{R}{2}.$$

[H] ✗ Substitute $t = 3T/2$ in (2) and in (6). ✗
This trap must be avoided by carefully reading the question

✓ When $t = 3T/2$, $t > T$.

$\therefore F = R$ and $v = v_m = \dfrac{(P - R)T}{2M}$. ✓

Chapter 5

Simple harmonic motion

5.1 GETTING STARTED

Any motion in which a particle moves backwards and forwards between points A and B along a fixed path (which need not be straight) in a regular fashion is called an **oscillation**. The period of an oscillation is the time taken for the completion of a single unit of the repeated pattern of the motion. The reciprocal of the period is called the frequency of the oscillation. The motion may be smooth or jerky; one kind of smooth oscillation is called simple harmonic motion because it corresponds to acoustic vibrations which are perceived by the human ear as very pure musical tones. This chapter treats the simple harmonic motion of a point moving along a straight line.

5.2 KINEMATICS OF SHM

ESSENTIAL FACTS

At time t let x be the displacement from O of a point P which moves on the x-axis. Thus x is the coordinate of P.

F1. When P moves so that $x = A\cos\omega t + B\sin\omega t$, where A, B and ω are constants and $\omega > 0$, it is said to move in **simple harmonic motion** (SHM) of **frequency** $\dfrac{\omega}{2\pi}$ about the **centre of oscillation** O.

The **period** of the oscillation is $\dfrac{2\pi}{\omega}$. The initial displacement is A and the initial velocity is ωB.

F2.

The equation $x = A\cos\omega t$ describes SHM with centre O, period $\dfrac{2\pi}{\omega}$, and **amplitude** $|A|$. The initial displacement is A and initial velocity is zero.

F3.

The equation $x = B\sin\omega t$ describes SHM with centre O, period $\dfrac{2\pi}{\omega}$, initial displacement 0 and initial velocity ωB. The amplitude is $|B|$.

F4.

The equation in F1 may be written $x = a\cos(\omega t - \varepsilon)$, where the amplitude $a = \sqrt{A^2 + B^2}$ and the **phase angle** ε is given by $a\cos\varepsilon = A$, $a\sin\varepsilon = B$. The angle ε is the **phase of the motion relative** to the motion of F2.

F5.

The equation $x = A\cos\omega(t - t_0) + B\sin\omega(t - t_0)$ describes SHM with displacement A and velocity ωB at time t_0.

F6. Differential equation of SHM

All the relations F1–F5 satisfy the equation $\ddot{x} + \omega^2 x = 0$. Conversely, every solution of this differential equation is of the form $x = A\cos\omega t + B\sin\omega t$, where A and B are constant.

F7.

For the motions F1–F5, $\dot{x}^2 = \omega^2(a^2 - x^2)$, where a is the amplitude. The maximum and minimum displacements, velocities and accelerations are $\pm a$, $\pm\omega a$ and $\pm\omega^2 a$ respectively.

F8.

The equation $x = a\cos(\omega t - \varepsilon) + c$, where $a > 0$, describes SHM with centre C (coordinate c), period $\dfrac{2\pi}{\omega}$, amplitude a and phase ε relative to the motion of F2. It satisfies the differential equation $\ddot{x} + \omega^2(x - c) = 0$.

F9.

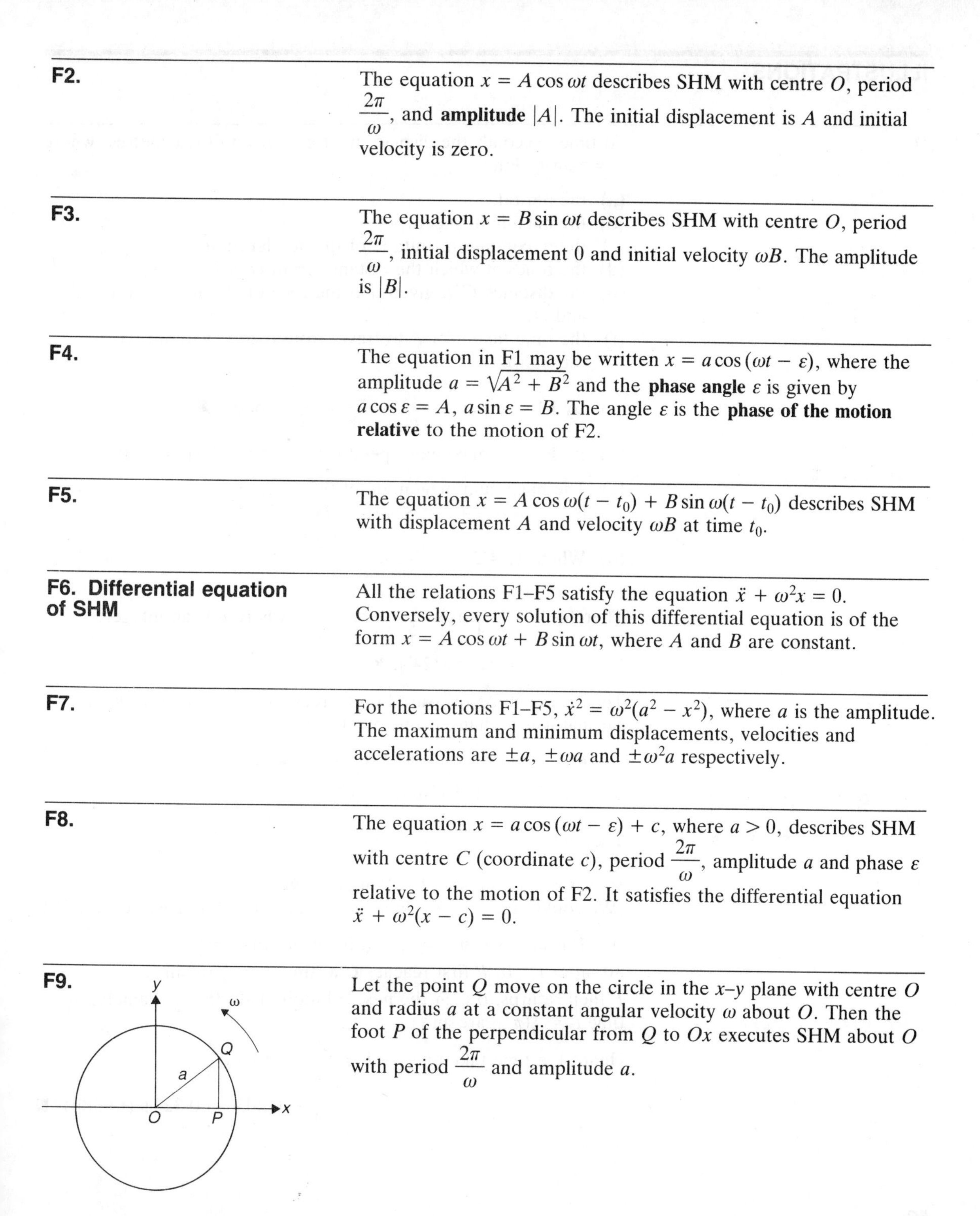

Let the point Q move on the circle in the x–y plane with centre O and radius a at a constant angular velocity ω about O. Then the foot P of the perpendicular from Q to Ox executes SHM about O with period $\dfrac{2\pi}{\omega}$ and amplitude a.

ILLUSTRATIONS

I1. At time t seconds the displacement of P from O is x metres, where $x = 5\sin 4t$. Find

(a) the period,
(b) the maximum speed,
(c) the maximum magnitude of the acceleration,
(d) the times at which the distance from O is 2 metres,
(e) the distance CD, given that the speed is $12\ \text{m s}^{-1}$ at both C and D,
(f) the time taken for P to travel from C to C without passing through D. □

(a) By F1, the period $\dfrac{2\pi}{\omega} = \dfrac{2\pi}{4} = \dfrac{\pi}{2}$ seconds. ■

(b) By F7, the maximum speed $\omega a = 4 \times 5 = 20\ \text{m s}^{-1}$. ■

(c) The maximum magnitude of the acceleration is $\omega^2 a = 4^2 5 = 80\ \text{m s}^{-2}$. ■

(d) When $|x| = 2$, $\quad |\sin 4t| = \dfrac{2}{5}$.

Therefore $t = \pm\dfrac{1}{4}\arcsin\dfrac{2}{5} + n\pi$, where n is an integer,

✗ $\quad = n\pi \pm \frac{1}{4}(24°)$. ✗

Time is not to be measured in degrees. Remember to set your calculator to radians at this point.

✓ $t = n\pi \pm 0{\cdot}103$ (to 3 d.p.). ✓ ■

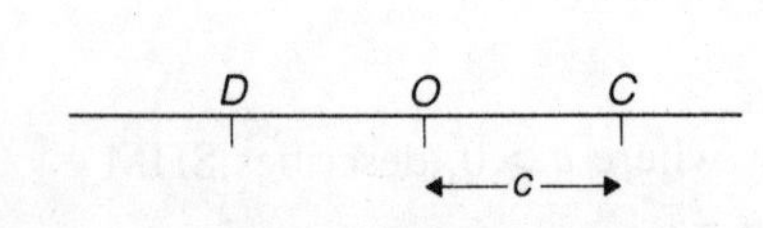

(e) By F7, $\quad 12^2 = \omega^2(a^2 - c^2)$.

Therefore $\quad c^2 = a^2 - \dfrac{144}{\omega^2} = 25 - \dfrac{144}{16} = 16$.

Hence $c = 4$, and so $CD = 8$ metres. ■
(We could take $c = -4$, but this would only interchange C and D.)

(f) P is at C when $x = c$, that is, when $\sin 4t = \frac{4}{5}$.
So, after $t = 0$, P first reaches C at time $t_1 = \frac{1}{4}\arcsin\frac{4}{5}$.
P then returns to C at the next solution of $\sin 4t = \frac{4}{5}$, namely, at time $t_2 = \frac{1}{4}(\pi - \arcsin\frac{4}{5})$.

Hence the time taken is $t_2 - t_1 = \dfrac{\pi}{4} - \dfrac{1}{2}\arcsin\dfrac{4}{5}$

$$= \frac{\pi}{4} - 0{\cdot}4636 = 0{\cdot}322 \text{ s (to 3dp).}$$ ■

12.

A particle P moves along a line in SHM with period 5 seconds. Initially its displacement from the centre of oscillation O is 2 metres in the direction OC and its velocity is 3 m s^{-1} in the same direction.

(a) Find the amplitude of the oscillation.
(b) Find the time at which the distance CP is first at a maximum.
(c) Answer part (b) in the case when the initial velocity is 3 m s^{-1} in the direction CO. □

Let the displacement of P from O (measured in the direction OC for convenience) be $x = A\cos\omega t + B\sin\omega t$, where $\omega = \dfrac{2\pi}{5}$.

By F1, $A = 2$ and $\omega B = 3$, from which $B = \dfrac{15}{2\pi}$.

(a) $\therefore$ the amplitude $a = \sqrt{2^2 + \left(\dfrac{15}{2\pi}\right)^2} = 3{\cdot}11$ metres (to 2 dp). ■

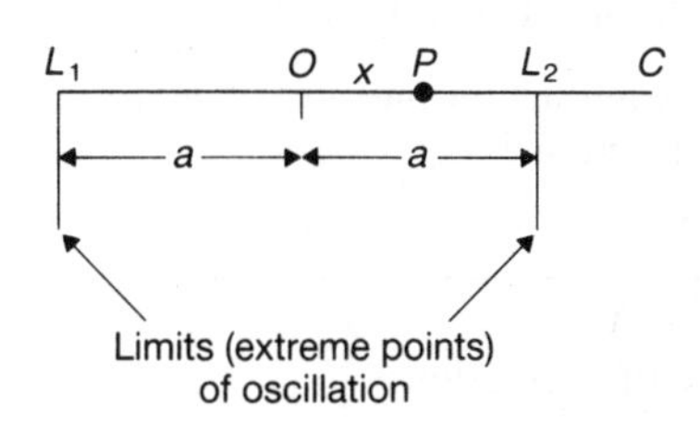

(b) $x = 2\cos\omega t + \dfrac{15}{2\pi}\sin\omega t = a\cos(\omega t - \varepsilon)$, using F4, where $a\cos\varepsilon = 2$ and $a\sin\varepsilon = \dfrac{15}{2\pi}$. Both of these are positive, so ε is acute, and so $\varepsilon = \arctan\left(\dfrac{15}{4\pi}\right)$.

From the diagram we see that CP is maximum when P is at L_1, that is, when $x = -a$. We need the least positive value of t for which this is so. This occurs when $\omega t - \varepsilon = \pi$. The time required is therefore $\dfrac{1}{\omega}(\pi + \varepsilon) = \dfrac{5}{2\pi}\left\{\pi + \arctan\left(\dfrac{15}{4\pi}\right)\right\} = 3{\cdot}20\ldots$ seconds. ■

(c) Now: $\omega B = -3$, so

$$x = 2\cos\omega t - \frac{15}{2\pi}\sin\omega t = a\cos(\omega t + \varepsilon),$$

where ε has the same value as before. The time at which CP is maximum is therefore given by $\omega t + \varepsilon = \pi$. Hence the time required is

$$\frac{5}{2\pi}\left\{\pi - \arctan\left(\frac{15}{4\pi}\right)\right\} = 1{\cdot}80\ldots \text{ seconds.} \ ■$$

Check: Since in (c) the particle is projected *towards* L_1, we expect the answer in (c) to be less than the answer in (b). It is.√

5.3 FORCE AND ENERGY IN SHM

ESSENTIAL FACTS

In the following Facts x is the displacement at time t of a particle P of mass m which moves on the x-axis in SHM with centre of oscillation O, frequency $\frac{\omega}{2\pi}$ and amplitude a.

F1.

The resultant force F acting on P is $-m\omega^2 x$.
$|F|$ is greatest when $|x| = a$ (P at an extreme point).
$|F|$ is least when $x = 0$ (P at the centre of oscillation O).

F2.

The motion of a particle of mass m which moves on a straight line subject to a force of attraction a constant k times its distance from a fixed point A on the line is SHM about A with period $2\pi\sqrt{\frac{m}{k}}$.
The centre of oscillation is also the **equilibrium position** of P, which is the point at which the force of attraction $F = 0$.

F3. Kinetic energy

The kinetic energy of P is $\frac{1}{2}m\omega^2(a^2 - x^2)$.
It is greatest when x is at the centre of oscillation.

F4. Potential energy

Taking O as the zero level, the potential energy of P is $\frac{1}{2}m\omega^2 x^2$.
The total energy $E = \frac{1}{2}m\omega^2 a^2$.

ILLUSTRATIONS

I1.

A particle P of mass m is suspended from a point A by a light inextensible string AP. The point of support A is made to move in a vertical line in SHM with amplitude a and period T, and the string remains taut. Find the tension in the string at time t after P is at the highest point of the motion. Show that the amplitude cannot exceed $\frac{1}{2\pi^2}$ times the distance through which a particle would fall freely during one period of the oscillation. □

At time t, let the depth of A below its centre of oscillation A_0 be x.
At the highest point $x = -a$.
Therefore, when $t = 0$, $\dot{x} = 0$ and $x = -a$.

$$\therefore x = -a\cos\omega t, \qquad \text{where } \omega = \frac{2\pi}{T}. \qquad \text{[From 5.2F2.]}$$

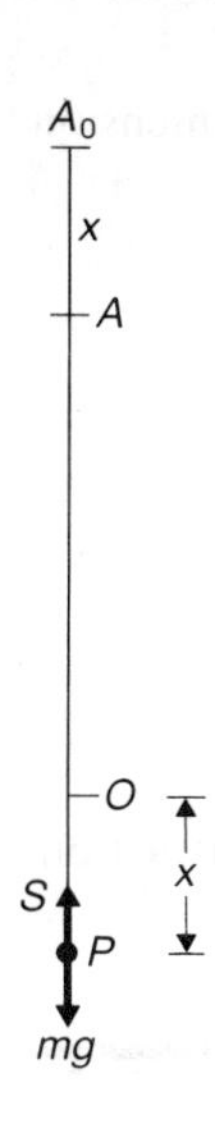

Because the string is **inextensible** and **taut**, AP is constant throughout the motion. Therefore P also moves in SHM so that its depth is x below a fixed point O, where A_0O is the length of the string. Let the tension in the string be S. Then the equation of motion of P is

$$m\ddot{x} = mg - S.$$

Now $\quad \ddot{x} = +\omega^2 a \cos \omega t.$

$$\therefore S = mg - m\omega^2 a \cos \omega t = m(g - \omega^2 a \cos \omega t). \blacksquare$$

Because S is the tension in a string, it can never be negative.

$\therefore g - \omega^2 a \cos \omega t \geqslant 0$ for all values of t.

The least value of the LHS occurs when $\cos \omega t$ takes its maximum value, which is 1. Therefore we require

$g - \omega^2 a \geqslant 0.$

$$\therefore \quad a \leqslant \frac{g}{\omega^2} = g\left(\frac{T}{2\pi}\right)^2 = \frac{1}{2\pi^2}\left(\frac{1}{2}gT^2\right),$$

which is the required result. ■

12.

A particle P of mass 2 kg is subject to an attraction towards O proportional to OP. When P is at rest at O it suddenly acquires kinetic energy 225 J, and thereafter executes SHM with amplitude 5 m. Find

(a) the period of the motion,
(b) the force of attraction when $OP = 3$ m,
(c) the rate of working of the attraction after (i) $\frac{\pi}{12}$ s (ii) $\frac{\pi}{4}$ s. □

(a) Let the displacement of P be $x = a \sin \omega t$.
When $t = 0$, $x = 0$, $\dot{x} = \omega a$, and the kinetic energy is $\frac{1}{2}m\omega^2 a^2$.

$$\therefore (\tfrac{1}{2})\, 2(\omega^2)5^2 = 225. \qquad \therefore \omega = 3.$$

$$\therefore \text{the period of the motion is } \frac{2\pi}{3} = 2{\cdot}09\ldots \text{ s. } \blacksquare$$

(b) Let F be the force acting on P.
Then the equation of motion of P is

$$F = m\ddot{x} = -\,m\omega^2 a \sin \omega t = -\,m\omega^2 x.$$

$\therefore$ when $OP = 3$, the force is of magnitude $2(3^2)3 = 54$ N and is directed along PO. ■

(c) The rate of working is $F\dot{x} = (-\,m\omega^2 a \sin \omega t)(\omega a \cos \omega t)$

$$= -\,m\omega^3 a^2 \sin \omega t \cos \omega t.$$

Dimensional check: ω has dimension T^{-1}, so $m\omega^3a^2$ has dimension $\mathrm{MT}^{-3}\mathrm{L}^2$; power. √

(i) When $t = \dfrac{\pi}{12}$,

$$\text{the power is} \quad -(2)(3^3)5^2 \sin\frac{3\pi}{12}\cos\frac{3\pi}{12} = -675 \text{ W}$$

(ii) When $t = \dfrac{\pi}{4}$,

$$\text{the power is} \quad -(2)(3^3)5^2 \sin\frac{3\pi}{4}\cos\frac{3\pi}{4} = +675 \text{ W}. \blacksquare$$

In case (ii) $x < 0$, so P is moving towards O, and so the attraction is working at a positive rate.

5.4 EXAMINATION QUESTIONS AND SOLUTIONS

Q1. A man of mass m stands on a horizontal platform which begins to move at time $t = 0$ and thereafter performs vertical oscillations such that its depth x below a fixed level is given by $x = a\cos\omega t$, where a and ω are positive constants. Find the force at time t which he exerts on the platform while he remains on it. Deduce that he never loses contact with the platform if $a\omega^2 < g$.

In the case when $a\omega^2 = 2g$ find, in terms of ω, the time at which the man first loses contact with the platform. (JMB 1983)

Q2. A particle P is free to move in a straight line OA, where O and A are fixed points on the line. The acceleration of P at time t seconds is $12\sin 2t$ m s^{-2} in the direction OA. When $t = 0$, the particle P passes through O moving with speed u m s^{-1} towards A. Find

(a) the velocity of P at time t seconds,
(b) the displacement of P from O at time t seconds,
(c) the value of u so that the motion is simple harmonic.

The acceleration of P in the direction OA is changed to $12|\sin 2t|$ m s^{-2}. Given also that, when $t = 0$, the particle P passes through O with speed 8 m s^{-1} in the direction OA, find the displacement of P from O when $t = \pi$. (AEB 1984)

Q3.

A ship arrives at noon at the entrance to a harbour. The depth of water over the harbour bar is 4 m at low tide and 10 m at high tide. Low tide is at 11.20 a.m. and high tide is at 5.40 p.m., and it is assumed that the water level moves in simple harmonic motion. The ship requires a depth of 9·4 m. Find the earliest time at which the ship can cross the bar so as to enter the harbour, and determine the rate, in centimetres per minute, at which the water will then be rising. ★

Q4.

(a) A particle moves in a straight line such that its acceleration is proportional to its distance x from a fixed point O on the line and is directed towards O. If t denotes time, the motion is described by the equation

$$\frac{d^2x}{dt^2} = -\omega^2 x$$

where ω is a positive constant.

Show, by substitution or otherwise, that this equation has solution $x = A\cos(\omega t + \varepsilon)$, where A and ε are constants.

Hence, or otherwise, show that the speed v of the particle is given by $v^2 = \omega^2(A^2 - x^2)$.

(b) A particle moves with simple harmonic motion such that its speed is 2 m s^{-1} when it is 0·1 m from the centre O of its motion and 1·75 m s^{-1} when it is 0·15 m from O. Find

(i) the amplitude of the motion,
(ii) the period of the motion,
(iii) the maximum speed and maximum acceleration of the particle.

If at time $t = 0$, the particle is at O, find the speed of the particle at time $t = 0·1$ seconds. (NI 1986)

Q5.

A particle describes simple harmonic motion with period $\frac{1}{2}\pi$ s about a point O as centre and the maximum speed in the motion is 20 m s^{-1}. Find

(a) the amplitude of the motion,
(b) the speed of the particle when at a point Q distance 2 m from O,
(c) the time taken to travel directly from O to Q.

Given that the particle is of mass 0·5 kg, find the rate at which the force acting on the particle is working at a time $(\pi/12)$ s after it has passed through O. Find also the maximum rate of working of this force. (AEB 1985)

SOLUTIONS

S1.

Let the force exerted on the platform by the man be R downwards. By Newton's Third Law the force exerted on the man by the platform is R upwards. [A]

The only other force acting on the man is his weight mg downwards.

The man is in contact with the platform, and so his motion, also, is described by the equation $x = a\cos\omega t$, where x is his depth below some fixed level.

$\therefore$ the downward acceleration of the man is given by $\ddot{x} = -\omega^2 a\cos\omega t$.

The equation of motion of the man is

$m\ddot{x} = mg - R$. [B]

Hence $R = mg - m\ddot{x} = mg + m\omega^2 a\cos\omega t$
$= m(g + \omega^2 a\cos\omega t)$.

$\therefore$ the force he exerts on the platform is $m(g + \omega^2 a\cos\omega t)$ downwards. ■

Since $\cos\omega t \geqslant -1$, $g + \omega^2 a\cos\omega t \geqslant g - \omega^2 a$. [C]
Hence R is always positive if $a\omega^2 < g$, and the man will not leave the platform. [D] ■

When $a\omega^2 = 2g$, $\quad R = mg(1 + 2\cos\omega t)$.

From the start, R decreases steadily, becoming zero when

$\cos\omega t = -\dfrac{1}{2}$, $\quad$ that is, $\quad \omega t = \dfrac{2\pi}{3}$.

Hence contact is lost at time $\dfrac{2\pi}{3\omega}$. $\quad$ *13m* ■

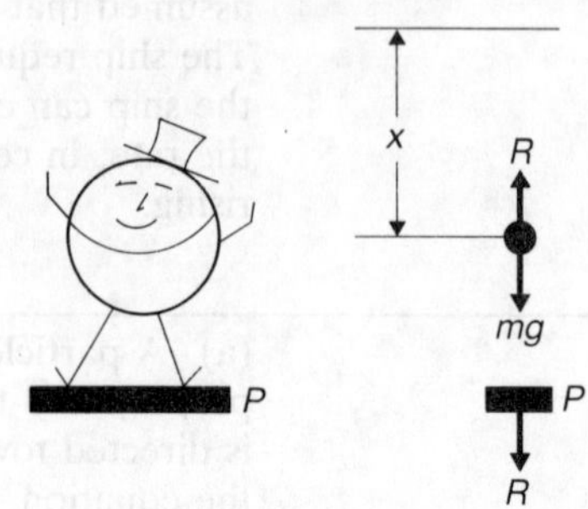

[A] This is an essential step. If you omit it you will obtain the reaction with the wrong sign.

[B] The equation has $m\ddot{x}$ on one side and the forces acting *on the man* on the other side. If you put either of the forces on the same side as the acceleration you will be very likely to make a sign error.

[C] Note that the property $\cos\theta \leqslant 1$ is of no use here.

[D] An argument like
Force $= ma\omega^2\cos\omega t$; maximum value $ma\omega^2$.
$\therefore$ Weight $mg > ma\omega^2$
✗ $\therefore$ he will not leave the platform ✗
is plausible on physical grounds but is not acceptable as a deduction. It would be better to say:
✓ If the man were falling freely his acceleration would be g downwards. But, if $a\omega^2 < g$, the maximum downward acceleration of the platform is less than g, and so contact will be maintained.✓,
but this would be dubious unless carefully worded. Actual calculation of R provides a much more reliable proof.

S2.

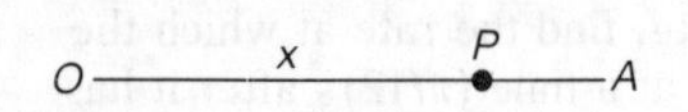

Let the displacement of P from O be x m. Then $\ddot{x} = 12\sin 2t$.

(a) Therefore the velocity $\dot{x} = -6\cos 2t + B$. When $t = 0$, $\dot{x} = u$.
$\therefore u = -6 + B$. Hence $\dot{x} = -6\cos 2t + 6 + u$. ■

(b) Then $x = -3\sin 2t + (6 + u)t + 0$, because $x = 0$ when $t = 0$. ■

(c) This is SHM if $x = -3\sin 2t$, that is, if $u = -6$. ■

We are now given $u = 8$. $\therefore \dot{x} = 14 - 6\cos 2t$ and $x = 14t - 3\sin 2t$.
This is true so long as

$12\sin 2t \geqslant 0$, that is, for $0 \leqslant t \leqslant \dfrac{\pi}{2}$.

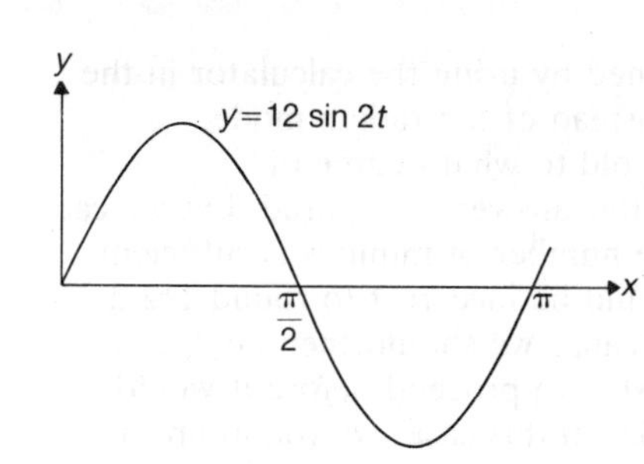

When $t = \frac{\pi}{2}$; $\quad \dot{x} = 20$ and $x = 7\pi$.

For $\frac{\pi}{2} < t < \pi$, $\quad \ddot{x} = -12\sin 2t$.

Therefore $\dot{x} = 6\cos 2t + C$, where $20 = 6\cos\pi + C$.
Hence $x = 6\cos 2t + 26$, which leads, on further integration, to
$x = 3\sin 2t + 26t + D$, where $7\pi = 3\sin\pi + 13\pi + D$.
Hence $x = 3\sin 2t + 26t - 6\pi$.
When $t = \pi$, $x = 0 + 26\pi - 6\pi = 20\pi$. *25m* ■

S3.

[A] The time from low tide to high tide is 6 hours 20 minutes = 380 min.
∴ the period of the motion is 760 min.
The centre O of the oscillation is at a height 7 m above the bar. O is the *half-tide* point, and the water is at O at 2.30 p.m. Let x m be the height of the water above O at a time t minutes after this half tide.
The amplitude is $10 - 7 = 3$ m.
Then, at time t min, the height of the water above the bar is $(7 + x)$ m, where

$$x = a\sin\omega t,\ a = 3,\ \text{and}\ \omega = \frac{2\pi}{760} = \frac{\pi}{380}.$$

The depth is 9·4 m when $x = 2{\cdot}4$.
Then $2{\cdot}4 = 3\sin\omega t$.
∴ $\sin\omega t = 0{\cdot}8$. [B]
As the tide is rising at noon, when the ship arrives, it will continue rising until high tide at $\omega t = \frac{\pi}{2}$. It will first reach the required level when $\omega t = \arcsin 0{\cdot}8$.
∴ the earliest time after half tide at which the ship can cross the bar is
$\frac{380}{\pi}\arcsin 0{\cdot}8 = 112{\cdot}2\ldots$ minutes. [C]
Hence the earliest time is 4.23 p.m. [D] ■
The rate of increase of the depth of the water is $\dot{x} = \omega a\cos\omega t$.
At the stated time the rate of rise is

$$\omega a\cos\omega t = \omega a\left(\frac{3}{5}\right) = \frac{\pi}{380}(3)\left(\frac{3}{5}\right)\ \text{metres min}^{-1}$$
$$= 1{\cdot}49\ldots\text{cm min}^{-1}.$$ *15m* ■

[A] A diagram is helpful here.

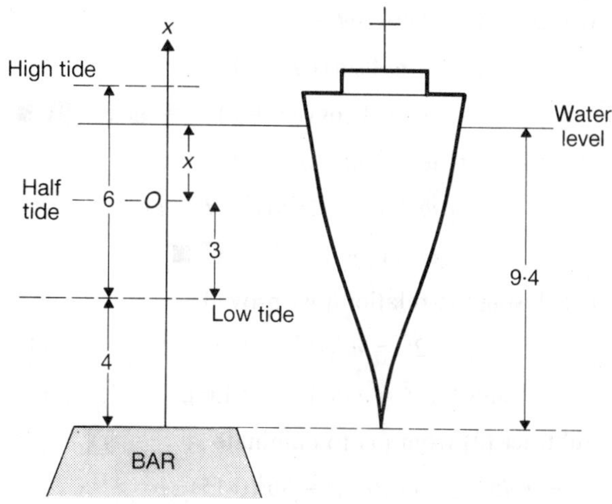

In the diagram the water is shown at the lowest level which permits the ship to cross the bar.

[B]

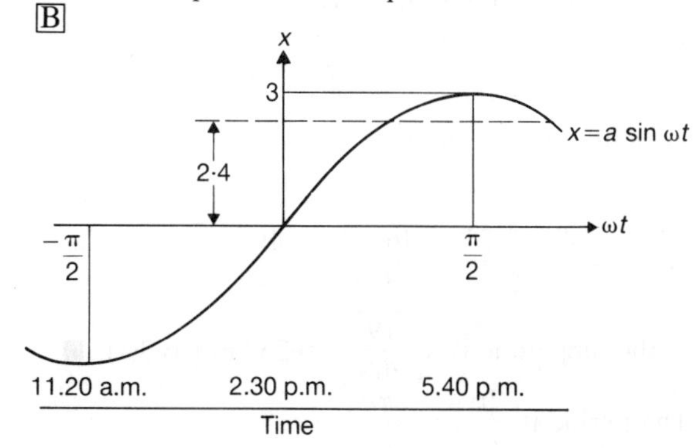

[C] If you obtain

$$\times\ \frac{380}{\pi}\arcsin 0{\cdot}8 = 6426\ \times$$

you should realise that something is wrong. This

answer is obtained by using the calculator in the degree mode instead of the radian mode.

[D] We are not told to what degree of approximation this answer is required, but we can assume a whole number of minutes is sufficient. However, it would be incorrect to round 112·2 down to 112 because we should then be permitting the ship to proceed before it would clear the bar. So, in this case, we round up to 113 minutes.

S4.

(a) [A] $x = A\cos(\omega t + \varepsilon)$.

$\therefore \dot{x} = -\omega A \sin(\omega t + \varepsilon)$.

$\therefore \ddot{x} = -\omega^2 A\cos(\omega t + \varepsilon) = -\omega^2 x$. [B] ■

$\therefore v^2 = \dot{x}^2 = \omega^2 A^2 \sin^2(\omega t + \varepsilon)$

$= \omega^2\{A^2 - A^2\cos^2(\omega t + \varepsilon)\}$

$= \omega^2(A^2 - x^2)$. [C] ■

(b) Using the relation just proved,

$$2^2 = \omega^2(A^2 - 0{\cdot}1^2) \quad (1)$$

and $$1{\cdot}75^2 = \omega^2(A^2 - 0{\cdot}15^2). \quad (2)$$

Subtract (2) from (1) to eliminate A. [D]

$2^2 - 1{\cdot}75^2 = -\omega^2(0{\cdot}1)^2 + \omega^2(0{\cdot}15)^2$.

$$\therefore \omega^2 = \frac{4 - (7/4)^2}{0{\cdot}0225 - 0{\cdot}01} = \frac{64 - 49}{16 \times 0{\cdot}0125}$$

$$= \frac{15}{0{\cdot}2} = 75. \quad \text{[E]}$$

Substitute for ω^2 in (1).

$4 = 75(A^2 - 0{\cdot}01)$.

$$\therefore A^2 = \frac{4 + 0{\cdot}75}{75} = \frac{16+3}{300} = \frac{19}{300}.$$

$\therefore$ the amplitude is $\sqrt{\dfrac{19}{300}} = 0{\cdot}252$ m (to 3sf). ■

The period is $\dfrac{2\pi}{\omega} = \dfrac{2\pi}{\sqrt{75}}$

$= 0{\cdot}726$ s (to 3sf). [F] ■

Maximum $|\dot{x}|$ is $\omega A = \sqrt{\omega^2 A^2} = \sqrt{4{\cdot}75}$.

$\therefore$ the maximum speed is $2{\cdot}18\ \text{m s}^{-1}$ (to 3sf).

[A] Read the question carefully. The phrase 'by substitution' has been put in to make it quite clear that you are *not* expected to go through the process of *solving* the differential equation, which would waste a lot of your time. Indeed, even if 'by substitution' is not mentioned in a question like this, substitution still provides a perfect answer.

[B] Do not give an 'otherwise' answer like:

✕ Differential equation and quoted solution both represent SHM; this proves the result. ✕

Here the examiner is demanding explicit substitution.

[C] This result could be obtained 'otherwise' by integrating the equation $\ddot{x} = -\omega^2 x$ with respect to x, but it is no easier, and you would have to identify your integration constant with A.

[D] This is easier than going straight for A by eliminating ω. We shall need ω for the rest of the question anyway.

[E] You could use a calculator for this work, but be careful to retain enough figures in your intermediate results, preferably by using the calculator memory.

[F] The question does not ask for results in decimal form, and so answers involving π and surds are acceptable. If you do give decimal results, state the degree of approximation.

Maximum $|\ddot{x}|$ is $\omega^2 A = \omega(\omega A) = \sqrt{75 \times 4{\cdot}75}$.
$\therefore$ the maximum acceleration is

$18{\cdot}9\ \mathrm{m\,s^{-1}}$ (to 3sf). ■

If $x = 0$ when $t = 0$ we may, using 5.2F3, write

$x = A \sin \omega t$.

$\therefore \dot{x} = \omega A \cos \omega t$,

where $\omega A = \sqrt{4{\cdot}75}$ and $\omega = \sqrt{75}$.

Hence, when $t = 0{\cdot}1$, the speed

is $|\dot{x}| = |\sqrt{4{\cdot}75} \cos(0{\cdot}1 \times \sqrt{75})|$
$= 1{\cdot}41\ \mathrm{m\,s^{-1}}$ (to 3sf). [G] *22m* ■

[G] Here it is really necessary to give the results decimally, because the magnitude of 'cos $\sqrt{0{\cdot}75}$' is not easily appreciated.

S5.

Let the displacement of the particle P be

$x = a \sin \omega t$. [A]

[A] We may choose any of the forms of SHM, but, noticing that part (b) demands the time to travel from the centre O, we expect the choice which makes $x = 0$ when $t = 0$ to be convenient.

The period is $\frac{1}{2}\pi$ s. $\quad \therefore \omega = \dfrac{2\pi}{\frac{1}{2}\pi} = 4$.

The maximum speed is $\omega a = 20$. $\quad \therefore 4a = 20$.
$\therefore$ the amplitude is 5 m.

$\dot{x}^2 = \omega^2(a^2 - x^2) = 16(25 - x^2)$. [B]

[B] Using 5.2F7

When $|x| = 2$, $\quad \dot{x}^2 = 16(25 - 4)$.

$\therefore$ the required speed is $4\sqrt{21}\ \mathrm{m\,s^{-1}}$.

When $x = 2$, $\quad 2 = 5 \sin 4t$. $\quad \therefore \sin 4t = 0{\cdot}4$. [C]
The earliest time after $t = 0$ when this is true is when $4t = \arcsin 0{\cdot}4$.
$\therefore$ the time to go directly from O to Q is
$0{\cdot}25 \arcsin 0{\cdot}4 = 0{\cdot}103$ s (to 3dp). ■

[C]

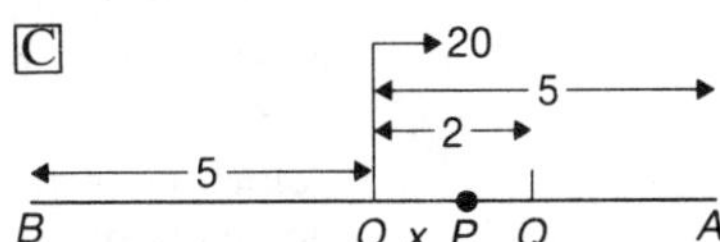

Even though this is very simple, a diagram may help here. The question says 'directly from O to Q', which we interpret as meaning that P is moving initially towards Q.
We therefore take $x = 2$ (not -2) at Q.

The force $F = m\ddot{x} \quad = -m\omega^2 a \sin \omega t$.

The velocity $v = \dot{x} \quad = \omega a \cos \omega t$.

$\therefore$ the power $P = Fv = -m\omega^3 a^2 \sin \omega t \cos \omega t$ [D]
$= -m\omega^3 a^2(\frac{1}{2} \sin 2\omega t)$. [E]

[D] Perform dimensional check (see 5.3I2)
[E] Noticing that the maximum value is demanded, we convert to $\sin 2\omega t$ which we know has maximum magnitude 1.

When $t = \dfrac{\pi}{12}$, $\quad P = -(0{\cdot}5)4^3 5^2\left(\dfrac{1}{2} \sin \dfrac{2\pi}{3}\right)$

$= -400\left(\dfrac{\sqrt{3}}{2}\right)$.

$\therefore$ the rate of working is $-200\sqrt{3}$ W.
The maximum rate of working
is $\frac{1}{2}m\omega^3 a^2 = 400$ W. *25m* ■

Chapter 6

Motion of two particles

6.1 GETTING STARTED

When we study the motion of two or more particles which move in the same straight line we can in principle treat the particles as separate, listing all the forces which act on each particle and then forming an equation of motion for each particle. In this way we obtain sufficient equations which can then be solved to obtain the acceleration of each particle. In practice it is often the case that the forces which the particles exert **on each other**, called the **internal forces of the system of particles**, are not given in the problem, and so the programme just suggested cannot be carried out. Other data must be used to determine the motion. In this chapter we first consider systems in which there is a known relation between the velocities of the particles (such as a car and a trailer, which normally have equal velocities). We then discuss the total momentum of two particles and continue to the methods of dealing with collision problems.

6.2 CONNECTED PARTICLES

ESSENTIAL FACTS

F1. The tension in a **light string** subject only to forces applied at its ends is the same at all points of the string (4.4F2). Consequently the forces exerted by the string on the bodies to which its ends are fastened are both of magnitude equal to the tension, and pull in the line of the string.

F2.

The forces exerted by a **smooth surface** on a string which is in contact with it are all **perpendicular to the string** by 4.4F4, and so the tension is the same everywhere in a light string passing round smooth pegs or pulleys (or light pulleys which can turn on smooth axles). So the forces exerted by the string at its ends are of magnitude equal to the tension and at each end pull in the line of the string.

F3. Total work done by tension

The total work done by the forces exerted at its ends by a light inextensible string at uniform tension is zero in any motion.

F4.

When a system of two particles connected by a light inextensible taut string passing over a smooth pulley is subject to gravitational forces only, the total gain of kinetic energy in any displacement is equal to the total loss of gravitational potential energy.

ILLUSTRATIONS

I1.

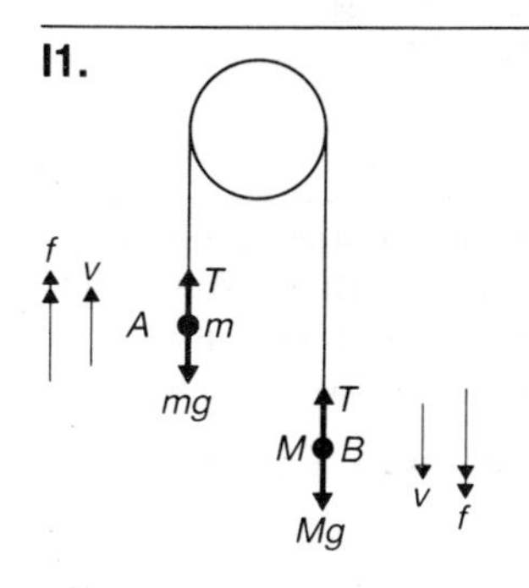

Two particles, A of mass m and B of mass M ($>m$), are connected by a light inextensible string passing over a smooth peg, and move vertically. Find the acceleration of each particle and the tension in the string. □

Because the string is **inextensible**, its length from A to B is constant. Consequently, if A undergoes an upward displacement x, B undergoes a downward displacement x. It follows that, if A has upward velocity v, B has downward velocity v, and similarly for the acceleration f, as shown in the diagram.

Because the string is **light** and the peg is **smooth**, the **tension is uniform** throughout the string. Let this tension be T. We can now write down the equations of motion:

$$A\uparrow: \qquad mf = T - mg$$

$$B\downarrow: \qquad Mf = Mg - T.$$

Add the equations to eliminate T.

$$(m + M)f = (M - m)g. \qquad \therefore f = \frac{M - m}{M + m}g.$$

$$\therefore T = mf + mg = m(f + g) = \frac{2Mmg}{M + m}. \ \blacksquare$$

12. One end of a light inextensible string is fixed. The string passes under a pulley B which carries a mass M and then over a fixed pulley C. A particle D of mass m is attached to the other end of the string. All parts of the string not in contact with the pulleys are vertical, and the pulleys are light and can turn on smooth axles. Find the acceleration of B and the tension in the string. □

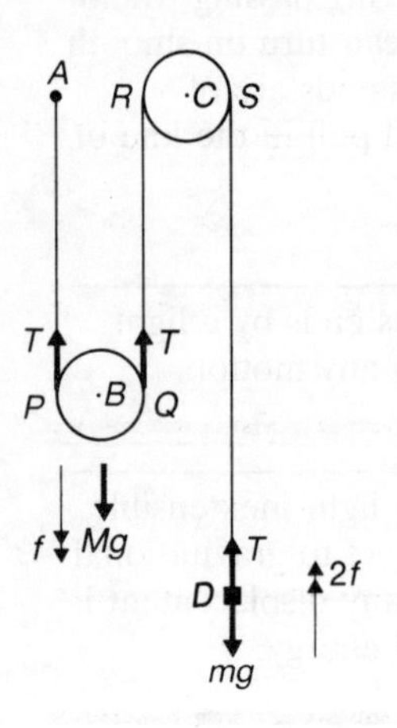

Let the vertical portions of the string be AP, QR and SD. Consider what happens when the pulley B undergoes a displacement x downwards. Both AP and QR increase by x, and so the increase in the length of string from A to R is $2x$. Since the string is inextensible, its length from A to D is constant. Therefore SD is reduced by $2x$, and so the displacement of D is $2x$ upwards. Consequently, if B has velocity v downwards, D has velocity $2v$ upwards, and similarly for the acceleration.

Let the acceleration of B be f downwards, so that the acceleration of D is $2f$ upwards.

Let the tension in the string be T. Then the equations of motion are:

$$B\downarrow: \quad Mf = Mg - 2T \quad \text{and} \quad D\uparrow: \quad 2mf = T - mg.$$

$$\therefore f = \frac{M - 2m}{M + 4m}g \quad \text{and} \quad T = \frac{3Mmg}{M + 4m}. \blacksquare$$

13. A lorry of mass M tows a trailer of mass m on a horizontal road. The resistance to the motion of the lorry is Q and the resistance to the motion of trailer is R. Find the acceleration and the tension in the coupling when the engine is exerting a tractive force F. In the case when $F = 0$ show that the coupling is in thrust if $\frac{Q}{M} > \frac{R}{m}$. □

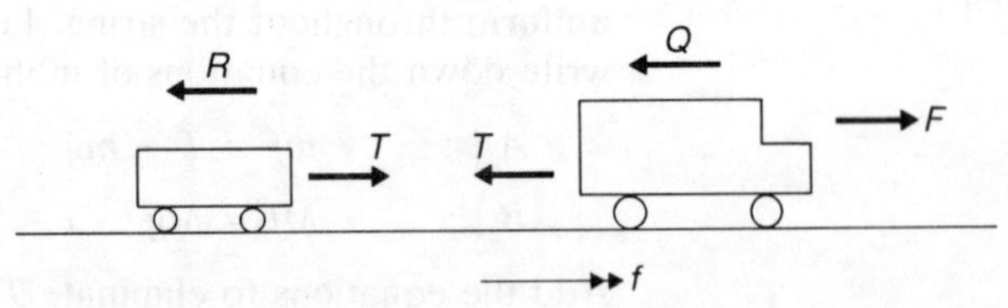

Because the lorry and trailer are moving on the same straight line with equal velocities, the system may be treated as a single particle with acceleration f. The equation of motion is

$$(M + m)f = F - Q - R.$$

$$\therefore \text{the acceleration is } f = \frac{F - Q - R}{M + m}.$$

For the trailer alone, $mf = T - R$, where T is the tension.

Hence $T = R + mf = \dfrac{m(F - Q) + MR}{M + m}$. ■

Check: For the lorry alone, $Mf = F - Q - T$.
Evaluate $Mf + T$ using the values found for f and T:

$$Mf + T = \frac{MF - MQ - MR + mF - mQ + MR}{M + m}$$
$$= F - Q. \checkmark$$

Putting $F = 0$, we see that T is negative when $mQ > MR$. ■

6.3 MOMENTUM OF TWO PARTICLES

ESSENTIAL FACTS

Two particles, P_1 of mass m_1 moving with velocity v_1 and P_2 of mass m_2 and velocity v_2, are subject to resultant external forces F_1 and F_2 respectively and to their mutual interaction. By Newton's Third Law (4.4F1), the forces they exert on each other are equal in magnitude and act in the line P_1P_2 in opposite directions ($\rightarrow R$ and $\leftarrow R$).

F1.

The **rate of change of the total momentum** of the two particles is equal to the sum of all the **external forces** acting on the particles.

$$\frac{d}{dt}(m_1v_1 + m_2v_2) = F_1 + F_2.$$

F2. Principle of conservation of momentum

When P_1 and P_2 are subject only to their mutual interaction, which means that **no external forces act**, the two particles are called an **isolated system**. In this case

$$m_1v_1 + m_2v_2 = \text{constant}.$$

In words:

> The total momentum of a system subject to no external forces remains constant.

Equivalently:

> Momentum is conserved in an isolated system.

F3. Energy in an isolated system

The increase in the total kinetic energy of an isolated system of two particles during a time interval is equal to the total work done in the interval by the forces of interaction between the particles. Note that kinetic energy is **not conserved** except when the work done by the internal forces is zero.

F4.

The impulses in any time interval of the forces exerted by two particles on each other are equal in magnitude and opposite in direction.

F5.

The increase in the total momentum of the two particles in any time interval is equal to the sum of the impulses (over the interval) of all the external forces acting on the particles.

ILLUSTRATIONS

I1.

From Newton's Laws of motion derive the principle of conservation of momentum for a system of two particles which move in the same straight line. □

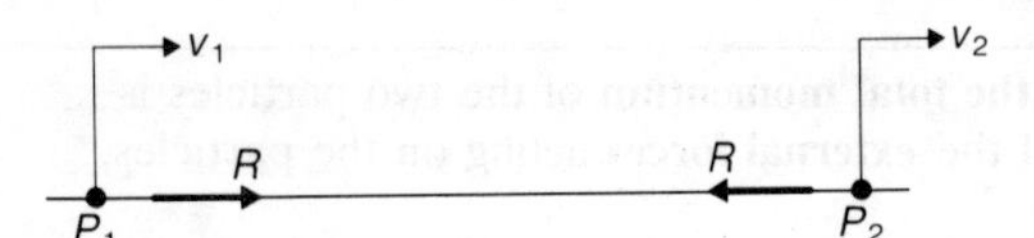

Let the two particles be P_1 of mass m_1 moving with velocity v_1 and P_2 of mass m_2 and velocity v_2. No forces external to the two-particle system act, but P_1 and P_2 exert forces on each other. Let the force exerted on P_1 by P_2 be R in the direction P_1P_2.

By Newton's **Third Law** (4.4F1), the force exerted on P_2 by P_1 is R in the direction P_2P_1.

By Newton's **Second Law**,

$$P_1 \rightarrow: \quad m_1\dot{v}_1 = R \quad \text{and} \quad P_2 \rightarrow: \quad m_2\dot{v}_2 = -R.$$

Add the equations:

$$m_1\dot{v}_1 + m_2\dot{v}_2 = 0.$$

$$\therefore \frac{\mathrm{d}}{\mathrm{d}t}(m_1v_1 + m_2v_2) = 0.$$

$$\therefore m_1v_1 + m_2v_2 = \text{constant}.$$

Hence, when no external forces act, the total momentum of P_1 and P_2 remains constant. ■

The principal of conservation of momentum applies when there are **no external** forces, and this **must be mentioned** in any statement of the principle.

The following argument is not a satisfactory derivation.

'Impulse = change in momentum (4.2F9).

∴ total increase in momentum during impact = (impulse on P_1) + (impulse on P_2)

✘ By Newton's Third Law:

Impulse exerted on P_2 by P_1 is equal and opposite to the impulse exerted on P_1 by P_2. ✘

∴ total increase in momentum is zero.'

Although the statement marked '✘' (which is 6.3F4) is true, it is *not* Newton's Third Law, for that Law (4.4F1) refers to forces, not impulses. The argument could be completed by *proving* 6.3F4, as follows:

✔ Sum of impulses on P_1 and P_2 in the duration, $t_1 \leqslant t \leqslant t_2$, of the impact is

$$\int_{t_1}^{t_2} R\,\mathrm{d}t + \int_{t_1}^{t_2} (-R)\,\mathrm{d}t \qquad \text{(using 4.2F8)}$$

$$= \int_{t_1}^{t_2} R\,\mathrm{d}t - \int_{t_1}^{t_2} R\,\mathrm{d}t = 0, \text{✔}$$

but this is less simple.

I2. Particles P and Q are subject to their mutual interaction and to no other forces. Initially Q is at rest and P is moving at a speed of $6\ \text{m s}^{-1}$ away from Q. When P comes to rest, Q is moving with speed $3\ \text{m s}^{-1}$. Given that the mass of P is 2 kg, find the mass of Q, and calculate the work done against the forces of interaction. □

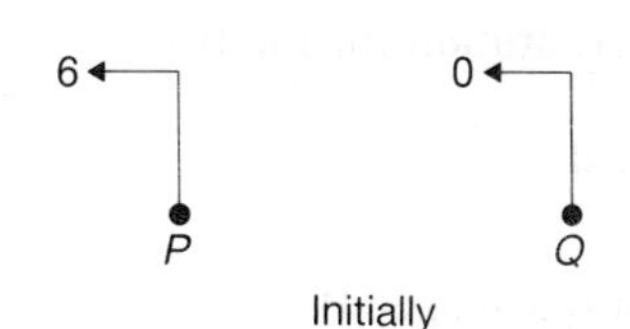

We do not know the interaction force, but we do know that the two particles form an **isolated system**, and so the principle of conservation of momentum applies. Let the mass of Q be m kg. Then

$$2 \times 6 + 0 = 0 + (m)3. \qquad \therefore m = 4.$$

$\therefore$ the mass of Q is 4 kg.

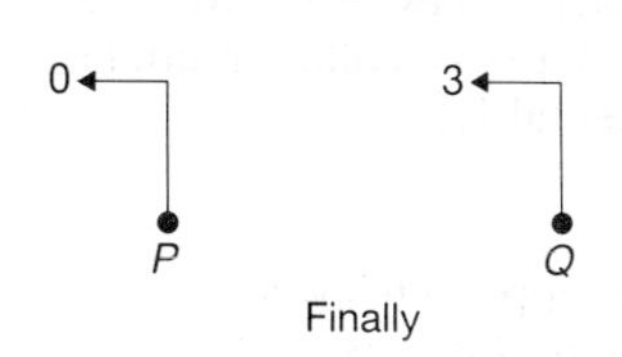

By F3, since the interactions are the only forces:

$$\begin{aligned}\text{Total work done by the interactions} &= \text{gain in kinetic energy} \\ &= \tfrac{1}{2}m(3^2) - \tfrac{1}{2}(2)6^2 \\ &= \tfrac{1}{2}(4)3^2 - \tfrac{1}{2}(2)6^2 = -18\ \text{J}.\end{aligned}$$

$\therefore$ the work done *against* the interaction forces is 18 J. ■

6.4 COLLISIONS

A collision between two bodies is an interaction which takes place within a limited time interval. During a collision, the force of interaction between the bodies is much larger than the forces exerted on them by all other bodies (external forces). The external forces (such as gravity, for example) are therefore treated as negligible in calculating the changes in the velocities which occur due to the collision.

Here we treat collisions between particles which move always in the same straight line, called **direct impact**. The same rules apply to collisions between uniform spheres when the centres of the spheres are moving on the same straight line.

ESSENTIAL FACTS

F1. In any collision between two particles the **total momentum is conserved**.

This follows from 6.3F2 because the neglect of all external forces amounts to taking the two particles as forming an isolated system.

F2. Restitution law

In a **direct impact** the magnitude v_S of the relative velocity of the particles after the collision is proportional to the magnitude v_A of the relative velocity before the collision.

$$v_S = ev_A.$$

The constant e is called the **coefficient of restitution** for the two particles; $0 \leqslant e \leqslant 1$.
For brevity, v_A is called the **approach speed** and v_S the **separation speed.** So:

$$\text{(Separation speed)} = e \times \text{(approach speed)}.$$

The law of restitution is not a fundamental law of nature, but it can be experimentally verified for some types of colliding bodies. It is sometimes called 'Newton's Experimental Law'.

F3. Perfectly elastic collision

When the separation speed is **equal** to the approach speed ($v_S = v_A$) the collision is said to be **perfectly elastic** (sometimes just 'elastic' is used).
When $e = 1$ the collision is perfectly elastic.

F4.

When the two particles **coalesce** (stick together, $v_S = 0$) the collision may be called **perfectly inelastic** (sometimes just 'inelastic').
When $e = 0$ the collision is perfectly inelastic.

F5. Energy in collisions

In an impact the work done by the interaction forces is normally negative, and so, by 6.3F3, kinetic energy is lost. Thus **kinetic energy is not necessarily conserved in a collision**. The lost kinetic energy is converted into some combination of sound with heat and other forms of energy within the bodies.

F6.

In a perfectly elastic collision no kinetic energy is lost.

An alternative definition of a perfectly elastic collision is that it is a collision in which kinetic energy is conserved.

When $e < 1$ some kinetic energy is lost. In a perfectly inelastic collision ($e = 0$) the greatest possible amount of kinetic energy, consistent with momentum conservation, is lost.

F7.

When a particle collides with a fixed smooth rigid body the component of its velocity tangential to the surface is unaltered. The magnitude of the normal component of the velocity of rebound is e times the magnitude of the normal component of the velocity of approach, where e is the coefficient of restitution between the particle and the surface.

I1.

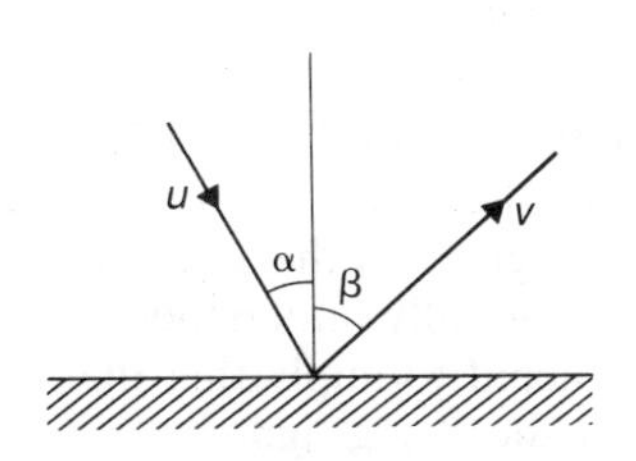

A particle collides with a fixed smooth plane surface. The coefficient of restitution between the particle and the plane is e. The particle approaches the plane with speed u along a line making an angle α with the normal to the plane. Find the velocity of rebound. □

Suppose the particle rebounds with speed v in a direction making an angle β with the outward normal to the plane. Then, by F7:

Tangential components: $v \sin \beta = u \sin \alpha$.

Normal components, restitution: $v \cos \beta = eu \cos \alpha$.

$\therefore\ e \tan \beta = \tan \alpha$ and

$$v^2 = v^2(\sin^2 \beta + \cos^2 \beta) = u^2 \sin^2 \alpha + e^2u^2 \cos^2 \alpha.$$

$\therefore$ the particle rebounds with speed $\sqrt{u^2 \sin^2 \alpha + e^2u^2 \cos^2 \alpha}$ at an angle to the normal of $\arctan(e^{-1} \tan \alpha)$. ■

I2.

A particle A of mass $2m$ moving with speed $5u$ impinges directly on a particle B of mass $4m$ moving with speed $2u$ in the same direction. The coefficient of restitution between the particles is e. Find the kinetic energy lost in the impact. □

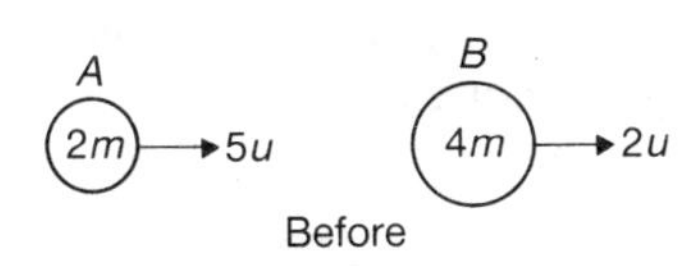

Let the velocities of A and B after the impact, measured in the same direction as the initial velocities, be v and w respectively.

Momentum conservation (F1): $2mv + 4mw = 2m(5u) + 4m(2u)$.

$$\therefore\ v + 2w = 9u. \qquad (1)$$

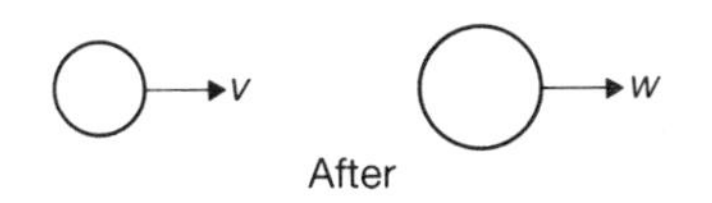

Restitution (F2): $w - v = e(5u - 2u) = 3eu. \qquad (2)$

Solve (1) and (2) for v and w.

$$v = (3 - 2e)u \quad \text{and} \quad w = (3 + e)u.$$

$\therefore$ the loss in kinetic energy is

$$\tfrac{1}{2}(2m)(5u)^2 + \tfrac{1}{2}(4m)(2u)^2 - \{\tfrac{1}{2}(2m)v^2 + \tfrac{1}{2}(4m)w^2\}$$

$$= \tfrac{1}{2}mu^2\{50 + 16 - 2(3 - 2e)^2 - 4(3 + e)^2\} = 6mu^2(1 - e^2). \ ■$$

Note that, in accordance with F6, the loss of kinetic energy is zero when $e = 1$ (perfectly elastic) and greatest when $e = 0$ (particles coalesce).

13. A particle P of mass $2m$ and a particle Q of mass $5m$ are connected by a light inextensible inelastic string and lie at rest on a smooth horizontal table with the string slack. P is then given a blow of impulse J in the direction QP. Find the speed of the particles after the string has become taut. Calculate the impulse of the tension in the string when the string tightens, and determine the loss of kinetic energy which then occurs. □

Let the speed of P before the string tightens be u. Since the table is **smooth** and **horizontal**, and there are no other forces with horizontal components, u is the speed initially given to P by the blow.

$$\therefore 2mu = J. \quad \boxed{A} \quad \therefore u = \frac{J}{2m}.$$

Because the string is **inextensible**, the distance PQ cannot exceed the fixed length of the string. Because the string is **inelastic**, there is **no rebound** when the string tightens; it will remain taut. Therefore P and Q have the same velocity after the string tightens; call it v. $\boxed{B}$
By the principle of conservation of momentum (6.3F2)

$$2mv + 5mv = 2mu + 0. \quad \boxed{C}$$

$$\therefore v = \frac{2u}{7} = \frac{J}{7m}. \blacksquare$$

The impulse K of the tension in the string is equal to the increase in the momentum of Q when the string tightens, as the diagram shows. $\boxed{D}$

$$K = 5mv - 0 = 5m\left(\frac{J}{7m}\right) = \frac{5J}{7}. \quad \boxed{E}$$

The loss of kinetic energy when the string tightens is

$$\frac{1}{2}(2m)u^2 - \left\{\frac{1}{2}(2m)v^2 + \frac{1}{2}(5m)v^2\right\}$$

$$= \frac{1}{2}\frac{J^2}{m}\left\{2\left(\frac{1}{2}\right)^2 - 7\left(\frac{1}{7}\right)^2\right\} = \frac{5J^2}{28m}. \blacksquare$$

$\boxed{A}$ Using the impulse momentum relation 4.2F9.

$\boxed{B}$ Very often only one of the words 'inextensible', 'inelastic' is given in the statement of a problem. It is conventionally understood that *either* of these words is meant to imply that the two particles move, after the string tightens, with the same velocity. Consequently the problem is similar to the case of particles which collide and coalesce (perfectly inelastic collision).

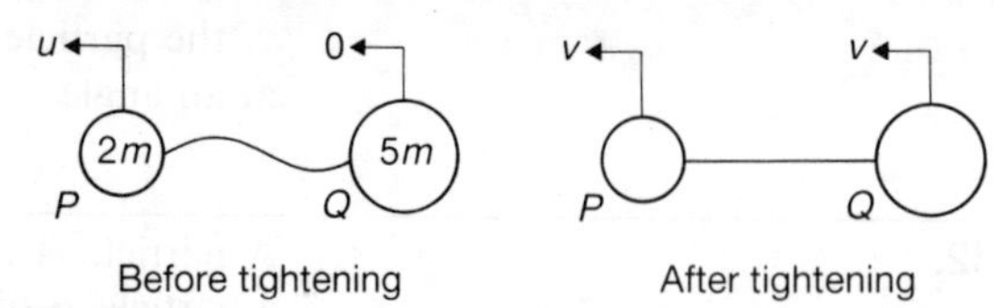

$\boxed{C}$ As the string tightens P and Q are subject to forces, due to the tension in the string, which are equal and opposite. No external forces act, and so the total momentum does not change.

$\boxed{D}$

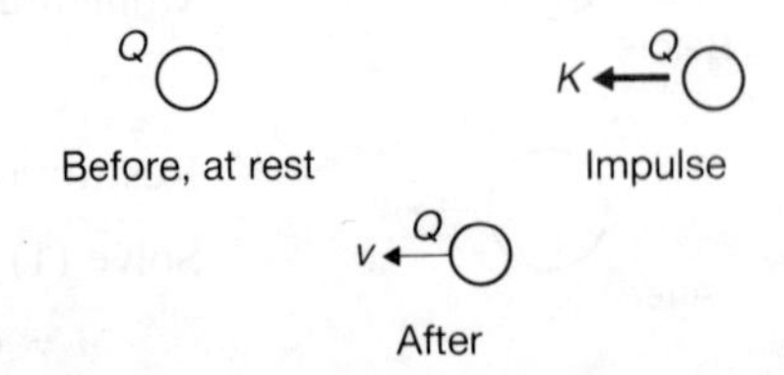

$\boxed{E}$ The phrase 'impulsive tension' is sometimes used to mean 'impulse of the tension'.

6.5 EXAMINATION QUESTIONS AND SOLUTIONS

Q1.

A smooth peg, of negligible diameter, is fixed at a height $3l$ above a horizontal table and a light inextensible string of length $4l$ hangs over the peg. A particle of mass m is attached to one end of the string and a particle of mass $2m$ is attached to the other end. The system is held at rest with the particles hanging at the same level and at a distance l from the table. The parts of the string not in contact with the peg are vertical. The system is then released from rest. Find, in terms of g and l, the speed u with which the particle of mass $2m$ strikes the table. Given that this particle rebounds from the table with speed $\frac{1}{2}u$, find, in terms of m, g and l, the magnitude of the impulse on the table. (JMB 1984)

Q2.

A small sphere of mass m is moving with speed u on a smooth horizontal plane along a straight line AO. At O the sphere collides with a smooth vertical wall, the normal to which makes an angle α with AO. Explain why the component of the velocity of the sphere parallel to the wall is unaltered by the impact.

Given that the sphere rebounds from the wall along a line OB perpendicular to OA, show that $e = \tan^2 \alpha$, where e is the coefficient of restitution between the sphere and the wall. Deduce the range of possible values of α.

Find, in terms of α, the fraction of the initial kinetic energy lost during the impact. Show that the magnitude of the impulse on the wall is $mu \sec \alpha$. (JMB 1986)

Q3.

A car of mass 1500 kg tows a caravan of mass 500 kg. The car and caravan move on a straight level road with the engine of the car exerting a constant pull of 4100 N. Given that there are frictional resistances of 800 N on the car and 300 N on the caravan, find the magnitude of

(a) the acceleration of the car,
(b) the tension in the tow-bar between the car and the caravan.

The car and caravan then go up a straight road which is inclined at an angle $\sin^{-1}(1/8)$ to the horizontal, the frictional resistances on the car and caravan remaining unchanged. The speed of the car and caravan increases from 10 m s^{-1} to 20 m s^{-1} in 16 seconds with the engine of the car now exerting a constant pull of P newtons. Find the value of P and the rate at which the engine is working when the speed is 15 m s^{-1}.

When the speed is 20 m s^{-1}, the tow-bar breaks. Find, to the nearest 10 m, the further distance travelled by the caravan before it comes momentarily to rest.

(Take g as 10 m s^{-2}.) (LON 1984)

Q4. Two small smooth spheres A and B of equal radius but of masses $3m$ and $2m$ respectively are moving towards each other so that they collide directly. Immediately before the collision sphere A has speed $4u$ and sphere B has speed u. The collision is such that sphere B experiences an impulse of magnitude $6mcu$, where c is a constant. Find

(a) in terms of u and c, the speeds of A and B immediately after collision,
(b) the coefficient of restitution, in terms of c,
(c) the range of values of c for which such a collision would be feasible,
(d) the value of c such that $\frac{9}{16}$ of the total kinetic energy would be destroyed by the collision. (AEB 1984)

Q5. Two particles of masses $4m$ and $3m$ respectively are attached one to each end of a light inextensible string which passes over a small smooth pulley. The particles move in a vertical plane with both the hanging parts of the string vertical. Write down the equation of motion for each of the particles and hence determine, in terms of m and/or g as appropriate, the magnitude of the acceleration of the particles and of the tension in the string.

When the particle of mass $3m$ is moving upwards with a speed V it picks up from rest at a point A an additional mass $2m$ so as to form a composite particle Q of mass $5m$. Determine

(a) the initial speed of the system,
(b) the impulsive tension in the string immediately the additional particle has been picked up,
(c) the height above A to which Q rises. (AEB 1983)

Q6. Two small smooth spheres A and B of equal mass moving in the same straight line (and in the same direction) with speeds $2u$ and u respectively, collide directly. After collision their directions of motion remain unchanged but their speeds are v and $1{\cdot}5v$ respectively. Show that the coefficient of restitution is $0{\cdot}6$.

Sphere A is set to move on a smooth horizontal floor and collides directly with B which is at rest at a distance of 2 m from a smooth vertical wall, B being the nearer to the wall. The motions before and after all possible collisions are perpendicular to the wall, and the coefficient of restitution for all collisions is 0·6.

(i) Show that when B collides with the wall for the first time, A is at a distance of 1·5 m from the wall.

(ii) Find the distance of the spheres from the wall when they collide for the second time. (WJEC 1985)

Q7.

Two small beads P and Q of masses 0·003 kg and 0·006 kg respectively are threaded on a smooth circular wire of radius 0·5 m which is maintained in a horizontal plane. Initially the beads P and Q are at rest at points A and B respectively, where A and B are at opposite ends of a diameter of the wire. The coefficient of restitution between the beads is 1/4. The bead P is then projected towards Q with a speed of $u\ \text{m s}^{-1}$. Find the velocities of the beads immediately after they first collide. Given that $u = 6$ find, as a multiple of π, the time that elapses between the first and second collision.

When the wire is rough and P is projected from A with a speed of $6\ \text{m s}^{-1}$ so as to collide with Q which is at rest at B, the speed of Q, immediately after the collision, is $1{\cdot}25\ \text{m s}^{-1}$. Find the work done by friction as P travels from A to collide with Q and the magnitude of the impulse acting on P during its collision with Q. (AEB 1985)

Q8.

Three small smooth spheres A, B and C, of masses $15m$, $5m$ and λm, are at rest in a straight line on a horizontal plane. Sphere A then moves along the plane and strikes directly sphere B which moves off with speed v. Sphere B goes on to strike sphere C directly. The coefficient of restitution between each pair of spheres is $\frac{1}{2}$.

(i) Find, in terms of v, the speed of A before and after impact with B.

(ii) Find, in terms of m and v, the change in

(*a*) the magnitude of the momentum of A,

(*b*) the magnitude of the momentum of B,

as a result of their collision.

Given that, after the impact of B with C, there are no further collisions, shown that $\lambda \leqslant 40/19$. (LON 1986)

Q9.

A particle of mass $4m$, which is at rest, explodes into two fragments, one of mass m and the other of mass $3m$. The explosion provides the fragments with total kinetic energy E. Find the velocities of the fragments just after the explosion. Hence find, in terms of m and E, the magnitude of the relative velocity of the fragments just after the explosion. (JMB 1982)

Q10. Two particles, P of mass $2m$ and Q of mass m, are subject to mutual forces of attraction, and no other force acts on them. At time $t = 0$, P is at rest at a fixed point O and Q is moving directly away from O with speed $5U$. At a later instant, when $t = T$, before any collision has taken place, Q is moving towards O with speed U. Find, in terms of m and U, the total work done by the forces of attraction during the time interval $0 \leqslant t \leqslant T$.
At the instant $t = T$ impulses of magnitudes J and K are applied to P and Q, respectively, bringing both of them to rest. Find J and K in terms of m and U. (JMB 1985)

SOLUTIONS

S1.

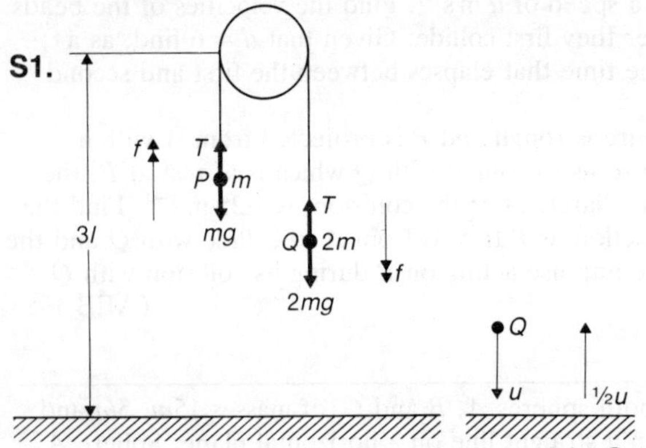

Let the particle P have mass m and the particle Q mass $2m$. Let the downward acceleration of Q be f; then f is also the upward acceleration of P. Let the tension in the string be T. The equations of motion of the particles are:

$P\uparrow$: $\quad mf = T - mg.$

$Q\downarrow$: $\quad 2mf = 2mg - T.$

$\therefore 3mf = mg.$ [A] $\quad \therefore f = \frac{g}{3}.$

The speed u with which Q strikes the table is the speed acquired in a distance l at constant acceleration f.

$\therefore u = \sqrt{2fl} = \sqrt{\frac{2gl}{3}}.$ [B] ■

The upward impulse exerted on Q by the table is equal to the increase in the upward momentum of Q, which is $2m(\frac{1}{2}u) - 2m(-u)$. [C]
Hence the magnitude of the impulse on the table is $3mu = m\sqrt{6gl}$. [D] 11m ■

[A] It is unwise to present the following derivation.
✕ Total mass in motion with acceleration f is

$$m + 2m = 3m.$$

Net force acting on this mass is

$$2mg - mg = mg.$$
$$\therefore \quad 3mf = mg. \text{ ✕}$$

Although not untrue, this might be considered incomplete and judged as evidence of defective understanding of Newton's Second Law, and thus lose marks. It is possible to elaborate an argument which justifies this method but it is not simple. The method is sometimes called 'resolving round the string'.
[B] Alternatively the principle of energy (4.3F6) may be applied to P and Q. From 6.2F3, the total work done by the string on P and Q is zero. Therefore the total increase in kinetic energy of P and Q is equal to the total work done by gravity. In the displacement in which Q falls to the table:

Total increase in K.E. $= \frac{1}{2}mu^2 + \frac{1}{2}(2m)u^2.$

Total work done by gravity $= -mgl + 2mgl.$

$\therefore \frac{3}{2}mu^2 = mgl$, giving the same value for u.
[C] This is positive, as expected, for the diagram shows Q losing downward momentum and gaining upward momentum. So the error of
✕ Impulse $= 2m(\frac{1}{2}u) - 2mu$ ✕ is clearly seen.
[D] By 6.3F4 the impulse *on* the table is downwards, and has the same magnitude as the impulse exerted *by* the table on Q.

S2.

The wall is smooth.
Therefore, at all times during the contact, the force acting on the sphere is along the normal ON. [A]
Therefore the component of the acceleration along the wall is always zero. [B]
Therefore the velocity component along the wall is unaltered. [C] ■
Let the speeds before and after the impact be u and v respectively.
Because the wall is smooth,

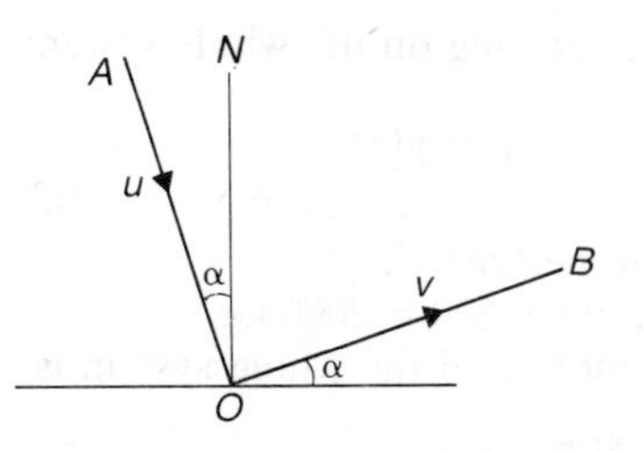

$$v \cos \alpha = u \sin \alpha \qquad (1)$$

Restitution: $$v \sin \alpha = eu \cos \alpha \qquad (2)$$

Divide (2) by (1): $\tan \alpha = e \cot \alpha$.

$\therefore$ $e = \tan^2 \alpha$. [D] ■

$0 \leqslant e \leqslant 1$. $\therefore 0 \leqslant \alpha \leqslant 45°$. ■

From (1), $v = u \tan \alpha$.
$\therefore$ the loss in kinetic energy is

$$\tfrac{1}{2}mu^2 - \tfrac{1}{2}mv^2 = \tfrac{1}{2}mu^2(1 - \tan^2 \alpha).$$

$\therefore$ the fraction lost is $1 - \tan^2 \alpha$. ■
The components of the momentum in the direction ON are:

Before impact: $-mu \cos \alpha$.
After impact: $mv \sin \alpha$.

$\therefore$ the impulse exerted by the wall on the sphere is

$$mv \sin \alpha - (-mu \cos \alpha) \qquad \text{[E]}$$

in the direction ON.
Its magnitude is $mv \sin \alpha + mu \cos \alpha$.
$\therefore$ the magnitude of the impulse exerted *on* the wall is

$$mv \sin \alpha + mu \cos \alpha \qquad \text{[F]}$$

$$= mu \tan \alpha \sin \alpha + mu \cos \alpha \qquad \text{[G]}$$

$$= mu\left(\frac{\sin^2 \alpha}{\cos \alpha} + \cos \alpha\right)$$

$$= mu \sec \alpha.$$ *21m* ■

[A] Using 4.4F4. Mention of smoothness is essential to the explanation.
[B] Resolving parallel to the plane (9.4I2 [A])
[C] **?** Smoothness ⇒ Impulse along ON **?**
This might just suffice, but really smoothness refers to forces, not impulses, and so it is better to argue via the acceleration.

[D] If you make the mistake of interchanging $\sin \alpha$ with $\cos \alpha$, or put e in the wrong place, you will not get the printed answer. When you have found the mistake, cross out all the work, clearly, and start again, writing down (1) and (2), so that the examiner can see that all the steps are correct.

[E] The sign must be very carefully checked here. ✖ $mv \sin \alpha - mu \cos \alpha$ ✖ is likely to be regarded not as an arithmetical slip, but as evidence of not understanding impulse, and so it may be heavily penalised.

[F] Using 6.3F4.

[G] This step, or the next one, must be inserted and the working must be very clear here, because the answer is printed.

S3.

The horizontal forces acting on the whole system are:

Pull of the engine $\rightarrow$ 4100 N.

Total resistance $\leftarrow$ (800 + 300) N. [A]

Let the acceleration be f m s^{-2}.

The total mass is 1500 + 500 = 2000 kg.

$\therefore$ the equation of motion of the whole system is

$$2000f = 4100 - 1100 = 3000.$$

$\therefore$ the acceleration is $1{\cdot}5$ m s^{-2}. ■

Let the tension in the tow-bar be T N.

The equation of motion of the caravan alone is

$$500f = T - 300.$$

$$\therefore T = 300 + 500f = 300 + 750.$$

$\therefore$ the tension is 1050 N. [B] [C] ■

The component of the weight of the whole system in the direction down the slope is a force Q N, where

$$Q = 2000g\left(\frac{1}{8}\right) = 2500. \quad \text{[D]}$$

The speed increases by an amount $20 - 10 = 10$ m s^{-1} in 16 seconds.

Therefore the acceleration up the slope is $\frac{10}{16}$ m s^{-2}.

The tractive force is now P N, and so the equation of motion of the system is

$$2000\left(\frac{10}{16}\right) = P - 1100 - 2500.$$

$$\therefore P = 3600 + 1250 = 4850.$$

When the speed is 15 m s^{-1}, the rate of working is $P(15) = 4850 \times 15 = 72\,750$ W. ■

When the tow-bar breaks, assuming that the caravan keeps rolling, the forces acting on the caravan are the resistance, 300 N down the slope, and the weight component S N, where

$$S = 500g\left(\frac{1}{8}\right) = 625.$$

Let the acceleration of the caravan up the slope be f_1 m s^{-2}. The equation of motion is

$$500f_1 = -300 - 625 = -925. \qquad \therefore f_1 = -\frac{37}{20}.$$

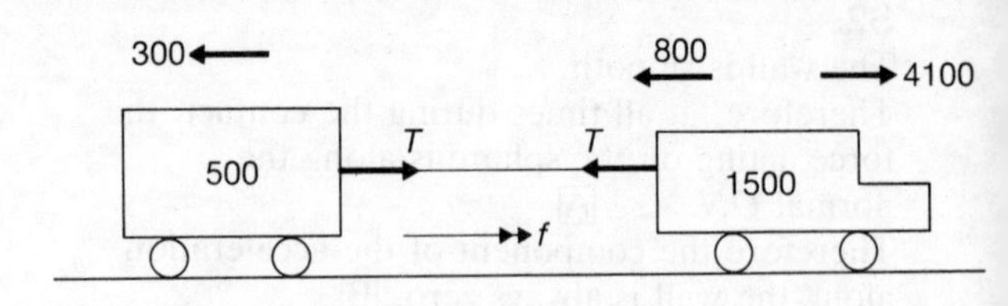

[A] Treating car-and-caravan as a single particle (see 6.2I3) leads directly to f without involving T.

[B] It is important to show $T\leftarrow$ acting on the car as well as $T\rightarrow$ acting on the caravan. Omission of the former force from the diagram easily leads to errors like

✗ Whole system:
$2000f = 4100 - 800 - 300 + T$ ✗

or

✗ Car alone: $1500f = 4100 - 800$ ✗

instead of ✓ $1500f = 4100 - 800 - T$ ✓ (1)

[C] *Check*: Write down equation (1) and substitute $f = 1{\cdot}5$, $T = 1050$.

LHS = 2250. RHS = 3300 − 1050 = 2250. √

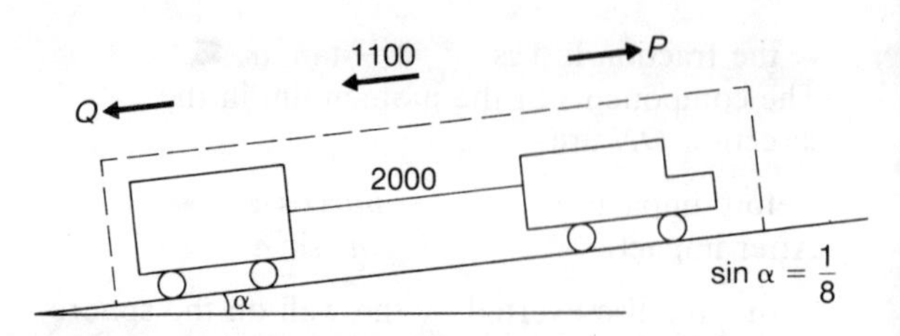

[D] The weight is $2000g$ vertically downwards. Its component in the direction down the slope is given by $Q = 2000g \sin \alpha$.

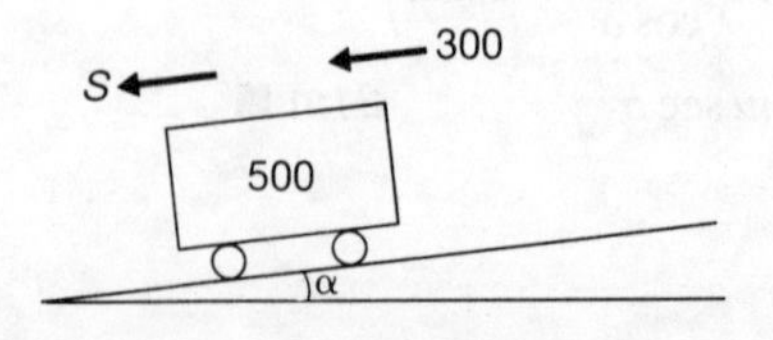

The distance travelled by the caravan starting with speed 20 m s^{-1} and retarded at a rate $\frac{37}{20}$ m s^{-2} is

$$\frac{20^2}{2(37/20)} = \frac{4000}{37} = 110 \text{ m (to 10 m).}$$ *25m* [E] ■

[E] Using 3.3F7D.

S4.

Let the velocities of A and B after the collision be v and w in the direction ⊕→ of the initial velocity of A.

The impulse–momentum equation for A is

$3mv - 12mu = -6mcu.$ [A]

$\therefore v = (4 - 2c)u.$ (1)

The impulse–momentum equation for B is

$2mw + 2mu = 6mcu.$ [B]

$\therefore w = (3c - 1)u.$ (2)

Therefore the speeds after the collision are

A: $|4 - 2c|u$ and B: $|3c - 1|u$. [C] ■

Let the coefficient of restitution be e.

Then $w - v = e(u + 4u) = 5eu$.

$\therefore 5e = 3c - 1 - (4 - 2c) = 5c - 5.$ [D]

$\therefore e = c - 1.$ ■

Since $0 \leqslant e \leqslant 1$, $\quad 1 \leqslant c \leqslant 2.$ ■

If $\frac{9}{16}$ of the kinetic energy is destroyed, $\frac{7}{16}$ remains. Therefore

$$\frac{1}{2}(3m)v^2 + \frac{1}{2}(2m)w^2 = \frac{7}{16}\left\{\frac{1}{2}(3m)(4u)^2 + \frac{1}{2}(2m)u^2\right\}.$$

Substitute for v and w using (1) and (2).

$$3(4 - 2c)^2 + 2(3c - 1)^2 = \frac{7}{16}(48 + 2).$$

$\therefore 8(30c^2 - 60c + 50) = 175.$

$\therefore 16c^2 - 32c + 15 = 0.$

$\therefore (4c - 3)(4c - 5) = 0.$

The root $c = \frac{3}{4}$ is outside the permitted range.

$\therefore c = \frac{5}{4}.$ [E] *25m* ■

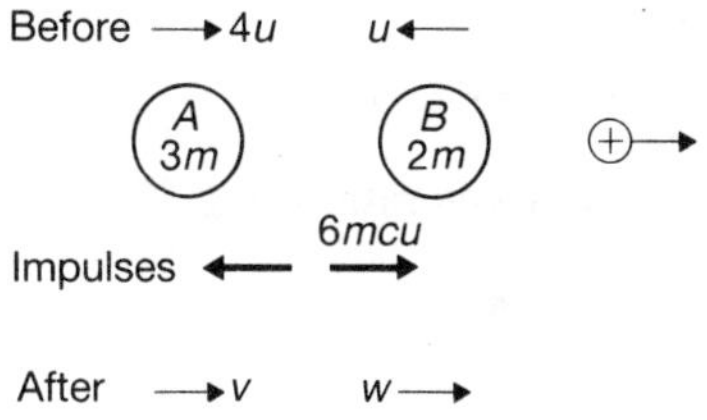

Before →4u u←

A 3m B 2m ⊕→

Impulses ← 6mcu →

After →v w→

[A] The impulse on A is equal and opposite to the impulse on B (6.3F4). In this question the impulse is given, and so v and w can be found directly from impulse–momentum relations (4.2F9).

[B] The increase in momentum is $2mw - (-2mu)$.

[C] As we do not know the values that c may take at this stage, the signs of v and w are unknown, so modulus signs are needed to ensure that the expressions for the speeds are positive.

[D] Since $6mcu$ is an impulse, c must have zero dimension. e is also dimensionless, being a ratio of speeds. So, if any terms involving u remain at this stage, the equation is dimensionally wrong, and you must look for an error.

[E] The question, in asking for the *value*, not the 'values' of c, hints that there is only one. You must take the hint, and explain why one root of the equation is to be rejected.

S5.

Let the particle B have mass $4m$ and the particle C mass $3m$.
Let the downward acceleration of B be f.
Then the upward acceleration of C is f.
Let the tension in the string be T.
The equations of motion of the particles are:

$B\downarrow$: $\quad 4mf = 4mg - T.$

$C\uparrow$: $\quad 3mf = T - 3mg.$

$\therefore\ 7mf = mg \quad$ and $\quad 7T = 24mg.$

$\therefore$ the magnitude of the acceleration is $\dfrac{g}{7}$

and the tension is $\dfrac{24mg}{7}$. ■

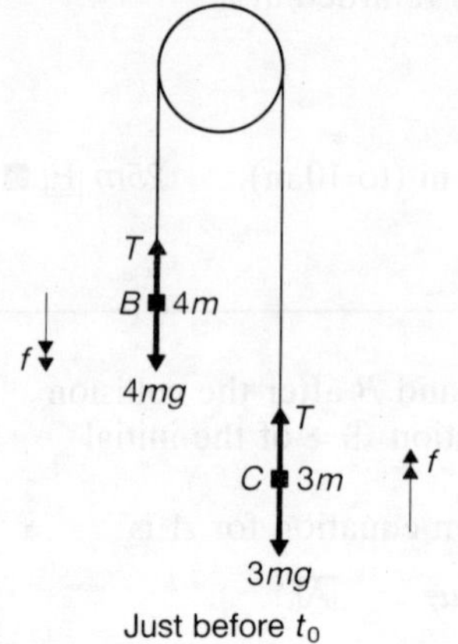

Just before t_0

Impulsive motion

Let the 'initial speed of the system' be v. [A]
Suppose the 'impulsive tension' at the instant t_0 when the particle D of mass $2m$ is picked up is J.
An impulse J is thus applied upwards both to B and to the 'composite particle' Q. [B]
$\therefore J$ is the increase in the upward momentum of each particle which suddenly occurs at time t_0 when D is picked up.

Momentum of B just before t_0 is $-4mV$.
Momentum of B just after t_0 is $-4mv$.

Therefore, by 4.2F9, $\quad -4mv - (-4mV) = J.$

Momentum of Q just before t_0 is
(momentum of C) + (momentum of D)
$3mV \quad + \quad 0.$

Momentum of Q just after t_0 is $5mv$.

$\therefore\ 5mv - 3mV = J.$ [C]

$\therefore\ 4mV - 4mv = 5mv - 3mV.$

Hence the 'initial speed' $\quad v = \dfrac{7V}{9}$. ■

The 'impulsive tension' $J = 4m(V - v)$

$= \dfrac{8mV}{9}$. ■

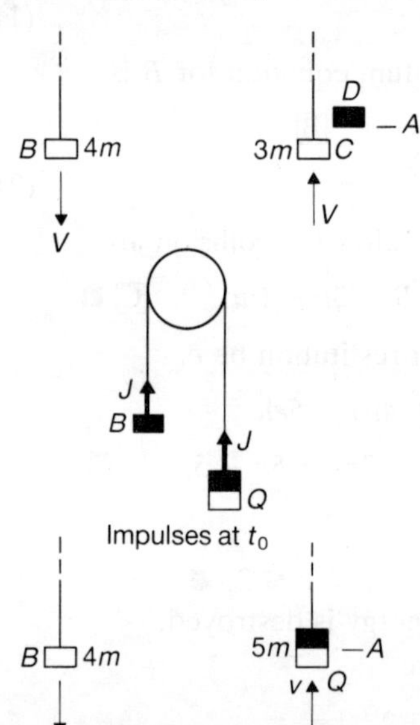

Just after t_0

[A] Meaning the speed just after the additional mass has been picked up.
[B] The 'impulsive tension' is the impulse of the tension in the string during the short time interval around t_0 in which D is picked up.
Do *NOT* represent J by a symbol like
✗ $\quad Ft \quad$ ✗.
The tension is large and not at all constant during the pick-up, so the notation is very misleading.
[C] In considering the impulsive motion which occurs at time t_0 we ignore the force of gravity, which does not have time to affect the velocities significantly during the pick-up.

Subsequent motion

Let the height above A to which Q rises be h.
When Q attains this height, the speed is zero.
From the pick-up to this position:

Loss in K.E. of Q is $\frac{1}{2}(5m)v^2$.
Loss in K.E. of B is $\frac{1}{2}(4m)v^2$.

Gain in gravitational P.E. of Q is $5mgh$.

As Q rises, B falls through a distance h.
$\therefore$ **Loss** in gravitational P.E. of B is $4mgh$.

Therefore, by 6.2F4,

$$5mgh - 4mgh = \frac{9}{2}mv^2 = \frac{9}{2}m\left(\frac{7V}{9}\right)^2.$$

Hence the distance risen by Q is $\dfrac{49V^2}{18g}$. *25m* ■

S6.

Before $\xrightarrow{2u}$ $\xrightarrow{u}$
A (m) (m) B
After $\xrightarrow{v}$ $\xrightarrow{1{\cdot}5v}$

Let the coefficient of restitution be e.

Momentum: $mv + m(1{\cdot}5v) = m(2u) + mu$. $\therefore 2{\cdot}5v = 3u$.

Restitution: (Separation speed) $= e \times$ (Approach speed)

$$1{\cdot}5v - v = e(2u - u).$$

$$\therefore e = \frac{0{\cdot}5v}{u} = \frac{3}{5}. \ ■$$

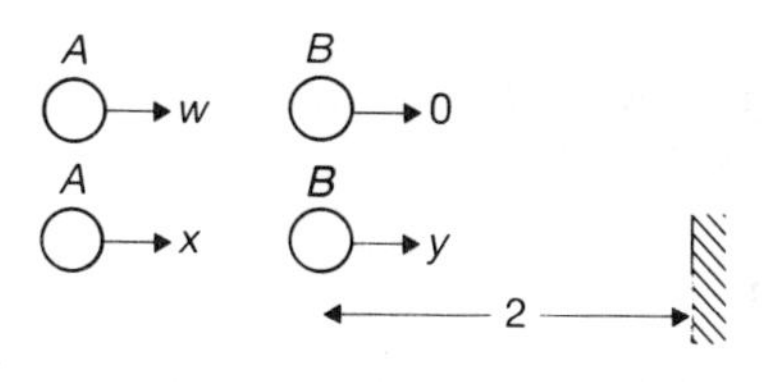

Suppose now that the initial speed of A is w and the velocities of A and B after collision are x and y towards the wall.

Momentum: $x + y = w$. Restitution: $y - x = 0{\cdot}6w$.

Eliminate w: $3x + 3y = 5y - 5x$. $\therefore y = 4x$.

$\therefore$ while B moves 2 m, A moves $\frac{1}{4}(2) = \frac{1}{2}$ m.
$\therefore$ distance of A from the wall is $2 - \frac{1}{2} = 1{\cdot}5$ m. ■
B then rebounds from the wall with speed $z = 0{\cdot}6y = 2{\cdot}4x$.
The speed of approach of A and B is then $z + x = 3{\cdot}4x$.

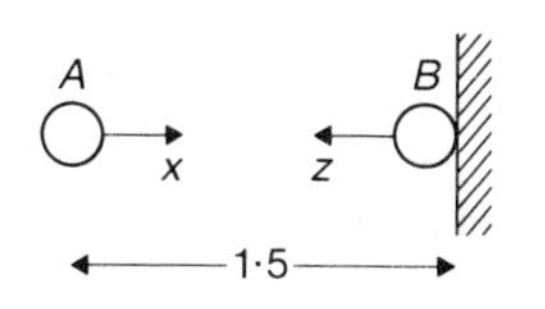

The time taken for AB to become zero is $\dfrac{1{\cdot}5}{3{\cdot}4x}$.

In this time B moves a distance $\left(\dfrac{1{\cdot}5}{3{\cdot}4x}\right)z = \dfrac{1{\cdot}5 \times 2{\cdot}4x}{3{\cdot}4x}$.

Hence the distance from the wall at the next collision is $\dfrac{18}{17}$ m.

27m ■

S7.

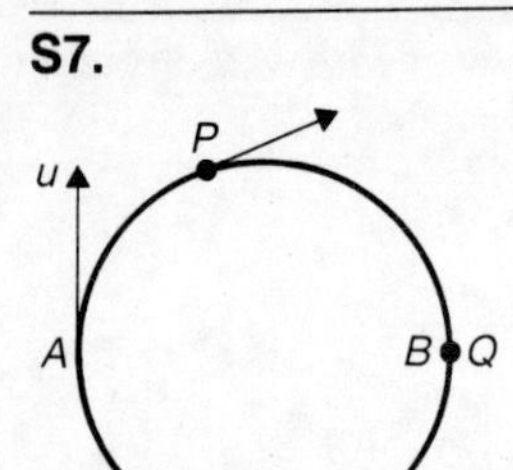

The hoop is smooth, and so P will move with constant speed v until it strikes Q.

After the collision, let the velocities of P and Q be v and w m s^{-1} respectively in the same direction as the approach velocity of P.

Momentum: $$0{\cdot}006w + 0{\cdot}003v = 0{\cdot}003u.$$

Restitution: $$w - v = 0{\cdot}25u.$$

$$\therefore 2w + v = u \qquad \text{and} \qquad 4w - 4v = u.$$

$$\therefore v = \frac{u}{6} \qquad \text{and} \qquad w = \frac{5u}{12}. \qquad (1)$$ ■

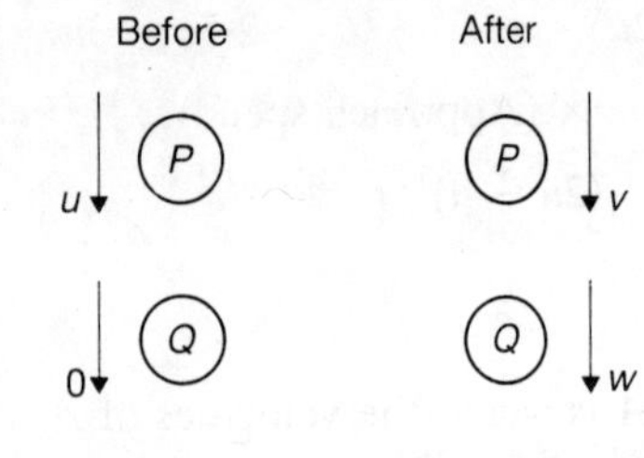

Q now moves away from P at a relative speed

$$w - v = \frac{u}{4} = \frac{6}{4} = 1{\cdot}5 \text{ m s}^{-1}.$$

They will collide again when Q has completed a revolution of the hoop *relative* to P, that is, Q laps P.

The circumference of the hoop is $2\pi(0{\cdot}5)$ m.

Therefore the time that elapses to the second collision is

$$\frac{2\pi(0{\cdot}5)}{1{\cdot}5} = \frac{2\pi}{3} \text{ seconds.}$$ ■

Suppose that, when the wire is rough, the speed of P just before the collision is x m s^{-1}. In this case $w = 1{\cdot}25$.

Using the result (1),

$$12 \times 1{\cdot}25 = 5x. \qquad \therefore x = 3.$$

The work done by friction is equal to the gain in kinetic energy as P goes from A to B, which is

$$\tfrac{1}{2}(0{\cdot}003)3^2 - \tfrac{1}{2}(0{\cdot}003)6^2 = -0{\cdot}0405 \text{ J}.$$

Notice that this is negative, as it should be, because friction always opposes the motion, thereby doing negative work.

The magnitude of the impulse acting on P is the same as the magnitude of the impulse exerted *by* P on Q, which is

$$0{\cdot}006 \times 1{\cdot}25 = 0{\cdot}0075 \text{ N s}.$$ *25m* ■

S8.

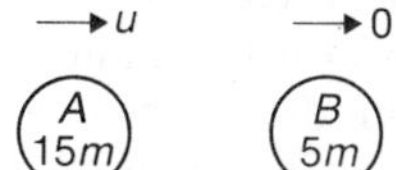

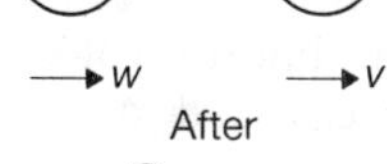

Suppose A approaches B with speed u and moves with velocity w in the same direction $\oplus\!\rightarrow$ after the collision.

Momentum: $15mw + 5mv = 15mu$.

Restitution: $v - w = \frac{1}{2}u$.

$\therefore 3u - 3w = v$ and $u + 2w = 2v$.

Solve for u and w. $u = \dfrac{8v}{9}$, $w = \dfrac{5v}{9}$.

Hence the speeds of A before and after the collision are $\dfrac{8v}{9}, \dfrac{5v}{9}$. ■

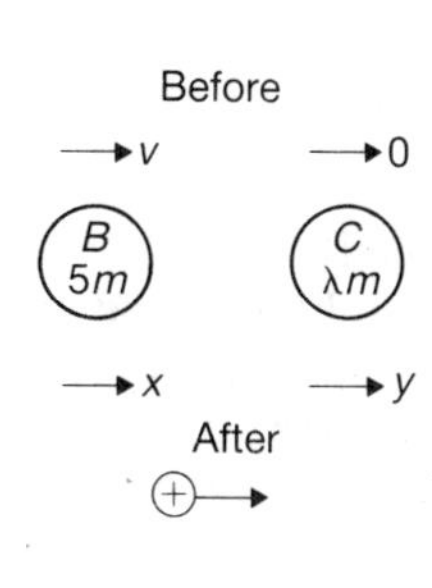

Momentum of A
Before collision: $15mu = \dfrac{40mv}{3}$. After collision: $15mw = \dfrac{25mv}{3}$.
Therefore the magnitude of the momentum of A is reduced by

$$\frac{40mv}{3} - \frac{25mv}{3} = 5mv. \blacksquare$$

Momentum of B
Before collision: 0. After collision: $5mv$.
$\therefore$ the magnitude of the momentum of B is increased by $5mv$. ■

Check: The momentum gained by B must be equal to the momentum lost by A. It is. √

Let the velocities of B and C after the second collision be x and y respectively in the positive direction.

Momentum: $5mx + \lambda my = 5mv$.

Restitution: $y - x = \frac{1}{2}v$.

Eliminate y: $5x + \lambda x = 5v - \frac{1}{2}\lambda v$.

The condition that there are no further collisions is that A does not strike B again, which means that $w \leqslant x$.

$$\therefore (5 + \lambda)w \leqslant (5 + \lambda)x = 5v - \frac{1}{2}\lambda v.$$

$$\therefore (5 + \lambda)\frac{5v}{9} \leqslant 5v - \frac{1}{2}\lambda v.$$

$$\therefore 2(5 + \lambda)5 \leqslant 90 - 9\lambda.$$

$$\therefore 50 + 10\lambda \leqslant 90 - 9\lambda.$$

$$\therefore 19\lambda \leqslant 40. \qquad \therefore \lambda \leqslant \frac{40}{19}.$$ 25m ■

S9.

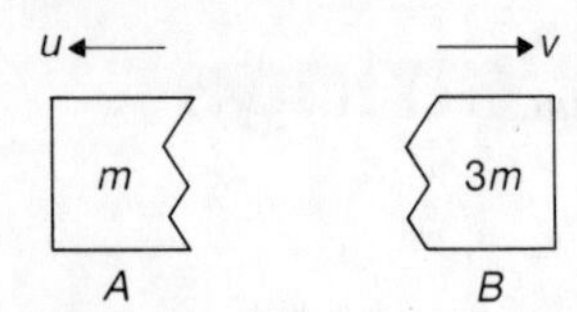

Suppose the speed of the fragment A of mass m is u and the speed of the fragment B of mass $3m$ is v. [A]

Momentum: $3mv - mu = 0$. $\therefore u = 3v$.

The kinetic energy acquired is therefore

$$E = \tfrac{1}{2}mu^2 + \tfrac{1}{2}3mv^2 = \tfrac{1}{2}m\,\{(3v)^2 + 3v^2\} = 6mv^2.$$

$$\therefore v = \sqrt{\frac{E}{6m}}, \quad \text{and so} \quad u = 3v = \sqrt{\frac{3E}{2m}}. \text{ [B]}$$

Hence the velocities of A and B after the explosion are $\sqrt{\frac{E}{6m}}$ and $\sqrt{\frac{3E}{2m}}$, in opposite directions. [C] ■

The magnitude of the relative velocity is $u + v$ [D]

$$= 3v + v = 4v = \sqrt{\frac{8E}{3m}}.$$ [E] *13m* ■

[A] As the initial momentum is zero, the momentum after the explosion is also zero, and so the fragments move in opposite directions. You can solve the problem by drawing them in the same direction in the diagram, but it involves unnecessary negative signs which can lead to errors.

[B] As expressing speed in terms of E and m is unusual, it is as well to conduct a dimensional check here. The dimension of energy is $\mathrm{ML^2T^{-2}}$. Therefore the dimension of

$$\sqrt{\frac{E}{m}} \text{ is } \sqrt{\frac{\mathrm{ML^2T^{-2}}}{\mathrm{M}}} = \sqrt{\mathrm{L^2T^{-2}}} = \mathrm{LT^{-1}} \text{ (speed).}$$

[C] We cannot give the velocities fully, because we do not know the direction of motion of either fragment, but we can say that they are opposed.

[D] Look at the diagram at this point, to avoid ✕ Relative velocity = $u - v$ ✕.

[E] Beware the very common algebraic error

$$✕ \quad \sqrt{\frac{E}{6m}} + \sqrt{\frac{9E}{6m}} = \sqrt{\frac{10E}{6m}} \quad ✕.$$

S10.

Let the velocity of P in the direction PQ when $t = T$ be v.

By the principle of conservation of momentum,

$2mv - mU = 5mU$. $\therefore v = 3U$. [A]

The total work done is equal to the increase in the total kinetic energy, which is

$$\tfrac{1}{2}(2m)v^2 + \tfrac{1}{2}mU^2 - \tfrac{1}{2}m(5U)^2$$

$$= mU^2\left(9 + \frac{1}{2} - \frac{25}{2}\right) = -\,3mU^2.$$ [B] ■

When $t = T$, the speed of P is $3U$.

$\therefore$ the magnitude of the impulse required to bring P to rest is $J = 2m(3U) = 6mU$.

Similarly the magnitude of the impulse required to bring Q to rest is $K = mU$. *13m* ■

t=0
→0 →5U
P 2m Q m
→v U←
t=T

[A] Although the question mentions 'forces of attraction', these are not important; the key fact is that 'no other force acts'. This is a clear signal that momentum conservation is likely to be useful.

[B] Using 6.3F3.

Chapter 7

Springs and elastic strings

7.1 GETTING STARTED

In this chapter we deal with the motion of particles which are attached to light strings or helical coiled springs which are **extensible** (in the case of the springs, also **compressible**). When no forces act on an extensible string, the length of the string is called its **natural length**, or unstretched length. When forces are applied to both ends of an extensible string, inducing tension in the string, the string extends beyond its natural length, the amount of the extension increasing as the tension increases.

If the **extension depends only on the tension** the string is said to be **elastic**. This means that, if the extension is found to be e_0 when the tension is T_0, you can change the applied forces in any way and then, when the tension is again T_0, the extension will again be e_0. No real string is so well behaved. If you pull any string hard enough and let go the string will suffer permanent deformation and will not return to its natural length. We say then that the string has been stretched beyond its **elastic limit**. If you pull still harder, the string will break.

All the strings and springs in this chapter are assumed to be within their elastic limits at all times, and further, they are assumed to be **linearly elastic**, which means that the extension is proportional to the tension at all times.

7.2 LINEAR ELASTICITY

ESSENTIAL FACTS

In the following Facts all strings are assumed to be light, and stretched only within the limits of linear elasticity.

The same Facts apply to helical springs which may be stretched or compressed in a straight line.

F1. Hooke's Law

When a string of natural (unstretched) length l is stretched to an **extension** x, so that its length becomes $l + x$, the tension is equal to αx, where α is a constant which is a property of the particular piece of string.

In the case of a spring, α is called the **stiffness**, or **spring constant**, of the spring. The extension x of a spring may be negative, in which case the positive quantity $(-x)$ is called the **compression** of the spring. Thus, in a spring of stiffness α at compression y, the **thrust** is αy.

F2. Modulus of a string

Hooke's Law is often stated as follows.
The tension T in a string of natural length l at extension x is

$$T = \lambda \frac{x}{l},$$

where the constant λ is the **modulus of the string**.
The dimension of the modulus is the same as the dimension of force (MLT^{-2}), and the modulus is measured in newtons (N).
In a **uniform string**, the modulus λ is a property of the string; any piece of the same string has the same modulus.

F3. Elastic energy

The work done in stretching an elastic string from its natural length l to an extension e is $\int_0^e T\,\mathrm{d}x = \frac{1}{2}\lambda \frac{e^2}{l}$. This is the work done *against* the tension. It is also called the **elastic energy** in the string.

ILLUSTRATIONS

I1.

A light helical spring is mounted in a vertical position with its lower end fixed. A particle P of mass m is placed on the upper end of the spring, and when P is at its equilibrium position O the compression of the spring is c. The particle is given a velocity $2\sqrt{gc}$ downwards when it is at O.

(a) Find the time taken for the spring to return to its natural length.

(b) Find the maximum height above O attained by P if

(i) P is fixed to the spring,

(ii) P is *not* fixed to the spring. □

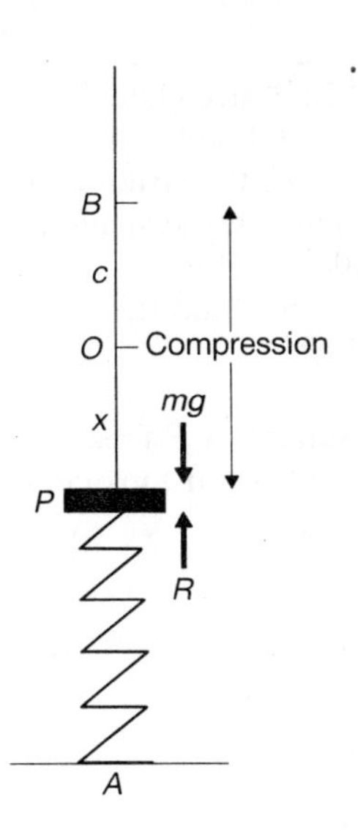

(a) Let the depth of P below O at time t be x, and let the thrust in the spring be R.

Then the equation of motion of P is $\qquad mg - R = m\ddot{x}$.

Let the stiffness of the spring be α.

Then, by Hooke's law (F1) $\qquad R = \alpha(c + x)$.

Eliminating R, we obtain $\qquad mg = m\ddot{x} + \alpha(c + x)$.

Since O is the equilibrium position, $\ddot{x} = 0$ when $x = 0$.

$\therefore\ mg = \alpha c$.

Hence the equation of motion becomes $\ddot{x} + \omega^2 x = 0$,

where $\omega = \sqrt{\dfrac{g}{c}}$.

Therefore P moves in SHM with centre O.

When $t = 0$, $x = 0$ and $\dot{x} = \sqrt{2gc}$.

Therefore (using 5.2F3) $\qquad x = \dfrac{\sqrt{2gc}}{\omega} \sin \omega t = 2c \sin \omega t$.

Hence the amplitude of the motion is $2c$.

[We shall assume that either the natural length (AB in the diagram) is greater than $3c$, or the spring is specially designed so that P can go below A.]

Let the time taken for the spring to return to its natural length, that is, for P to reach B, be t_1. When $t = t_1$, $x = -c$.

$$\therefore\ -c = 2c \sin \omega t_1. \qquad \therefore\ \sin \omega t_1 = -\frac{1}{2}.$$

This first occurs when $\omega t_1 = \dfrac{7\pi}{6}$.

Hence the time taken is $\dfrac{7\pi}{6}\sqrt{\dfrac{c}{g}}$. ■

(b) (i) If P is fixed to the spring the maximum height attained above O is equal to the amplitude, that is, $2c$. ■

(ii) If P is not fixed to the spring it will lose contact when the thrust R becomes zero, that is at B.

The velocity is $x = 2\omega c \cos \omega t = 2\sqrt{gc} \cos \omega t$.

So, when $t = t_1$, P is at B and its downward velocity is

$$2\sqrt{gc} \cos \omega t_1 = 2\sqrt{gc} \cos \frac{7\pi}{6} = -\sqrt{3gc}.$$

Under gravity, starting at B with a speed $\sqrt{3gc}$ and moving upwards, P will reach a further height $\dfrac{3gc}{2g} = \dfrac{3c}{2}$.

$\therefore$ the maximum height attained above O is $\dfrac{5c}{2}$. ■

I2. A light uniform elastic string of natural length $2a$ is stretched between two points A and B on a smooth horizontal table. A particle P of mass m is attached to the mid-point of the string and lies at rest. The distance AB is $4a$ and the modulus of the string is $2man^2$. The particle is projected towards B with speed u, and in the ensuing motion both parts of the string remain taut. Find the position of P at time t after its projection, and show that $u \leqslant 2an$. □

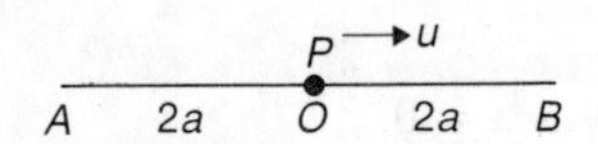

Let O be the mid-point of AB. Let the displacement of P from O in the direction OB (⊕→) be x at time t, and let the tensions be T_1 in AP and T_2 in BP. The equation of motion of P is

$$m\ddot{x} = T_2 - T_1.$$

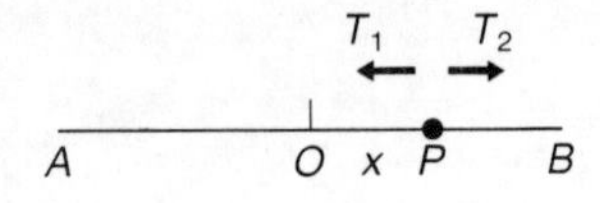

Now P was attached to the mid-point of the string when the tension was uniform, so that the string was uniformly stretched.
Therefore the natural length of each portion, AP and BP, is half the natural length of the whole string $= \frac{1}{2}(2a) = a$.
By Hooke's Law (F2), since it is given that both strings are taut:

$$T_1 = 2man^2\,\frac{(2a + x) - a}{a} = 2mn^2x + 2man^2$$

$$T_2 = 2man^2\,\frac{(2a - x) - a}{a} = -2mn^2x + 2man^2.$$

$$\therefore\ m\ddot{x} = -4mn^2x. \qquad \therefore\ \ddot{x} + (2n)^2x = 0;\ \text{which gives SHM.}$$

The general solution of this equation is, by 5.2F6,

$$x = A\cos 2nt + B\sin 2nt.$$

When $t = 0$, $x = 0$. $\quad\therefore A = 0.\quad \therefore x = B\sin 2nt.$

$$\therefore\ \dot{x} = 2nB\cos 2nt.$$

When $t = 0$, $\dot{x} = u$. $\quad\therefore 2nB = u.$

Hence the position of P is given by

$$x = \frac{u}{2n}\sin 2nt. \qquad \text{This is SHM with amplitude } \frac{u}{2n}.$$

Since both parts of the string remain taut, P does not come within a distance a of either end. Therefore the distance from O to P does not exceed a at any time, which means that the amplitude is not greater than a, and so $u \leqslant 2an$. ■

I3. A particle P of mass m is connected to a fixed point A by a light elastic string of natural length l and modulus $4mg$. It is released from rest at A. Find

(a) the point O at which the speed of P is maximum,
(b) the amplitude of the SHM part of the motion,
(c) the time taken for P to return to A. □

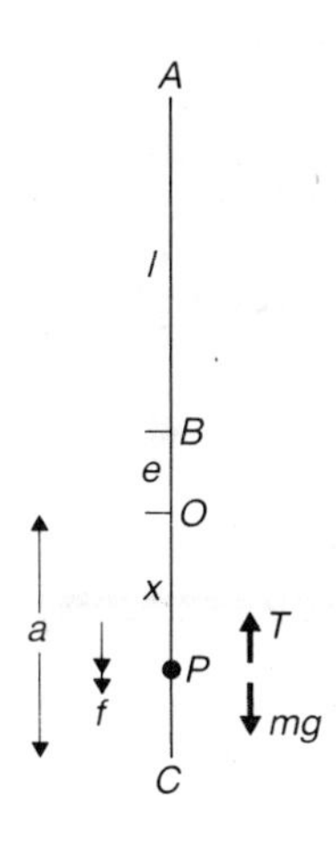

(a) At first P has downward acceleration g, so the speed increases. When P reaches B, at a depth l below A, the string becomes taut and exerts an upward force on P due to the tension T. The downward acceleration f of P is then given by

$$mf = mg - T.$$

The speed of P continues to increase while $f > 0$, that is, while $T < mg$. As P descends, T increases due to the extension, and when T passes through the value mg the acceleration becomes negative and the speed begins to decrease. Therefore the point O is the point at which $mg - T = 0$.

Let $BO = e$. Then, at O, $T = 4mg\frac{e}{l}$. $\therefore e = \frac{l}{4}$.

Hence the point O is at a depth $\frac{5l}{4}$ below A. ■

Note that O is also the *position of equilibrium*, which is the point at which P could be permanently at rest.

(b) Let P be at depth x below O at time t. The equation of motion of P is

$$m\ddot{x} = mg - T = mg - 4mg\frac{x + e}{l} = -4mg\frac{x}{l}.$$

which is SHM about O with $\omega = 2\sqrt{\frac{g}{l}}$.

Let C be the lowest point reached and let $OC = a$. Then a is the amplitude of the SHM.
C can be found by considering the energy changes which occur in the displacement from A to C. The kinetic energy is zero both when P is at A and when it is at C.

$\therefore$ Loss of gravitational P.E. – gain in elastic energy = 0.

$$\therefore mg\left(\frac{5l}{4} + a\right) = \frac{1}{2}(4mg)\frac{(l/4 + a)^2}{l}.$$

$$\therefore \frac{5mgl}{4} + mga = \frac{2mg}{l}\left(\frac{l^2}{16} + \frac{la}{2} + a^2\right).$$

$$\therefore \frac{5l^2}{4} = \frac{l^2}{8} + 2a^2. \qquad \therefore a^2 = \frac{9l^2}{16}. \qquad \therefore \text{amplitude} = \frac{3l}{4}.$$ ■

(c) In the first part of the motion P falls freely from A to B in time t_1, where $\frac{1}{2}gt_1^2 = l$.
Let the time taken from B to C be t_2. The speed of P at any point in AC is the same both when P is descending and when it is ascending, and so the time taken from C to B is also t_2, and the time from B to A is t_1.

Taking $t = 0$ at C, the SHM is given by $x = \frac{3l}{4}\cos\omega t$.

At B, $x = -\frac{l}{4}$. $\quad \therefore \cos \omega t = -\frac{1}{3}$.

$\therefore$ the time t_2 from C to B is given by $\omega t_2 = \pi - \arccos\frac{1}{3}$.

$$\therefore \text{ the total time is } 2(t_1 + t_2) = 2\left\{\sqrt{\frac{2l}{g}} + \frac{1}{2}\sqrt{\frac{l}{g}}\left(\pi - \arccos\frac{1}{3}\right)\right\}$$

$$= \sqrt{\frac{l}{g}}\left(2\sqrt{2} + \pi - \arccos\frac{1}{3}\right). \blacksquare$$

7.3 EXAMINATION QUESTIONS AND SOLUTIONS

Q1.

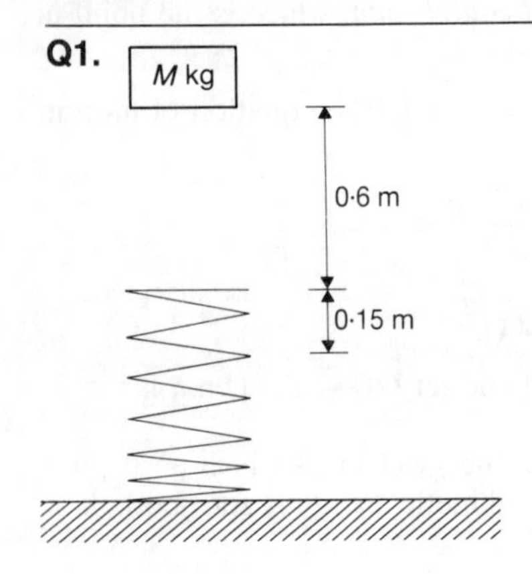

The figure illustrates a body of mass M kg falling vertically from rest through a distance of 0.6 m onto a spring. The body comes to rest after the spring has been compressed through 0.15 m.

If the modulus of elasticity divided by the natural length of the spring has value 7000 N m^{-1}, find M. (Neglect the mass of the spring and take the acceleration due to gravity to be 10 m s^{-2}.)

(NI 1986)

Q2.

A light elastic string is stretched between two fixed points A and B on a smooth horizontal table. The string is of natural length a and $AB = b$, where $b > a$. A particle P is attached to the string at the point which is initially at a distance kb from A. The particle is displaced and then released so that it oscillates on the line AB. Throughout the motion the portions AP and PB of the string are both in tension. Show that the motion is simple harmonic.

Show also that the periods of oscillation when $k = \frac{1}{2}$ and when $k = \frac{1}{3}$ are in the ratio $3{:}2\sqrt{2}$.

(JMB 1985)

Q3.

Prove that the elastic energy of a light spring of natural length a and modulus of elasticity λ, stretched by an amount x, is $\lambda x^2/(2a)$.

A trolley of mass m runs down a smooth track of constant inclination $\pi/6$ to the horizontal, carrying at its front a light spring of natural length a and modulus mga/c, where c is constant. When the spring is fully compressed it is of length $a/4$, and it obeys Hooke's law up to this point. After the trolley has travelled a distance b from rest the spring meets a fixed stop. Show that, when the spring has been compressed a distance x, where $x < 3a/4$, the speed v of the trolley is given by

$$cv^2/g = c(b + x) - x^2.$$

Given that $c = a/10$ and $b = 2a$, find the total distance covered by the trolley before it momentarily comes to rest for the first time.

(LON 1982)

Q4.

A light spring obeys Hooke's Law. A force of 20 N extends the spring by 0·01 m. Show that the work done in extending the spring by b m from the unstretched state is $10^3 b^2$ J.

This spring is placed in a long smooth straight cylindrical tube with one end fixed to the tube. The tube is fixed in a vertical position with the free end of the spring uppermost. The dimensions of the tube and of the spring are such that the spring can only move vertically and the spring always remains inside the tube. A particle of mass 4 kg is firmly attached to the free end of the spring. The particle is held so that the spring is compressed a distance of 0·1 m from its uncompressed state. The particle is then released. Show that subsequently

$$v^2 = 3 + 20y - 500y^2$$

where v m s^{-1} is the speed of the particle and y m is the compression of the spring.

Find (a) v^2 when the *extension* of the spring is 0·01 m,
(b) the value of y when the speed is a maximum,
(c) the maximum extension of the spring.

[Take g to be 10 m s^{-2}.] (AEB 1983)

Q5. A light elastic string of modulus mln^2 and natural length l has one end fixed at O and carries at its other end two particles, P and Q, of masses m and $2m$, respectively. Find, in terms of g and n, the extension of the string when the particles hang in equilibrium. When the system is in equilibrium the particle Q is gently removed. In the subsequent motion of P, let x denote the displacement at time t of P below the point at which it would hang in equilibrium. Show that, whilst the string is taut,

$$\ddot{x} + n^2 x = 0.$$

Find the time that elapses after the particle Q is removed before the string slackens.

Given that when P reaches its greatest height it is still below O, show that

(i) in the upward motion of P the ratio of the time for which the string is taut to the time for which it is slack is $2\pi : 3\sqrt{3}$,

(ii) $l > \dfrac{3g}{2n^2}$. (JMB 1986)

Q6. 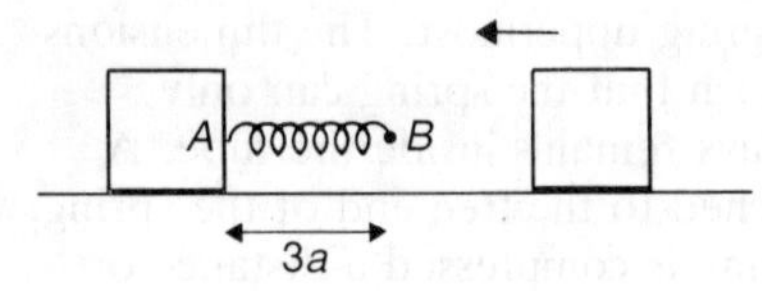

A small cubical block of mass $8m$ is attached to one end A of a light elastic spring AB of natural length $3a$ and modulus of elasticity $6mg$. The spring and block are at rest on a smooth horizontal table with AB equal to $3a$ and lying perpendicularly to the face to which A is attached. A second block of equal physical dimensions, but of mass m, moving with a speed $(2ga)^{1/2}$ in the direction parallel to BA impinges on the free end B of the spring.

Assuming that the heavier block is held fixed and that AB remains straight and horizontal in the subsequent motion, determine

(i) the maximum compression of the spring,
(ii) the time that elapses between impact and the lighter block first coming to instantaneous rest.

Assuming now that the heavier block is also free to move determine, at the instant when the blocks are first moving instantaneously with the same velocity, the values of

(iii) the common velocity of the blocks,
(iv) the compression in the spring. (AEB 1982)

Q7. A spring that obeys Hooke's Law in both extension and compression is extended by 0.1 m when one end is attached to a fixed point and a mass of 2 kg is fixed to the other end and the system is allowed to hang vertically. If the spring constant is defined as the modulus of elasticity divided by the natural length of the spring, what is its value? (Take the acceleration due to gravity to be 10 m s^{-2}).

The spring and the 2 kg mass are now placed on a smooth horizontal table. One end of the spring is attached to the table and the other end (to which the 2 kg mass is attached) is pulled so that the spring is extended by an amount 0.05 m. If, at this moment ($t = 0$), the 2 kg mass is projected with a speed of 1 m s^{-1} in the direction of increasing extension of the spring, find the extension (or compression) of the spring at any later time t.

What is the maximum extension of the spring and at what time will this occur?

At what time will the speed first reach its maximum value?

(NI 1984)

SOLUTIONS

S1.

Let the initial position, O, of the body P be the zero of gravitational potential energy.
During the motion, the total energy is

K.E. + gravitational P.E. + elastic P.E. [A]

The K.E. is zero, both at O and at A, where the body comes to rest.
At A, the gravitational P.E. is

$- Mg(0{\cdot}6 + 0{\cdot}15)$ J.

The spring has stiffness α N m^{-1}: $\alpha = 7000$. [B]
Therefore the elastic energy when the spring is compressed to A is

$\frac{1}{2}\alpha(0{\cdot}15)^2 = 3500(0{\cdot}15)^2$. [C]

Hence, by conservation of energy,
Total energy when P is at A = energy at O.

$$-Mg(0{\cdot}75) + 3500(0{\cdot}15)^2 = 0 + 0.$$

$$\therefore M = \frac{(3500)(0{\cdot}15)^2}{10 \times 0{\cdot}75} = 70 \times 0{\cdot}15 = 10{\cdot}5. \quad \text{[D]}$$

11m ■

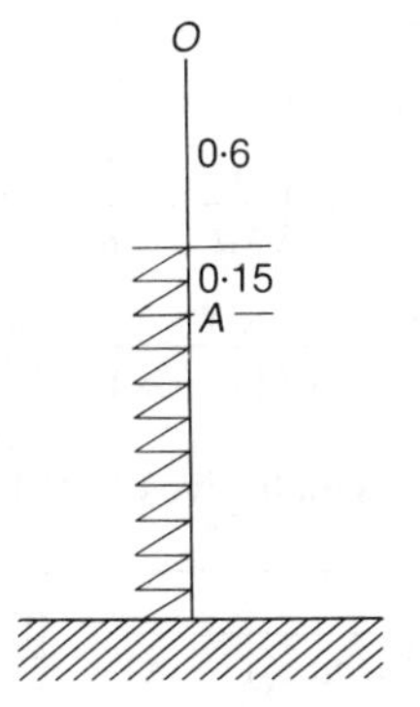

[A] The question involves only speeds and displacements, and therefore we expect the energy argument to be the simplest method.
[B] From 7.2F1 and F2. The stiffness of a spring of natural length l and modulus λ is λ/l.
[C] From 7.2F3. In the case of a spring, the extension e can be negative, but the formula for the elastic energy remains valid.
[D] Note that there is no impact when the body meets the spring because of the instruction 'neglect the mass of the spring'.

S2.

Let the mass of P be m and the modulus of the string λ. Let the equilibrium position of P be O, and the displacement of P from O be x in the direction OB. Let the tension in AP be T_1 and the tension in BP be T_2.

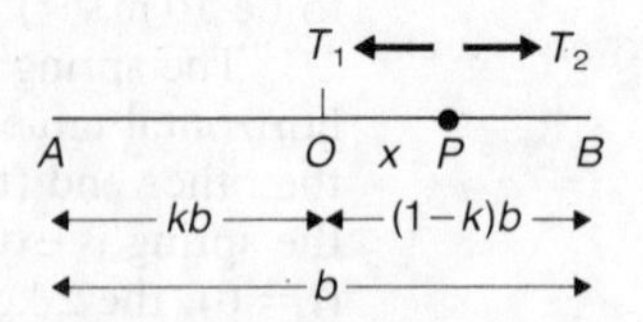

The equation of motion of P is

$$m\ddot{x} = T_2 - T_1. \quad \boxed{A}$$

The string AP has natural length ka and its extension is $kb + x - ka$. $\boxed{B}$

$$\therefore T_1 = \lambda \frac{x + kb - ka}{ka} = \lambda\left(\frac{x}{ka} + \frac{b}{a} - 1\right). \quad \boxed{C}$$

The string BP has natural length $(1 - k)a$ and its extension is $(1 - k)b - x - (1 - k)a$.

$$\therefore \quad T_2 = \lambda \frac{(1 - k)b - x - (1 - k)a}{(1 - k)a}$$

$$= \lambda\left\{-\frac{x}{(1 - k)a} + \frac{b}{a} - 1\right\}.$$

$$\therefore \frac{m\ddot{x}}{\lambda} = -\frac{x}{(1 - k)a} + \frac{b}{a} - 1 - \left(\frac{x}{ka} + \frac{b}{a} - 1\right)$$

$$= -\frac{x}{a}\left(\frac{1}{1 - k} + \frac{1}{k}\right) = -\frac{x}{ak(1 - k)}. \quad \boxed{D}$$

$$\therefore \quad \ddot{x} = -\frac{\lambda x}{mak(1 - k)}, \qquad \text{which gives SHM. } \blacksquare$$

$$\ddot{x} + \omega^2 x = 0, \qquad \text{where } \omega = \sqrt{\frac{\lambda}{mak(1 - k)}}. \quad \boxed{E}$$

The period is $\dfrac{2\pi}{\omega} = 2\pi\sqrt{\dfrac{ma}{\lambda}k(1 - k)}$.

The ratio of the periods when $k = \frac{1}{2}$ and $k = \frac{1}{3}$

is $\dfrac{\sqrt{\frac{1}{2}(1 - \frac{1}{2})}}{\sqrt{\frac{1}{3}(1 - \frac{1}{3})}} = \dfrac{3}{2\sqrt{2}}$. *13m* ■

$\boxed{A}$ Always write down the equation of motion before using Hooke's Law to find the tensions. This makes clear what is required and also shows the examiner that you have applied Newton's Second Law.

$\boxed{B}$ Questions of this type normally do not say that the string is uniform, but you are expected to assume that λ is the same at all points of the string. So the natural length of AP follows by the argument in 7.2I2.

$\boxed{C}$ It helps to separate the constant term from those involving x, because it is then easier to see that they cancel out later.

$\boxed{D}$ If the constant terms have *not* cancelled out at this stage, then you must look for the mistake, because the acceleration must be zero at the equilibrium position.

$\boxed{E}$ *Dimension check*: k is a pure number, and λ is a force, so the dimension of this expression is $\sqrt{\dfrac{MLT^{-2}}{ML}} = T^{-1}$, which is the correct dimension of ω.

S3.

Let the tension be T when the extension is z. By Hooke's Law,

$$T = \lambda \frac{z}{a}. \qquad (1)$$

The elastic energy is $\displaystyle\int_0^x T\,dz$ (2) [A]

$$= \int_0^x \lambda \frac{z}{a}\,dz = \frac{\lambda}{a}\left[\frac{z^2}{2}\right]_0^x = \frac{\lambda x^2}{2a}. \quad ■$$

The diagram shows the trolley at the instant when the compression of the spring is x. The front, P, of the trolley has then moved a distance $b + x$ from its starting point A. [B]

The forces which act in the line of motion are:

the component of the weight, $mg \sin \frac{\pi}{6} = \frac{mg}{2}$,

the thrust of the spring, R. [C]

In the motion from A to P, the **loss** of gravitational P.E. is $\frac{mg}{2}(b + x)$.

The **gain** in elastic P.E. is $\frac{1}{2}\frac{mga}{c}\frac{x^2}{a}$. [D]

Therefore, by conservation of energy,

$$\frac{1}{2}mv^2 = \frac{mg}{2}(b + x) - \frac{mgx^2}{2c}. \qquad \text{[E]}$$

$$\therefore \frac{cv^2}{g} = c(b + x) - x^2. \quad ■$$

Substitute the given values.

$$\frac{av^2}{10g} = \frac{a}{10}(2a + x) - x^2. \qquad (3)$$

$$\therefore \frac{av^2}{g} = 2a^2 + ax - 10x^2 = (2a + 5x)(a - 2x).$$

As x increases from zero the RHS stays positive until $x = \frac{1}{2}a$. [F]

Therefore the total distance covered before the trolley comes to rest is $b + \frac{1}{2}a = \frac{5}{2}a$. *25m* ■

[A] Equations (1) and (2) are essential ingredients in the proof, and should be written down. You might get less than full marks if you omit either. The argument ✖ 'We may assume Energy = (Average tension) × (extension)' ✖ is likely to be viewed unfavourably, because the validity of this statement depends on the *linear relation* between the tension and the extension. Unless this linearity is fully explained, the statement is worthless. Integration is much quicker and clearer than attempting such an explanation.

Also, avoid writing ✖ $\displaystyle\int_0^x T\,dx$ ✖, because it is confusing to use the same symbol x both as a variable of integration and as a limit in the definite integral.

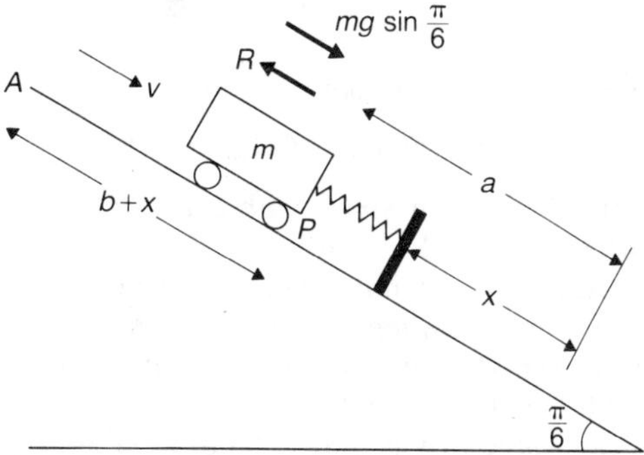

[B] A clear diagram which leads you to the correct total displacement is needed.

[C] *Thrust* is defined in 4.4F3. A spring also can exert a thrust.

[D] The energy method is to be chosen here because only v–x relations are involved; moreover the first part of the question, used here, can be expected to be relevant.

[E] Using 4.3F9.

[F] **?** Substitute $v = 0$ in (3) and solve for x. **?** Though correct in this case, because there is only one positive root of the equation for x, it is generally better to keep v^2 undetermined so as to see how it behaves as x varies.

S4.

Let the stiffness of the spring be α N m^{-1}.
Then $20 = \alpha(0{\cdot}01)$. $\quad \therefore \alpha = 2000$. [A]
$\therefore$ the force required to produce extension x m is $2000x$ N. Therefore the work done is

$$\int_0^b 2000x \, dx = \left[1000x^2\right]_0^b = 10^3 b^2 \text{ J.} \quad \text{[B]} \ ■$$

When the compression is y, the *extension* is $-y$. Then the elastic P.E. is $10^3(-y)^2$ J. The gravitational potential energy of the particle P of mass 4 kg is $4g(-y)$ J, taking the position A, when the spring is uncompressed, as the zero level. By conservation of energy,

$$\tfrac{1}{2}(4v^2) + 10^3y^2 - 4gy = \text{constant}$$
$$= 0 + 10^3(0{\cdot}1)^2 - 4g(0{\cdot}1).$$
$$\therefore v^2 = 3 + 20y - 500y^2. \quad (1) \ ■$$

When the extension is 0·01 m, $\quad y = -0{\cdot}01$.
Then $\quad v^2 = 3 - 0{\cdot}2 - 0{\cdot}05 = 2{\cdot}75$. ■

$$\ddot{y} = \frac{d}{dy}\left(\frac{1}{2}\dot{y}^2\right) = 10 - 500y$$
$$= -500(y - 0{\cdot}02). \quad \text{[C]}$$

Therefore P moves in SHM with centre O at the point where $y = 0{\cdot}02$. [D]
The speed of P is maximum when it passes through O, that is, when $y = 0{\cdot}02$. ■
As P is at rest when $y = 0{\cdot}1$, the amplitude of the SHM is $0{\cdot}1 - 0{\cdot}02 = 0{\cdot}08$.
$\therefore$ the least value of y is $0{\cdot}02 - 0{\cdot}08 = -0{\cdot}06$. [E]
Hence the maximum extension is 0·06 m. *25m* ■

[A] Using 7.2F1.
[B] Since you have to *show* that the result is true,
× Quoting $\frac{1}{2}\lambda\frac{e^2}{l}$ × from 7.2F3 will not suffice.
You must do the integration (see also S3 [A]).

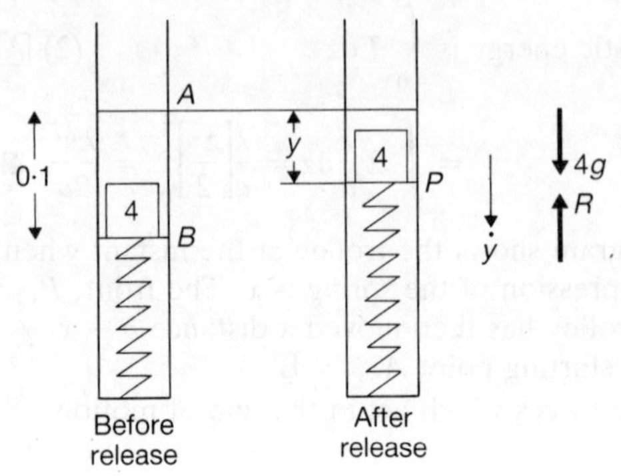

[C] The numbers in this problem are uncomfortable, and so the method of completing the square to obtain the maximum of v^2 and the point where it is zero is not preferred. It is better to find the acceleration and then use the known properties of SHM.
[D] Using 5.2F8.

[E] *Check*: Substitute $y = -0{\cdot}06$ in (1) and verify that the equation then gives $v^2 = 0$.

S5.

Let the equilibrium tension be T_0 and the extension e. Then

$$3mg = T_0 = mln^2\frac{e}{l} = mn^2e.$$

$$\therefore \text{extension } e = \frac{3g}{n^2}. \ ■$$

P alone would hang in equilibrium at C,

where the extension is $\dfrac{e}{3} = \dfrac{g}{n^2}$.

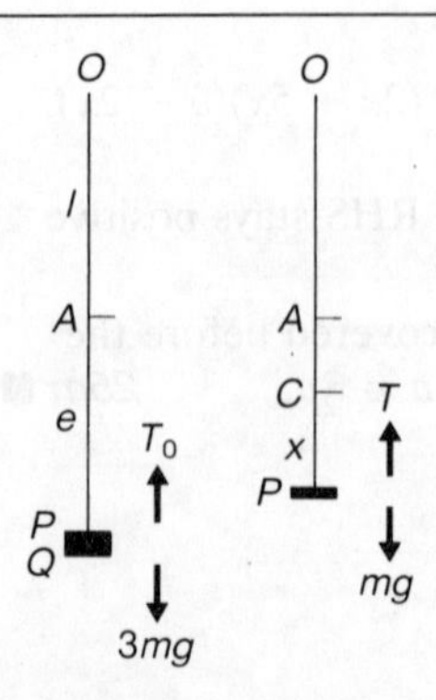

The equation of motion of P is

$$m\ddot{x} = mg - T, \quad (1)$$

where the tension $T = \dfrac{mln^2(g/n^2 + x)}{l}$

$$= mg + mn^2x.$$

$\therefore m\ddot{x} = mg - (mg + mn^2x) = -mn^2x.$

$\therefore \ddot{x} + n^2x = 0.$ [A] [B] ■

When $t = 0$, $\dot{x} = 0$ and $x = e - \dfrac{e}{3} = \dfrac{2g}{n^2}.$

$\therefore x = \dfrac{2g}{n^2}\cos nt.$

The string slackens when P is at A, where

$$x = -\frac{g}{n^2}, \text{ that is, when } \cos nt = -\frac{1}{2}. \quad (2)$$

As t increases from 0, $\cos nt$ decreases from 1, and it first becomes $-\dfrac{1}{2}$ when $nt = \dfrac{2\pi}{3}$. [C]

$\therefore$ the time before slackening is $\dfrac{2\pi}{3n}$. ■

When $t = \dfrac{2\pi}{3n}$, $\dot{x} = -\dfrac{2g}{n}\sin nt = -\dfrac{2g}{n}\sin\dfrac{2\pi}{3}$

$$= -\frac{g}{n}\sqrt{3}.$$

The particle now moves freely under gravity, starting with an upward velocity of $\dfrac{g}{n}\sqrt{3}$.

It comes to rest in time $\left(\dfrac{g}{n}\sqrt{3}\right)\Big/ g = \dfrac{\sqrt{3}}{n}$. [D]

$\therefore$ the ratio of times is $\left(\dfrac{2\pi}{3n}\right):\left(\dfrac{\sqrt{3}}{n}\right) = 2\pi:3\sqrt{3}$. ■

The height reached above A is

$$\frac{1}{2}\frac{\sqrt{3}}{n}\left(\frac{g}{n}\sqrt{3} + 0\right) = \frac{3g}{2n^2}. \quad \text{[E]}$$

Since P is still below O, $l > \dfrac{3g}{2n^2}$. *29m* ■

[A] Always write down a proper equation of motion like (1), showing clearly all the forces acting.

✗ 'Extension for mg is $\dfrac{g}{n^2}$.

$\therefore$ restoring force $= mg\left(\dfrac{x}{g/n^2}\right) = mn^2x$.

$\therefore \ddot{x} + n^2x = 0$' ✗, although not untrue, is not an acceptable demonstration because it does not start from mechanical principles, but uses the special notion of 'restoring force'.

[B] If you get $2\ddot{x} + n^2x = 0$, you have gently removed the wrong particle.

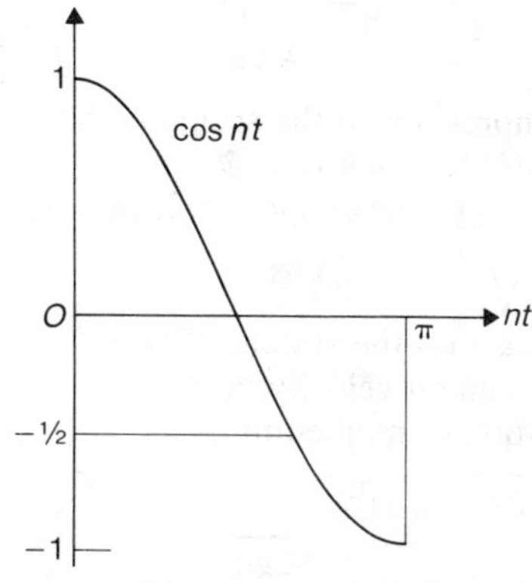

[C] Great care is generally needed in selecting the correct root of a trigonometrical equation like (2). In this case even a pocket calculator would have found the right one, but this is not always so.

[D] Using 3.3F7A.

[E] Using 3.3F7B.

S6.

Suppose that the compression of the spring is x at time t after the block has impinged on the spring. Let the thrust in the spring be R.

By Newton's Second Law, $m\ddot{x} = -R$. [A]

By Hooke's Law, $R = 6mg\dfrac{x}{3a}$.

Therefore $\ddot{x} = -\omega^2 x$, where $\omega = \sqrt{\dfrac{2g}{a}}$.

When $t = 0$, $x = 0$.

$\therefore x = B\sin\omega t$, where B is constant. [B]

When $t = 0$, $\dot{x} = \sqrt{2ga}$.

$\therefore \omega B = \sqrt{2ga}$. $\therefore B = \sqrt{2ga}\sqrt{\dfrac{a}{2g}} = a$. [C]

The maximum compression of the spring is the amplitude of the SHM, which is a. ■

Instantaneous rest occurs when $\omega t = \pi/2$, that is,

after time $\dfrac{\pi}{2\omega} = \pi\sqrt{\dfrac{a}{8g}}$. [D] ■

No external forces act on the system of two blocks. Let their common velocity be v. Then, by conservation of momentum,

$8mv + mv = m\sqrt{2ga}$. [E]

Hence the common velocity is $\dfrac{\sqrt{2ga}}{9}$ in the original direction of motion of the lighter block. [F]

Let the compression of the spring be y. Then the elastic energy is

$\dfrac{1}{2}(6mg)\dfrac{y^2}{3a} = \dfrac{mgy^2}{a}$. [G]

Gain in P.E. = Loss of K.E.

$\dfrac{mgy^2}{a} = \tfrac{1}{2}m(2ga) - \tfrac{1}{2}(9m)v^2$. [H]

$\therefore y^2 = a^2 - \dfrac{9a}{2g}\dfrac{2ga}{81} = \dfrac{8a^2}{9}$.

$\therefore$ the compression in the spring is $\dfrac{\sqrt{8}}{3}a$. 25m ■

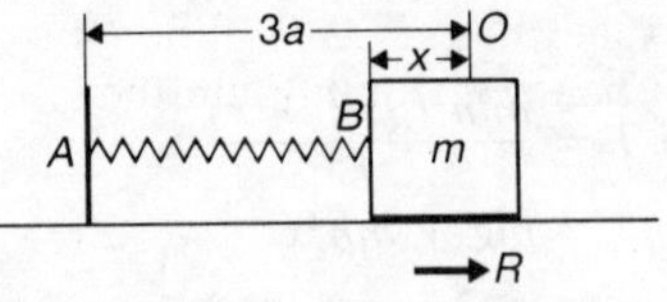

[A] The end A of the spring is fixed, and so $\ddot{x}$ is the acceleration.

[B] Using 5.2F3.

[C] This shows that it was wise to denote the constant by B at first, instead of using the symbol 'a', which would be quite natural, but would lead to the embarrassing result '$a = a$', which would be difficult for the examiner to assess.

[D] Alternatively, by the principle of energy,

$$\frac{1}{2}m\dot{x}^2 + \frac{1}{2}(6mg)\frac{x^2}{3a} = \frac{1}{2}m(2ga).$$

$$\therefore \dot{x}^2 = \frac{2g}{a}(a^2 - x^2).$$

By 5.2F7, this motion is SHM with period $2\pi\sqrt{\dfrac{a}{2g}}$ and amplitude a.

Hence the maximum compression is a, and the block comes to rest after a quarter of the period, which is a time $\pi\sqrt{\dfrac{a}{8g}}$.

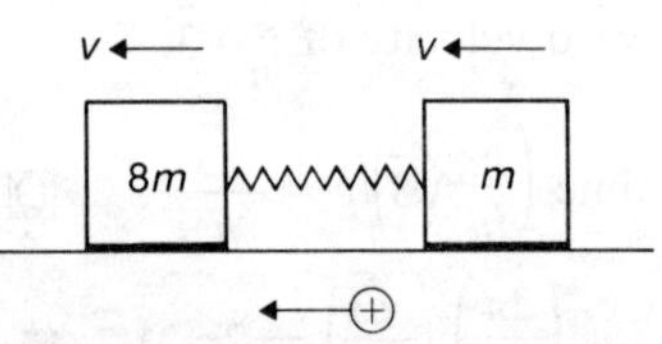

[E] Using 6.3F2.

[F] Remember to state the direction as well as the magnitude, or indicate it clearly on your diagram.

[G] Using 7.2F3.

[H] Using 6.3F3. Although there is no direct hint that the energy method should be tried we notice that (iv) asks for a displacement after giving information about velocity, and *time* is not mentioned, and this points to energy.

It is possible to solve this problem by setting up the equation of motion of each block, but it takes longer.

S7.

Let the spring constant be α N m^{-1}. When a particle of mass m hangs in equilibrium the tension is mg. In this case

$\alpha(0{\cdot}1) = 2 \times 10.$ $\quad \therefore \alpha = 200.$ [A] ■

[A] Using 7.2F1.

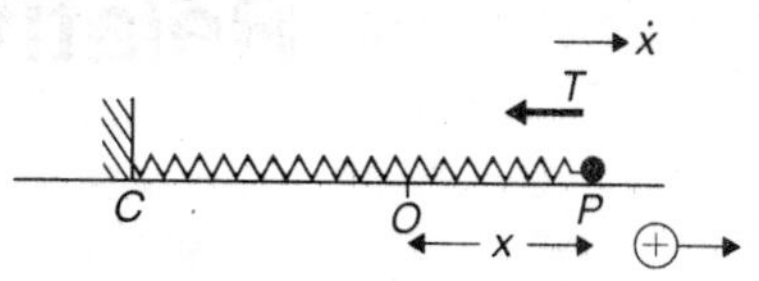

Suppose the extension is x m and the tension T N at time t s.
The equation of motion of the mass, P, is

$2\ddot{x} = -T.$

By Hooke's Law (7.2F1), $\quad T = \alpha x = 200x.$

Therefore $\ddot{x} = -100x = -10^2 x.$

$\therefore x = A\cos 10t + B\sin 10t.$ [B]

$\therefore \dot{x} = -10A\sin 10t + 10B\cos 10t.$ [C]

When $t = 0$, $\quad x = 0{\cdot}05 \quad$ and $\quad \dot{x} = 1.$

$\therefore 0{\cdot}05 = A \quad$ and $\quad 1 = 10B.$

$\therefore$ the extension is $0{\cdot}05\cos 10t + 0{\cdot}1\sin 10t$ m. ■

[B] Using 5.2F6. In Mechanics questions it is always permissible to quote the general solution $x = A\cos\omega t + B\sin\omega t$ of the equation $\ddot{x} + \omega^2 x = 0$. Never waste time trying to derive the solutions by integration methods.
[C] Preparing to use the given initial conditions.

$$20x = \cos 10t + 2\sin 10t \quad \text{[D]}$$
$$= \sqrt{1^2 + 2^2}\cos(10t - \beta),$$
where $\beta = \arctan 2$.

[D] Getting rid of awkward decimal fractions. Now use the formula on page 244.

Therefore the maximum extension is

$\dfrac{\sqrt{5}}{20} = 0{\cdot}112$ m (to 3dp), and it occurs at

time $\dfrac{\beta}{10} = \dfrac{\arctan 2}{10} = 0{\cdot}111$ s (to 3dp). ■

$20\dot{x} = (-10)\sqrt{5}\sin(10t - \beta).$

As t increases from zero, $-\sin(10t - \beta)$ is at first positive, and then falls to zero and continues to fall to a minimum when

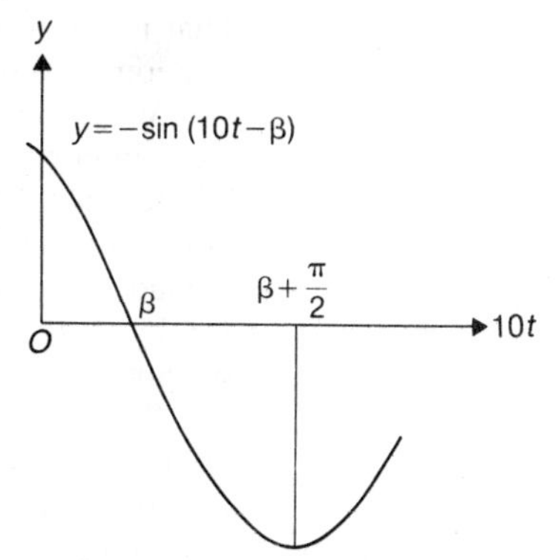

$10t = \beta + \dfrac{\pi}{2}.$ [E]

[E] It is helpful to sketch a graph at this stage. The velocity x in the ⊕→ direction is continually falling due to the tension T←. The mass reaches its maximum negative velocity (and hence maximum speed) when it is at O. After that, the spring will be in compression and will push against the mass and slow it down.

The mass is then travelling, at its maximum speed, towards the fixed end, C, of the spring.
$\therefore$ the maximum speed is reached when

$t = \dfrac{\pi + 2\arctan 2}{20} = 0{\cdot}268$ s (to 3 dp). *23m* ■

Chapter 8 Relative motion

8.1 GETTING STARTED

When considering the motion of a particle P we normally select a point O, which is fixed on earth, and specify the position of P by the displacement $\overrightarrow{OP}$. As the earth is rotating about the sun and also about its polar axis such a point O is clearly not fixed in space. What we are doing is to consider the motion of P relative to O. That is, we are describing the position of P as it would appear to someone standing at O.

When there are two moving particles P and Q it is sometimes more important to describe the motion of P as viewed from Q than it is to describe the motion as viewed from O. For instance, a pilot in one aeroplane may wish to take action in order to avoid a second aeroplane. In cases of this type we investigate the motion of P relative to Q; that is, we consider the displacement $\overrightarrow{QP}$ as this specifies the position of P as viewed by an observer moving with Q.

The technique used in problems of this type is vector algebra, although sometimes we will see that some manipulation may be circumvented by applying trigonometry to a vector diagram.

8.2 DISPLACEMENT, VELOCITY AND ACCELERATION

ESSENTIAL FACTS

F1. Position vector

The **position vector** of a point A relative to a point O is the displacement $\overrightarrow{OA}$.

When an origin O has been fixed, it is often convenient to denote the position vectors of points A, B, C, . . . by the symbols $\mathbf{a}$, $\mathbf{b}$, $\mathbf{c}$, . . . so that $\mathbf{a} = \overrightarrow{OA}$, $\mathbf{b} = \overrightarrow{OB}$, $\mathbf{c} = \overrightarrow{OC}$, . . .

F2. Velocity

When a point P has position vector $\mathbf{r}$ relative to an origin O, the velocity $\mathbf{v}$ of P relative to O is given by

$$\mathbf{v} = \dot{\mathbf{r}} = \frac{\mathrm{d}\mathbf{r}}{\mathrm{d}t}.$$

The **speed** of P is the magnitude, $|\mathbf{v}|$, of the velocity.

Thus, if $\mathbf{r} = x\mathbf{i} + y\mathbf{j} + z\mathbf{k}$, $\mathbf{v} = \dot{x}\mathbf{i} + \dot{y}\mathbf{j} + \dot{z}\mathbf{k}$ and the speed is $\sqrt{\dot{x}^2 + \dot{y}^2 + \dot{x}^2}$

F3. Acceleration

When a point P has position vector $\mathbf{r}$ and velocity $\mathbf{v}$ relative to an origin O, the **acceleration** $\mathbf{f}$ of P relative to O is given by

$$\mathbf{f} = \dot{\mathbf{v}} = \frac{\mathrm{d}\mathbf{v}}{\mathrm{d}t} = \ddot{\mathbf{r}} = \frac{\mathrm{d}^2\mathbf{r}}{\mathrm{d}t^2}.$$

Thus, if $\mathbf{r} = x\mathbf{i} + y\mathbf{j} + z\mathbf{k}$, $\mathbf{f} = \ddot{x}\mathbf{i} + \ddot{y}\mathbf{j} + \ddot{z}\mathbf{k}$.

F4.

When a point P has a velocity $\mathbf{v}$ relative to an origin O, the position vector of P at time t is given by

$$\mathbf{r} = \int \mathbf{v}\,\mathrm{d}t + \mathbf{a},$$

where $\mathbf{a}$ is a constant vector whose value may be determined from a knowledge of the position vector of P at some specific time.

F5.

When a point P has an acceleration $\mathbf{f}$ relative to an origin O, the velocity of P relative to O at time t is given by

$$\mathbf{v} = \int \mathbf{f}\,\mathrm{d}t + \mathbf{b},$$

where $\mathbf{b}$ is a constant vector whose value may be determined from a knowledge of the velocity of P at some specific time.

F6.

The statement 'the position vector of P relative to O' is often abbreviated to 'the position vector of P'.
Whenever unqualified statements are made, such as

'the position vector of P' or 'the velocity of P'

it is conventional to interpret them as meaning that the relevant quantity is measured relative to the origin O, which is assumed to be a fixed point.

ILLUSTRATIONS

I1.

Find, relative to an origin O, the velocity, speed and acceleration of a point P whose position vector $\mathbf{r}$ m relative to O at time t s is given by

(a) $\mathbf{r} = 5(1 + \cos t)\mathbf{i} + 5(2 - \sin t)\mathbf{j} + 3\mathbf{k}$.
(b) $\mathbf{r} = (t^3 - 3t + 1)\mathbf{i} + (3 - 2t^2)\mathbf{j} + (\sqrt{5}t^2 + 4)\mathbf{k}$. □

(a) $\dot{\mathbf{r}} = -5\sin t\mathbf{i} - 5\cos t\mathbf{j}$.

$|\dot{\mathbf{r}}| = 5(\sin^2 t + \cos^2 t) = 5$.

$\ddot{\mathbf{r}} = -5\cos t\mathbf{i} + 5\sin t\mathbf{j}$.

$\therefore$ P has velocity $5(-\sin t\mathbf{i} - \cos t\mathbf{j})$ m s^{-1}, speed 5 m s^{-1} and acceleration $5(-\cos t\mathbf{i} + \sin t\mathbf{j})$ m s^{-2}. ■

(b) $\dot{\mathbf{r}} = 3(t^2 - 1)\mathbf{i} - 4t\mathbf{j} + 2\sqrt{5}t\mathbf{k}$.

$|\dot{\mathbf{r}}| = \sqrt{9(t^2 - 1)^2 + 16t^2 + 20t^2} = 3(t^2 + 1)$.

$\ddot{\mathbf{r}} = 6t\mathbf{i} - 4\mathbf{j} + 2\sqrt{5}\mathbf{k}$.

$\therefore$ P has velocity $3(t^2 - 1)\mathbf{i} - 4t\mathbf{j} + 2\sqrt{5}t\mathbf{k}$ m s^{-1},
speed $3(t^2 + 1)$ m s^{-1}
and acceleration $6t\mathbf{i} - 4\mathbf{j} + 2\sqrt{5}\mathbf{k}$ m s^{-2}. ■

I2.

A particle P is moving with constant speed u m s^{-1} in the direction of the vector $\mathbf{i} + 4\mathbf{j} - 8\mathbf{k}$. Initially the position vector of P is $(\mathbf{i} - 8\mathbf{j})$ m. After 1 second P is at a point P_1 on the plane $y = 0$. Find u and the position vector of P_1. Show that P passes through the point with position vector $(2\mathbf{i} - 4\mathbf{j} - 8\mathbf{k})$ m and find when this occurs. □

Let $\mathbf{v}$ m s^{-1} be the constant velocity of P.

Then, $$\mathbf{v} = u\,\frac{(\mathbf{i} + 4\mathbf{j} - 8\mathbf{k})}{|\mathbf{i} + 4\mathbf{j} - 8\mathbf{k}|} = \frac{u}{9}(\mathbf{i} + 4\mathbf{j} - 8\mathbf{k}).$$

Using F4, the position vector of P at time t s is $\mathbf{r}$ m, where

$$\mathbf{r} = \frac{ut}{9}(\mathbf{i} + 4\mathbf{j} - 8\mathbf{k}) + \mathbf{a}$$

and $\mathbf{a}$ is a constant vector.
Initially, that is, when $t = 0$, $\mathbf{r} = \mathbf{i} - 8\mathbf{j}$

$\therefore$ $\mathbf{a} = \mathbf{i} - 8\mathbf{j}$.

$$\therefore \mathbf{r} = \frac{ut}{9}(\mathbf{i} + 4\mathbf{j} - 8\mathbf{k}) + \mathbf{i} - 8\mathbf{j}. \qquad (1)$$

When $t = 1$, P is on the plane $y = 0$; that is, the $\mathbf{j}$ component of $\mathbf{r}$ is zero.

Then $\frac{u(1)}{9}4\mathbf{j} - 8\mathbf{j} = \mathbf{0}$. $\therefore u = 18$. ■

Then $\vec{OP}_1 = \frac{18(1)}{9}(\mathbf{i} + 4\mathbf{j} - 8\mathbf{k}) + \mathbf{i} - 8\mathbf{j} = (3\mathbf{i} - 16\mathbf{k})$ m.

Substitute $u = 18$ in equation (1).

$$\mathbf{r} = (2t + 1)\mathbf{i} + (8t - 8)\mathbf{j} - 16t\mathbf{k}.$$

If $\mathbf{r} = 2\mathbf{i} - 4\mathbf{j} - 8\mathbf{k}$, then equating coefficients gives

$$2t + 1 = 2, \qquad 8t - 8 = -4, \qquad -16t = -8.$$

Each of these equations is satisfied by $t = \frac{1}{2}$, and so it follows that P passes through the point with position vector $(2\mathbf{i} - 4\mathbf{j} - 8\mathbf{k})$m at $\frac{1}{2}$ second after the initial instant. ■

---

13. The game of 'crash' is played on a smooth horizontal surface. In the game, a child A rolls a ball, at a constant speed of $3\sqrt{2}$ m s^{-1}, in a straight line in the direction NE. A second child B, standing 6 m due east of A, also rolls a ball which moves in a straight line with constant speed u m s^{-1}. The balls are released at the same instant and a 'crash' occurs if they collide.

(i) Show that a crash can only occur if $u \geqslant 3$.

(ii) Given that $u = 6$, and that a 'crash' occurs, find the direction in which B rolls the ball and the time that elapses before the balls collide.

(iii) Given that $u = 2\sqrt{3}$, and that a 'crash' occurs, find the two possible directions in which B may roll the ball. □

Let the crash occur after time t s and assume that the balls meet at a point C.

Then $AC = 3\sqrt{2}t$ m and $BC = ut$ m.

So, if B rolls the ball in the direction Wθ N, we have

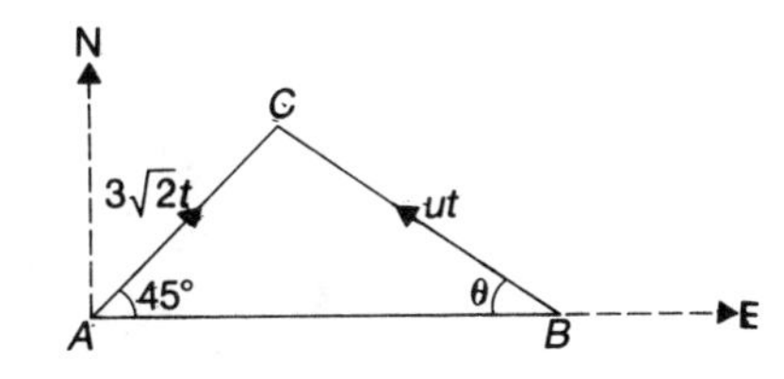

$$\frac{3\sqrt{2}t}{\sin\theta} = \frac{ut}{\sin 45^\circ}.$$

$$\therefore \frac{3}{\sin\theta} = u. \qquad (1)$$

(i) As $\sin\theta \leqslant 1$ if follows that $u \geqslant 3$. ■

(ii) Substituting $u = 6$ in equation (1):

$$\sin\theta = \tfrac{1}{2}, \text{ that is } \theta = 30^\circ \text{ or } 150^\circ.$$

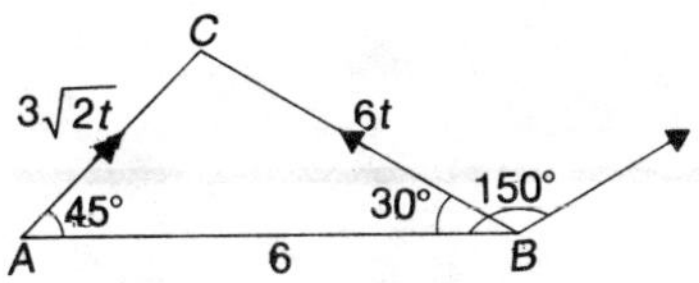

As $\theta = 150^\circ$ corresponds to a direction N60° E the paths of the balls would diverge.

Hence $\theta = 30^\circ$, that is, B projects the ball in the direction W30° N.■

Now $AB = AC\cos 45^\circ + BC\cos 30^\circ$.

$$\therefore 6 = 3\sqrt{2}t\,\frac{\sqrt{2}}{2} + \frac{6t\sqrt{3}}{2} = 3t + 3\sqrt{3}t,$$

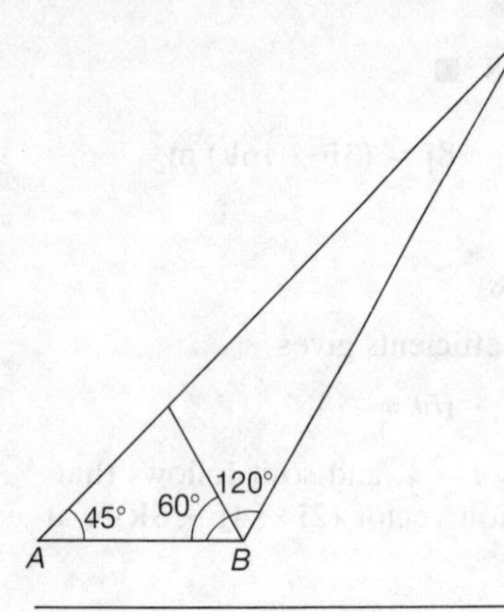

that is $t = \dfrac{2}{1 + \sqrt{3}} = 0{\cdot}73$ to 2 dp.

So the balls collide after 0·73 seconds have elapsed. ■

(iii) Substituting $u = 2\sqrt{3}$ in equation (1):

$$\sin\theta = \frac{\sqrt{3}}{2}, \text{ that is } \theta = 60° \text{ or } 120°.$$

The paths of the balls collide for both values of θ, hence B may roll the ball in either of the directions E60° N or W60° N. ■

I4. A particle P has acceleration $\mathbf{f}\ \text{m s}^{-1}$ at time t s, where

$$\mathbf{f} = 2\sin t\mathbf{i} + 2\cos t\mathbf{j} + (6t - 4)\mathbf{k}.$$

Given that P is initially at rest at the origin, find the velocity and the position vector of P at time t seconds. □

Let the velocity and position vector of P be $\mathbf{v}\ \text{m s}^{-1}$ and $\mathbf{r}$ m respectively. Using F5:

$$\mathbf{v} = \int \mathbf{f}\,dt + \mathbf{a}$$

$$= -2\cos t\mathbf{i} + 2\sin t\mathbf{j} + (3t^2 - 4t)\mathbf{k} + \mathbf{a}.$$

When $t = 0$, $\mathbf{v} = \mathbf{0}$; therefore $\mathbf{0} = -2\mathbf{i} + \mathbf{a}$.

$$\therefore \mathbf{v} = 2(1 - \cos t)\mathbf{i} + 2\sin t\mathbf{j} + (3t^2 - 4t)\mathbf{k}. \ ■$$

Using F4:

$$\mathbf{r} = \int \mathbf{v}\,dt + \mathbf{b}$$

$$= 2(t - \sin t)\mathbf{i} - 2\cos t\mathbf{j} + (t^3 - 2t^2)\mathbf{k} + \mathbf{b}.$$

When $t = 0$, $\mathbf{r} = \mathbf{0}$; therefore $\mathbf{0} = -2\mathbf{j} + \mathbf{b}$.

$$\therefore \mathbf{r} = 2(t - \sin t)\mathbf{i} + 2(1 - \cos t)\mathbf{j} + t^2(t - 2)\mathbf{k}. \ ■$$

8.3 RELATIVE DISPLACEMENT, VELOCITY AND ACCELERATION

ESSENTIAL FACTS

F1. Relative displacement

The following terms all mean the same thing.

Displacement of Q relative to P

Displacement of Q from P

Displacement from P to Q
Displacement $\overrightarrow{PQ}$
Displacement undergone by moving from P to Q
Position vector of Q relative to P

Let the position vectors of P and Q with respect to an origin O be $\mathbf{r}_P$ and $\mathbf{r}_Q$. Then

the position vector of Q relative to P is $\mathbf{r}_Q - \mathbf{r}_P$.

The distance between P and Q is $|\mathbf{r}_Q - \mathbf{r}_P|$.

F2. Relative velocity

The velocity of a point Q relative to a point P is the rate of change of the displacement $\overrightarrow{PQ}$.
Let the points P, Q have position vectors $\mathbf{r}_P$, $\mathbf{r}_Q$ and velocities $\mathbf{v}_P$, $\mathbf{v}_Q$ relative to an origin O at time t.
Then the velocity of Q relative to P is

$$\frac{\mathrm{d}}{\mathrm{d}t}(\mathbf{r}_Q - \mathbf{r}_P) = \dot{\mathbf{r}}_Q - \dot{\mathbf{r}}_P = \mathbf{v}_Q - \mathbf{v}_P.$$

The magnitude of the relative velocity, $|\mathbf{v}_Q - \mathbf{v}_P|$, is sometimes called the **relative speed**.

F3. Relative acceleration

Relative acceleration is the rate of change of relative velocity. Let the points P, Q have position vectors $\mathbf{r}_P$, $\mathbf{r}_Q$, velocities $\mathbf{v}_P$, $\mathbf{v}_Q$ and accelerations $\mathbf{f}_P$, $\mathbf{f}_Q$ with respect to an origin O at time t. Then the acceleration of Q relative to P is

$$\frac{\mathrm{d}}{\mathrm{d}t}(\mathbf{v}_Q - \mathbf{v}_P) = \dot{\mathbf{v}}_Q - \dot{\mathbf{v}}_P = \mathbf{f}_Q - \mathbf{f}_P = \ddot{\mathbf{r}}_Q - \ddot{\mathbf{r}}_P$$

$$= \frac{\mathrm{d}^2}{\mathrm{d}t^2}(\mathbf{r}_Q - \mathbf{r}_P).$$

F4. Maximum and minimum separations

Let $\mathbf{r}$ be the position vector of a point Q relative to a point P. The distance $PQ = r = |\mathbf{r}|$ and, as $r^2 = \mathbf{r}\cdot\mathbf{r}$, maximum and minimum values of the distance PQ may occur when

$$\frac{\mathrm{d}}{\mathrm{d}t}(r^2) = \frac{\mathrm{d}}{\mathrm{d}t}(\mathbf{r}.\mathbf{r}) = 0, \quad \text{that is,} \quad \text{when } \mathbf{r}.\dot{\mathbf{r}} = 0.$$

As $\dot{\mathbf{r}}$ is the velocity of Q relative to P, maximum and minimum separations of P and Q may occur when the position vector of Q relative to P is perpendicular to the velocity of Q relative to P.

F5.

Path of Q relative to P

$\dot{\mathbf{r}} = \mathbf{v}_Q - \mathbf{v}_P$

Shortest distance

Q

r

P

In the special case when $\dot{\mathbf{r}}$, the velocity of Q relative to P, is constant the equation $\mathbf{r}.\dot{\mathbf{r}} = 0$ has only one solution which always corresponds to the position of minimum separation of P and Q.

ILLUSTRATIONS

I1.

A speedboat S is travelling with speed $6\sqrt{3}\ \mathrm{m\,s^{-1}}$ in a direction N60° E and a yacht Y is travelling with speed $4\ \mathrm{m\,s^{-1}}$ in a direction N30° E. Initially Y is 52 m due east of S. Find the least distance between S and Y and the time at which they are nearest to one another. □

Take the initial position of S as the origin and let $\mathbf{i}$ and $\mathbf{j}$ be unit vectors pointing east and north respectively.

The velocities of S and Y are $\mathbf{v}_S\ \mathrm{m\,s^{-1}}$ and $\mathbf{v}_Y\ \mathrm{m\,s^{-1}}$, where

$$\mathbf{v}_S = 6\sqrt{3}\sin 60°\mathbf{i} + 6\sqrt{3}\cos 60°\mathbf{j} = 9\mathbf{i} + 3\sqrt{3}\mathbf{j}$$

and $\mathbf{v}_Y = 4\cos 60°\mathbf{i} + 4\sin 60°\mathbf{j} = 2\mathbf{i} + 2\sqrt{3}\mathbf{j}$.

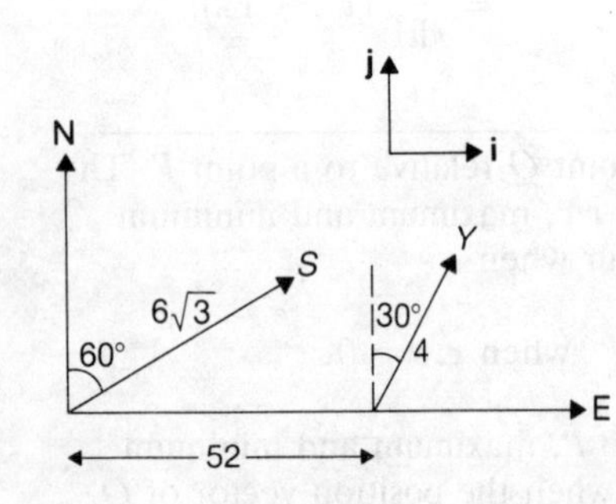

The initial position vectors of S and Y are $\mathbf{0}$ and $52\mathbf{i}$ m.

The position vectors of S and Y at time t s are $\mathbf{r}_S$ and $\mathbf{r}_Y$ m, where

$$\mathbf{r}_S = 9t\mathbf{i} + 3\sqrt{3}t\mathbf{j}$$

and $\mathbf{r}_Y = 2t\mathbf{i} + 2\sqrt{3}t\mathbf{j} + 52\mathbf{i}$.

Let $\mathbf{r} = \overrightarrow{YS}$

Then $\mathbf{r} = \mathbf{r}_S - \mathbf{r}_Y = (7t - 52)\mathbf{i} + \sqrt{3}t\mathbf{j}$.

$$\begin{aligned}\text{Hence } |\mathbf{r}|^2 &= (7t - 52)^2 + (\sqrt{3}t)^2 \\ &= 52(t^2 - 14t + 52) \\ &= 52\,[(t - 7)^2 + 3].\end{aligned}$$

As $(t - 7)^2 \geqslant 0$, the minimum value of $|\mathbf{r}|^2$ is $52 \times 3 = 156$ and it occurs when $t = 7$.

Hence the least distance between S and Y is $\sqrt{156}$ m and they are closest after 7 seconds. ■

Note: Instead of finding $|\mathbf{r}|^2$ and completing the square, we may use F4.

As $\mathbf{r} = (7t - 52)\mathbf{i} + \sqrt{3}t\mathbf{j}$, $\quad \dot{\mathbf{r}} = 7\mathbf{i} + \sqrt{3}\mathbf{j}$.

When $\mathbf{r}.\dot{\mathbf{r}} = 0$, $\quad 7(7t - 52) + 3t = 0$, $\quad$ that is, $\quad t = 7$.

Then substituting $t = 7$ gives $\mathbf{r} = -3\mathbf{i} + 7\sqrt{3}\mathbf{j}$, and
$|\mathbf{r}| = \sqrt{9 + 147} = \sqrt{156}$.

Alternative solution using trigonometry

We first draw a vector velocity triangle OCD in which $\vec{OC}$, $\vec{OD}$ represent the velocities of S and Y. Then $\vec{CD} = \vec{OD} - \vec{OC}$, represents the velocity, $\mathbf{v}\ \mathrm{m\,s^{-1}}$, of Y relative to S.

Then $|\mathbf{v}|^2 = v^2 = (4)^2 + (6\sqrt{3})^2 - 2 \times 4 \times 6\sqrt{3} \times \cos 30°$

$$= 16 + 108 - 72 = 52.$$

$$\therefore v = 2\sqrt{13}.$$

Let angle OCD be α. Then

$$(4)^2 = v^2 + (6\sqrt{3})^2 - 2v \times 6\sqrt{3}\cos\alpha.$$

$$\therefore \cos\alpha = \frac{2\sqrt{3}}{\sqrt{13}} \quad \text{and} \quad \sin\alpha = \frac{1}{\sqrt{13}}.$$

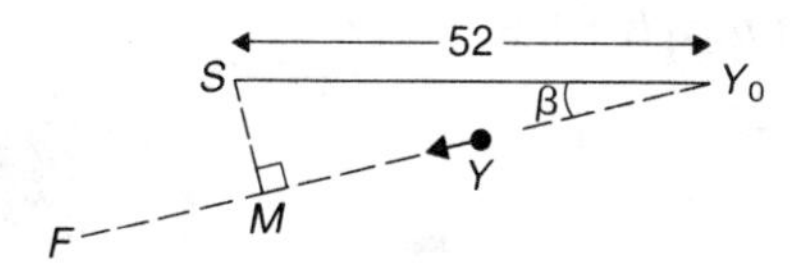

Now draw a diagram showing the motion of Y relative to S. Mark Y_0, the initial position of Y relative to S, and draw the line Y_0F parallel to CD.

Relative to S, the yacht Y moves along Y_0F with speed $v = 2\sqrt{13}\ \mathrm{m\,s^{-1}}$.

Then, if M is the foot of the perpendicular from S to Y_0F, the minimum distance between S and Y is SM and it occurs at time MY_0/v.

Let the angle SY_0M be $\beta = 30° - \alpha$.

Then $\sin\beta = \sin(30° - \alpha) = \frac{1}{2}.\frac{2\sqrt{3}}{\sqrt{13}} - \frac{\sqrt{3}}{2}.\frac{1}{\sqrt{13}} = \frac{\sqrt{3}}{2\sqrt{13}}$.

Hence $SM = 52\sin\beta = 2\sqrt{39}$, $\quad$ and

$$\frac{MY_0}{v} = \frac{\sqrt{52^2 - SM^2}}{v} = \frac{\sqrt{52^2 - 4 \times 39}}{2\sqrt{13}} = \frac{2 \times \sqrt{13} \times \sqrt{49}}{2\sqrt{13}} = 7.$$

---

12. A man cycling on level ground finds that when he travels due east with speed $20\ \mathrm{km\,h^{-1}}$ the wind appears to be blowing from a direction EθS, where $\tan\theta = \frac{3}{4}$. When the man travels with speed $20\ \mathrm{km\,h^{-1}}$ in a direction EθS the wind appears to be blowing from the south. Find

(i) the apparent speed of the wind on both occasions

(ii) the true velocity of the wind. □

Let $\mathbf{v}_m$ km h^{-1} and $\mathbf{v}_w$ km h^{-1} be the velocities of the man and the wind and let $\mathbf{v}$ km h^{-1} be the velocity of the wind relative to the man. That is, $\mathbf{v}$ km h^{-1} is the velocity of the wind as it appears to the cyclist.
Then $\mathbf{v} = \mathbf{v}_w - \mathbf{v}_m$. (1)
Let $\mathbf{i}$ and $\mathbf{j}$ be unit vectors in the east and north directions.

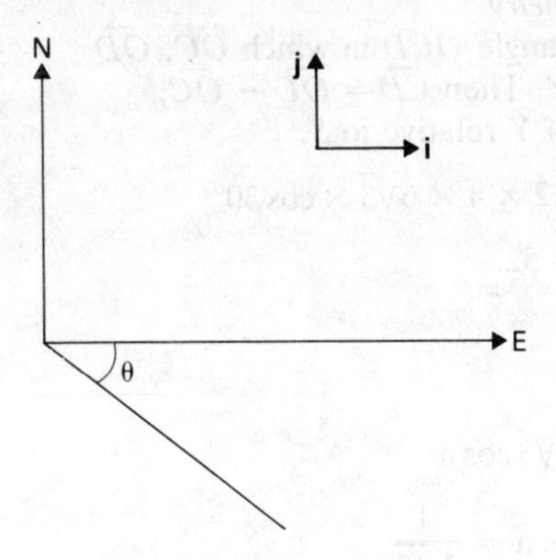

When the cyclist travels due east.
$\mathbf{v}_m = 20\mathbf{i}$ and $\mathbf{v}$ is in the direction $-4\mathbf{i} + 3\mathbf{j}$.
So $\mathbf{v} = \lambda(-4\mathbf{i} + 3\mathbf{j})$, where λ is a constant.
Substituting in (1),

$$-4\lambda\mathbf{i} + 3\lambda\mathbf{j} = \mathbf{v}_w - 20\mathbf{i}. \quad (2)$$

When the cyclist travels EθS
A unit vector in the direction of $\mathbf{v}_m$ is

$$\cos\theta\mathbf{i} - \sin\theta\mathbf{j} = \tfrac{4}{5}\mathbf{i} - \tfrac{3}{5}\mathbf{j}.$$

$$\therefore \mathbf{v}_m = 20(\tfrac{4}{5}\mathbf{i} - \tfrac{3}{5}\mathbf{j}) = 16\mathbf{i} - 12\mathbf{j}.$$

Also $\mathbf{v} = \mu\mathbf{j}$, where μ is a constant.
Substituting in (1),

$$\mu\mathbf{j} = \mathbf{v}_w - (16\mathbf{i} - 12\mathbf{j}). \quad (3)$$

Eliminating $\mathbf{v}_w$ from (2) and (3),

$$(20 - 4\lambda)\mathbf{i} + 3\lambda\mathbf{j} = 16\mathbf{i} + (\mu - 12)\mathbf{j}.$$

Equating coefficients of $\mathbf{i}$ and $\mathbf{j}$,

$$20 - 4\lambda = 16 \text{ and } 3\lambda = \mu - 12.$$

$$\therefore \lambda = 1 \text{ and } \mu = 15.$$

Substituting in (2) or (3) then gives

$$\mathbf{v}_w = 16\mathbf{i} + 3\mathbf{j}.$$

Alternative Solution
Define $\mathbf{v}_m$, $\mathbf{v}_w$ and $\mathbf{v}$ as in the first solution, so that

$$\mathbf{v} = \mathbf{v}_w - \mathbf{v}_m. \quad (1)$$

Let $\mathbf{v}_w$ km h^{-1} consist of a speed W km h^{-1} in a direction EψN.
We now use equation (1) to draw a vector velocity triangle for each of the two occasions.
When the cyclist travels due east

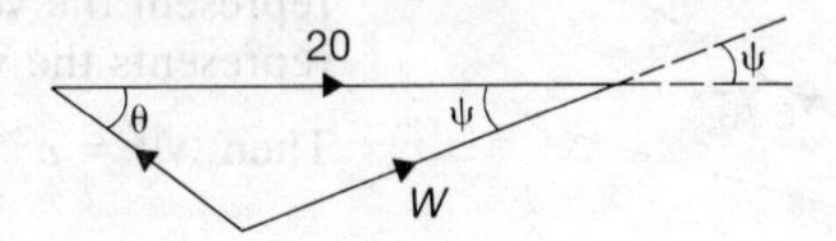

Applying the sine rule

$$\frac{20}{\sin(\theta + \psi)} = \frac{W}{\sin\theta}$$

$$\therefore \quad 20 \times \frac{3}{5} = W\left(\frac{3}{5}\cos\psi + \frac{4}{5}\sin\psi\right)$$

$$\therefore \quad 60 = W(3\cos\psi + 4\sin\psi). \quad (2)$$

When the cyclist travels EθS

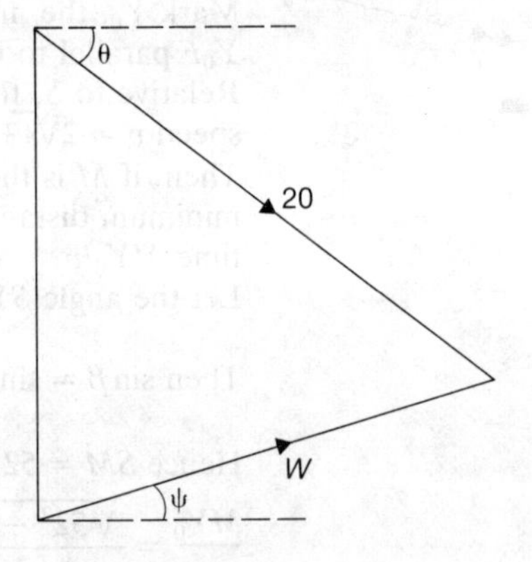

Applying the sine rule

$$\frac{20}{\cos\psi} = \frac{W}{\cos\theta}.$$

$$\therefore 16 = W\cos\psi. \quad (3)$$

Hence the apparent speed of the wind on the first occasion is

$|\lambda(-4\mathbf{i} + 3\mathbf{j})| = 5 \text{ km h}^{-1}$

and on the second occasion it is

$|\mu\mathbf{j}| = 15 \text{ km h}^{-1}$. ■

Finally, the velocity of the wind is $\sqrt{16^2 + 3^2} = \sqrt{265} \text{ km h}^{-1}$ in a direction EϕN, where $\tan\phi = \frac{3}{16}$. ■

Eliminating W from (2) and (3),

$$\frac{60}{16} = 3 + 4\tan\psi.$$

$$\therefore \tan\psi = \frac{3}{16}.$$

Hence, using (3), $W = \sqrt{265}$.
The apparent speed of the wind is equal to the length of the third side in the vector triangles. Once W and ψ have been found the lengths of these sides may be determined by using the sine rule or the cosine rule on each of the triangles. Hence the solution may be continued.

13.

A river has straight parallel banks 100 m apart and the water flows at a constant speed of $0{\cdot}8 \text{ m s}^{-1}$. A man stands at a point A on one bank of the river, directly opposite a point B on the other bank. The man enters the water at A and swims at a speed, relative to the water, of $1{\cdot}6 \text{ m s}^{-1}$. He reaches the opposite bank at a point C. Given that he chose his direction such that the crossing was completed in the shortest possible time, find the time taken and the distance BC. □

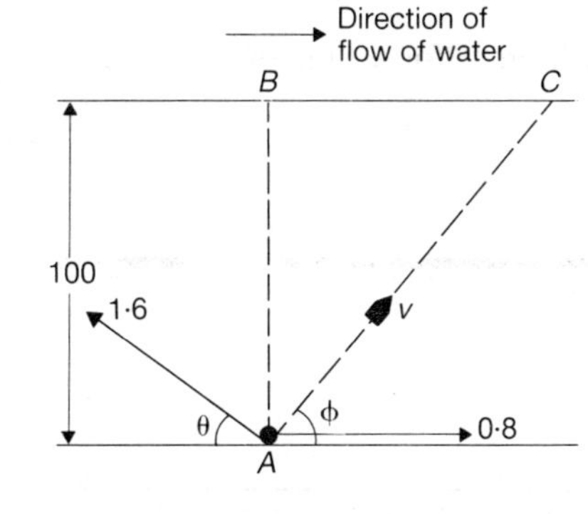

The velocity of the man relative to the water equals his actual velocity minus the velocity of the water.
So the man's actual velocity equals the sum of his velocity relative to the water and the velocity of the water.
Let the man's actual velocity be $v \text{ m s}^{-1}$ in a direction ϕ with the downstream direction and let his velocity relative to the water be $1{\cdot}6 \text{ m s}^{-1}$ in a direction θ with the upstream direction.
Taking components parallel and perpendicular to the bank.

$$v\cos\phi = 0{\cdot}8 - 1{\cdot}6\cos\theta \quad \text{and} \quad v\sin\phi = 1{\cdot}6\sin\theta.$$

The crossing is completed in time T s, where

$$T = \frac{100}{v\sin\phi} = \frac{100}{1{\cdot}6\sin\theta}.$$

This is a minimum when $\theta = \frac{1}{2}\pi$, that is, when $T = 62{\cdot}5$.
With $\theta = \frac{1}{2}\pi$, $v\cos\phi = 0{\cdot}8$ and $v\sin\phi = 1{\cdot}6$.

$$\therefore \tan\phi = 2.$$

Now $BC = 100\cot\phi = 50$.
Therefore the crossing is completed in 62·5 s and $BC = 50$ m. ■

14.

At time $t = 0$ a ship A leaves a harbour O and moves with a constant speed of 13 km h^{-1} in the direction of the vector $5\mathbf{i} + 12\mathbf{j}$. At the same time a ship B, which is moving with constant velocity at 5 km h^{-1} in the direction of the vector $-3\mathbf{i} + 4\mathbf{j}$, is at the point H with position vector $16\mathbf{j}$ km referred to O. Given that visibility is 12 km find the length of time during which the ships are in sight of each other. □

$|5\mathbf{i} + 12\mathbf{j}| = 13$ so, using 8.2F2, the velocity of A is $\mathbf{v}_A$ km h^{-1}, where

$$\mathbf{v}_A = 5\mathbf{i} + 12\mathbf{j}.$$

Similarly, as $|-3\mathbf{i} + 4\mathbf{j}| = 5$, the velocity of B is $\mathbf{v}_B$ km h^{-1}, where

$$\mathbf{v}_B = -3\mathbf{i} + 4\mathbf{j}.$$

Hence, at time t hours, the position vectors of A and B referred to O are $\mathbf{r}_A$ m and $\mathbf{r}_B$ m, where

$$\mathbf{r}_A = 5t\mathbf{i} + 12t\mathbf{j}$$

and $$\mathbf{r}_B = -3t\mathbf{i} + 4t\mathbf{j} + 16\mathbf{j}.$$

The position vector of A relative to B is $\mathbf{r}$ m, where

$$\mathbf{r} = \mathbf{r}_A - \mathbf{r}_B = 8t\mathbf{i} + (8t - 16)\mathbf{j}.$$

The ships are in sight of each other when $|\mathbf{r}| \leqslant 12$.
That is, when $|\mathbf{r}|^2 \leqslant 144$.

$$\therefore (8t)^2 + (8t - 16)^2 \leqslant 144.$$

$$\therefore 8(t^2 - 2t) + 7 \leqslant 0.$$

$$\therefore 8(t-1)^2 \leqslant 1.$$

$$\therefore t \leqslant 1 + \frac{1}{2\sqrt{2}} \text{ and } t \geqslant 1 - \frac{1}{2\sqrt{2}}.$$

Hence the ships are in sight of each other for a time of

$2 \times \frac{1}{2\sqrt{2}} = \frac{1}{\sqrt{2}}$ hours. ■

8.4 EXAMINATION QUESTIONS AND SOLUTIONS

Q1.

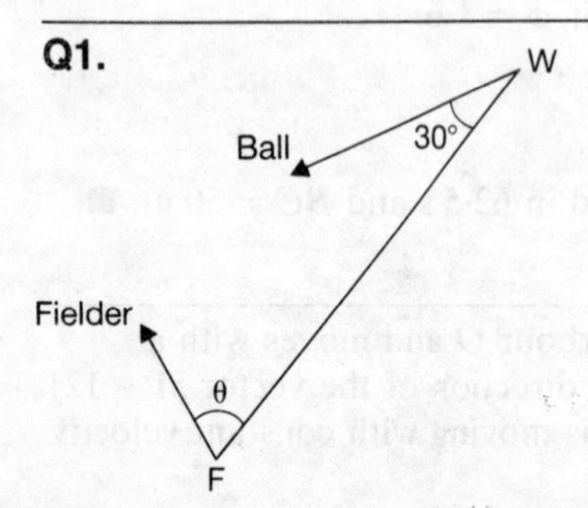

A batsman is at the wicket W and the fielder is in the outfield at point F. The batsman strikes the ball along the ground in a direction making 30° with the line WF, as shown. The ball travels at constant speed $\frac{3}{2}V$ and the fielder starts running with constant speed V at the instant at which the ball is struck. The fielder runs in a direction making an angle θ with FW.

Find θ so that the fielder stops the ball as soon as possible. If the fielder has run a distance of 20 metres to stop the ball, find the distance WF. (NI 1985)

Q2.

(a) A particle moves in a plane so that at time t ($\geqslant 0$) its position vector relative to a set of Cartesian axes is

$$\mathbf{r} = \tfrac{1}{2}(t^2 - 3t)\mathbf{i} + 2t\mathbf{j}.$$

Find $\mathbf{v}$ and $\mathbf{a}$, its velocity and acceleration vectors, respectively and show that its minimum speed is 2.

(b) A radar station is located at the origin of a coordinate system, in which the unit of distance is 1 km, and relative to it airport P has position vector $60\mathbf{i} + 90\mathbf{j}$.

At noon, on a day when the wind has a constant velocity vector $\mathbf{w} = c\mathbf{i} + d\mathbf{j}$, where the constants c and d are measured in km h^{-1}, helicopters A and B leave airport P flying at the same *speed* of v km h^{-1} relative to the air. Relative to the air, A's velocity is in the direction of the unit vector $\frac{3}{5}\mathbf{i} + \frac{4}{5}\mathbf{j}$ and B's velocity is in the direction of the unit vector $\frac{7}{25}\mathbf{i} - \frac{24}{25}\mathbf{j}$.

One hour later, the radar station records the position vectors of A and B to be $132\mathbf{i} + 175\mathbf{j}$ and $100\mathbf{i} - \mathbf{j}$ respectively. Find the speeds v and $|\mathbf{w}|$ of the helicopters and the wind, respectively. Find also the direction of the wind.

Use vector methods to compute the angle between the paths, relative to the ground, of the helicopters. (WJEC 1984)

Q3.

In a wind blowing from the south with constant speed w a helicopter flies horizontally with constant velocity in a direction θ east of north from a point A to a point B. The speed of the helicopter relative to the air is λw, where $\lambda > 1$. Find the speed of the helicopter along AB.

The helicopter returns from B to A with constant velocity and the same speed λw relative to the air, and in the same wind. Show that the total time for the two journeys is

$$\frac{2c\sqrt{\lambda^2 - \sin^2\theta}}{w(\lambda^2 - 1)},$$

where $c = AB$. (JMB 1984)

Q4.

At time $t = 0$, a particle A, which moves with constant velocity $(3\mathbf{i} + u\mathbf{j} + 5\mathbf{k})\ \text{m s}^{-1}$, has position vector $(4\mathbf{i} + 9\mathbf{j} - 10\mathbf{k})$ m relative to a fixed origin O. At time $t = 0$, a particle B which has constant velocity $2\mathbf{i}\ \text{m s}^{-1}$ is at O.

(i) Given that $u = 2$, find the value of t at the instant when the distance between A and B is least.

(ii) At time t s, a third particle C has position vector

$$[10\mathbf{i} + 5\mathbf{j}\sin(\pi t/4) + 5\mathbf{k}\cos(\pi t/4)]\ \text{m}$$

relative to the fixed origin O. Find the value of u in this case, given that A and C collide. (LON 1986)

Q5. A ship is steaming at 15 knots due east, while the wind speed is 20 knots from due north. Find the magnitude and direction, to the nearest degree, of the wind velocity relative to the ship. Find also the course, between east and south, along which the ship would have to steer at 16 knots for the wind velocity relative to the ship to be at right angles to the course of the ship. Obtain the magnitude of the velocity of the wind relative to the ship in this case. (LON 1982)

Q6. At time t two points P and Q have position vectors $\mathbf{p}$ and $\mathbf{q}$ respectively, where

$$\mathbf{p} = 2a\mathbf{i} + (a\cos\omega t)\mathbf{j} + (a\sin\omega t)\mathbf{k},$$
$$\mathbf{q} = (a\sin\omega t)\mathbf{i} - (a\cos\omega t)\mathbf{j} + 3a\mathbf{k}$$

and a, ω are constants. Find $\mathbf{r}$, the position vector of P relative to Q, and $\mathbf{v}$, the velocity of P relative to Q. Find also the values of t for which $\mathbf{r}$ and $\mathbf{v}$ are perpendicular.

Determine the smallest and greatest distances between P and Q. (LON 1982)

Q7. The map of a holiday resort shows that the point A is 1600 m north of a hotel O, B is 800 m east of O and C is 600 m north of B. The map also shows a straight horizontal road between O and C and a ski-lift carrying ski-buckets in a straight line between A, which is 150 m higher than O, to B, which is 50 m lower than O. Taking O as origin, with unit vectors $\mathbf{i}$ due east, $\mathbf{j}$ due north and $\mathbf{k}$ vertically upwards, write down the position vectors, relative to O, of A, B and C. Hence find unit vectors parallel to the directions of $\overrightarrow{AB}$ and $\overrightarrow{OC}$.

At a given instant a ski-bucket leaves A and travels towards B at a uniform speed of 9 km h^{-1} and, at the same time, a man leaves O and travels towards C at a uniform speed of 5 km h^{-1}. Write down the position vectors of the ski-bucket and the man at a subsequent time t hours. Hence find the position vector of the bucket relative to the man at time t hours.
Safety regulations require that the man and the bucket are never less than 50 m apart. By considering the distance between the man and the bucket at a time $\frac{1}{7}$ hours after their departures, or otherwise, show that the regulations are broken. (LON 1984)

S1.

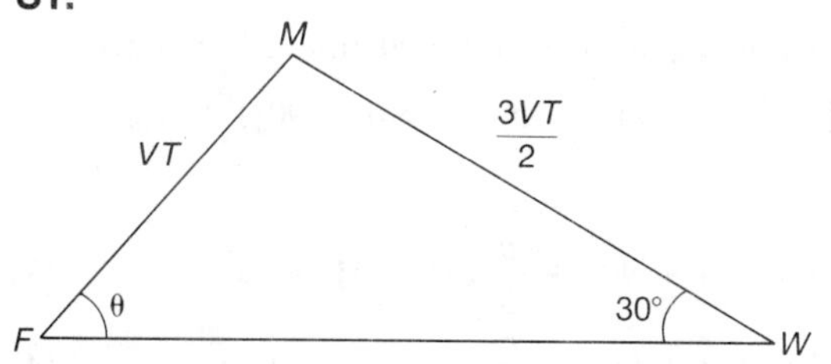

For the fielder to stop the ball as soon as possible he must meet the ball whilst he is still running. [A]
Let the fielder and the ball meet at a point M and after time T.

Then $FM = VT$ and $MW = \dfrac{3VT}{2}$.

Hence $\dfrac{3VT}{2\sin\theta} = \dfrac{VT}{\sin 30°}$. [B]

$\therefore \sin\theta = \frac{3}{4}$.

So $\theta = 48{\cdot}9°$ or $(180 - 48{\cdot}9)°$ to 1dp.
Rejecting the larger value of θ, we see that the fielder stops the ball in the shortest time when $\theta = 48{\cdot}9°$ to 1dp. [C] ■
Now $FW = FM\cos\theta + MW\cos 30°$
and when $FM = VT = 20$ m,

$$FW = 20 \times \frac{\sqrt{7}}{4} + 20 \times \frac{3}{2} \times \frac{\sqrt{3}}{2}$$
$$= 39{\cdot}2 \text{ m to 1dp.}$$ *11m* ■

[A] If the fielder reaches the line of motion of the ball and then has to wait for the ball to reach him he has obviously not minimized the time taken to stop the ball. As a consequence, it follows that the shortest time is when he meets the ball whilst he is still running.

[B] Here the equation for $\sin\theta$ has been obtained by using the sine rule. However it may also be obtained by the other methods. Firstly, we could say that the velocity component of the fielder perpendicular to FW must equal the velocity component of the ball perpendicular to FW.
That is, $V\sin\theta = \dfrac{3V}{2}\sin 30°$.
Secondly, we could, equivalently, say that for the fielder to intercept the ball the velocity of the fielder relative to the ball must be parallel to FW.
[C] The larger value of θ corresponds to a point of intersection which is more distant from W than M.
The time taken for the ball to cover this larger distance will then be greater than the time for it to travel to M.

S2.

(a) $\mathbf{r} = \frac{1}{2}(t^2 - 3t)\mathbf{i} + 2t\mathbf{j}$.
$\therefore$ the velocity $\mathbf{v} = \dot{\mathbf{r}} = (t - \frac{3}{2})\mathbf{i} + 2\mathbf{j}$
and the acceleration vector $\mathbf{a} = \ddot{\mathbf{r}} = \mathbf{i}$. ■

$$|\dot{\mathbf{r}}|^2 = (t - \tfrac{3}{2})^2 + 4 \geqslant 4. \qquad \therefore |\dot{\mathbf{r}}| \geqslant 2.$$

Hence the minimum speed of the particle is 2. ■
(b) The velocities of A and B relative to the air are

$$\frac{v}{5}(3\mathbf{i} + 4\mathbf{j})\text{ km h}^{-1} \quad \text{and} \quad \frac{v}{25}(7\mathbf{i} - 24\mathbf{j})\text{ km h}^{-1} \quad \text{respectively.}$$

Let $\mathbf{v}_A$ km h^{-1}, $\mathbf{v}_B$ km h^{-1} be the velocities of A and B relative to the fixed origin.
As the velocity of a helicopter relative to the air equals the velocity of the helicopter minus the velocity of the air, we have that

$$\frac{v}{5}(3\mathbf{i}+4\mathbf{j}) = \mathbf{v}_A - \mathbf{w} \tag{1}$$

and $$\frac{v}{25}(7\mathbf{i}-24\mathbf{j}) = \mathbf{v}_B - \mathbf{w}. \tag{2}$$

The position vectors $\mathbf{r}_A$ m, $\mathbf{r}_B$ m of A and B at time t hours are

$$\mathbf{r}_A = (60\mathbf{i}+90\mathbf{j}) + \mathbf{v}_A t \quad \text{and} \quad \mathbf{r}_B = (60\mathbf{i}+90\mathbf{j}) + \mathbf{v}_B t.$$

$\therefore$ when $t = 1$,

$$132\mathbf{i} + 175\mathbf{j} = 60\mathbf{i} + 90\mathbf{j} + \frac{v}{5}(3\mathbf{i}+4\mathbf{j}) + \mathbf{w} \tag{3}$$

and $$100\mathbf{i} - \mathbf{j} = 60\mathbf{i} + 90\mathbf{j} + \frac{v}{25}(7\mathbf{i}-24\mathbf{j}) + \mathbf{w} \tag{4}$$

Subtracting (4) from (3),

$$32\mathbf{i} + 176\mathbf{j} = \frac{v}{25}(8\mathbf{i}+44\mathbf{j}).$$

$$\therefore \quad 800\mathbf{i} + 4400\mathbf{j} = v\,(8\mathbf{i}+44\mathbf{j}).$$

$$\therefore \quad v = 100.$$

Substituting $v = 100$ in equation (3),

$$\mathbf{w} = (132 - 60 - 60)\mathbf{i} + (175 - 90 - 80)\mathbf{j} = 12\mathbf{i} + 5\mathbf{j}.$$

$$\therefore \ |\mathbf{w}| = 13.$$

The wind velocity is $12\mathbf{i} + 5\mathbf{j}$ and this vector also gives the direction of the wind. ■

From (1) and (2),

$$\mathbf{v}_A = (60\mathbf{i} + 80\mathbf{j}) + (12\mathbf{i} + 5\mathbf{j}) = 72\mathbf{i} + 85\mathbf{j}$$

and $\mathbf{v}_B = (28\mathbf{i} - 96\mathbf{j}) + (12\mathbf{i} + 5\mathbf{j}) = 40\mathbf{i} - 91\mathbf{j}.$

Let θ be the angle between the paths, relative to the ground, of the helicopters.

Then $\cos\theta = \dfrac{\mathbf{v}_A.\mathbf{v}_B}{|\mathbf{v}_A|\,|\mathbf{v}_B|} = \dfrac{-4855}{111.39 \times 99.40}.$

$\therefore \ \theta = 116.0°$ to 1 dp. *25m* ■

Note: The symbols for the components c and d of the wind velocity were not needed. They were included in the question as an assurance that the wind had no vertical component.

S3.

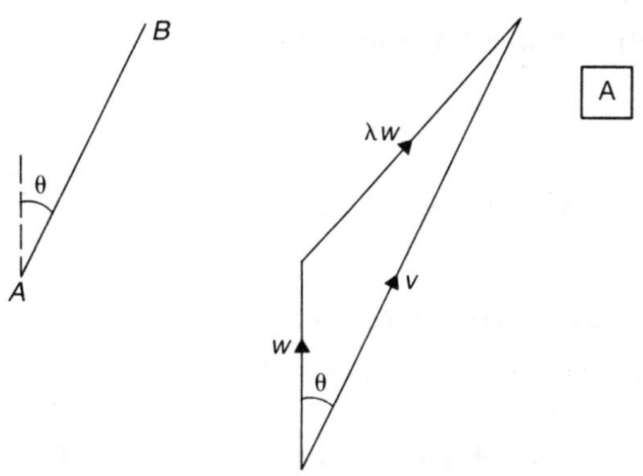

[A] We first draw a vector velocity triangle representing the fact that the velocity of the helicopter relative to the air is equal to the velocity of the helicopter (direction from A to B) minus the velocity of the air (direction from south to north).
[B] Using the cosine rule.
[C] As $\lambda \geqslant 1$, $\lambda^2 - \sin^2\theta \geqslant 1 - \sin^2\theta = \cos^2\theta$. Hence, the root $w\cos\theta - w\sqrt{\lambda^2 - \sin^2\theta}$ is negative, and so may be rejected.

Motion from A to B
Let v be the speed of the helicopter along AB.

$\lambda^2 w^2 = w^2 + v^2 - 2wv\cos\theta.$ [B]

$\therefore\ v^2 - 2vw\cos\theta = w^2(\lambda^2 - 1).$

$\therefore\ (v - w\cos\theta)^2 = w^2(\lambda^2 - 1) + w^2\cos^2\theta$

$= w^2(\lambda^2 - \sin^2\theta).$

$\therefore\ v = w\cos\theta \pm w\sqrt{\lambda^2 - \sin^2\theta}.$

Hence, as $v > 0$, the speed of the helicopter is

$v = w\cos\theta + w\sqrt{\lambda^2 - \sin^2\theta}.$ [C] ■

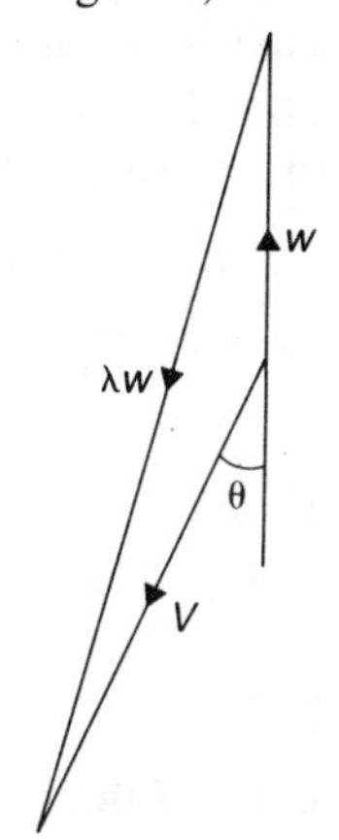

Motion from B to A [D]
Let V be the speed of the helicopter along BA.

$\lambda^2 w^2 = w^2 + V^2 + 2wV\cos\theta.$ [E]

$\therefore\ V = -w\cos\theta + w\sqrt{\lambda^2 - \sin^2\theta}.$ [F]

The total time for the two journeys is T, where

$$T = \frac{c}{v} + \frac{c}{V} = \frac{c(v + V)}{vV}$$

$$= \frac{2cw\sqrt{\lambda^2 - \sin^2\theta}}{w^2(\lambda^2 - \sin^2\theta) - w^2\cos^2\theta}$$

$$= \frac{2cw\sqrt{\lambda^2 - \sin^2\theta}}{w^2(\lambda^2 - 1)}.$$

19m ■

[D] Draw a second vector velocity triangle in which the velocity of the helicopter has direction from B to A.
[E] Using the cosine rule and remembering that $\cos(\pi - \theta) = -\cos\theta$.
[F] This may be written down immediately by comparison with the first part of the question. That is, we replace $\cos\theta$ by $-\cos\theta$.

S4.

Let $\mathbf{r}_A$ m, $\mathbf{r}_B$ m be the position vectors of A and B at time t s.

Then $\mathbf{r}_A = 4\mathbf{i} + 9\mathbf{j} - 10\mathbf{k} + (3\mathbf{i} + u\mathbf{j} + 5\mathbf{k})t$

and $\mathbf{r}_B = 2t\mathbf{i}$.

(i) When $u = 2$,

$$\mathbf{r} = \mathbf{r}_A - \mathbf{r}_B = (t + 4)\mathbf{i} + (2t + 9)\mathbf{j} + (5t - 10)\mathbf{k}$$

and $\dot{\mathbf{r}} = \mathbf{i} + 2\mathbf{j} + 5\mathbf{k}$.

$\mathbf{r}.\dot{\mathbf{r}} = 0$ when $(t + 4) + 2(2t + 9) + 5(5t - 10) = 0$.

(Using 8.3 F5 to find the time of nearest approach).

That is, when $t = \frac{14}{15}$.

Therefore, the distance between A and B is least when $t = \frac{14}{15}$. ■

(ii) C has position vector $\mathbf{r}_C$ m, where

$$\mathbf{r}_C = 10\mathbf{i} + 5\sin(\pi t/4)\mathbf{j} + 5\cos(\pi t/4)\mathbf{k}.$$

$\mathbf{r}_A = \mathbf{r}_C$ when

$$3t + 4 = 10,\ ut + 9 = 5\sin(\pi t/4) \text{ and } 5t - 10 = 5\cos(\pi t/4).$$

Thus $t = 2$ and $u = -2$.

So if A and C collide, the value of $u = -2$. *25m* ■

S5.

Let $\mathbf{V}_S$ knots, $\mathbf{V}_W$ knots be the velocities of the ship and the wind. Then, $\mathbf{V}$ knots, the velocity of the wind relative to the ship is given by

$$\mathbf{V} = \mathbf{V}_W - \mathbf{V}_S$$

and this equation may be used to construct a vector velocity diagram.

When the ship steams at 15 knots due east:

Assuming $\mathbf{V}$ knots to be a speed v_1 knots in a direction Wθ S, the vector velocity triangle is

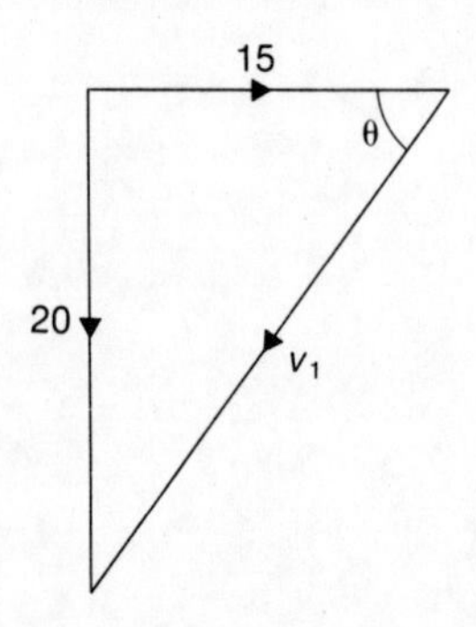

Alternative solution.

Define $\mathbf{V}_S$, $\mathbf{V}_W$ and $\mathbf{V}$ as in the first solution, and let $\mathbf{i}$, $\mathbf{j}$ be unit vectors in the east and north directions.

When the ship steams at 15 knots due east:

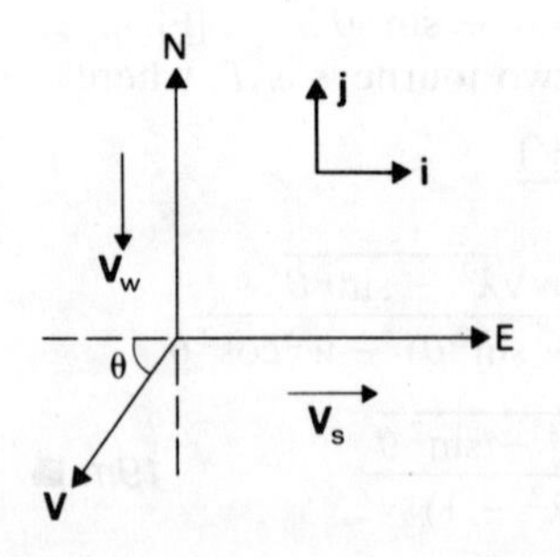

$\mathbf{V}_W = -20\mathbf{j}$,

$\mathbf{V}_S = 15\mathbf{i}$.

Hence $\mathbf{V} = \mathbf{V}_W - \mathbf{V}_S = -15\mathbf{i} - 20\mathbf{j}$.

Thus $|\mathbf{V}| = \sqrt{15^2 + 20^2} = 25$ and $\mathbf{V}$ has direction Wθ S, where $\tan\theta = \frac{4}{3}$. That is, $\theta = 53°$.

Hence $v_1 = \sqrt{15^2 + 20^2} = 25$
and $\tan\theta = \frac{4}{3}$ that is, $\theta = 53°$ to the nearest degree.
So the wind velocity relative to the ship has magnitude 25 knots and direction W53° S.
Note: The result could be presented equivalently as '25 knots from the direction E53° N'.
When the ship steams at 16 knots:
Let $\mathbf{V}_S$ be in the direction Eϕ S and let $|\mathbf{V}| = v_2$.
Then the vector velocity triangle is

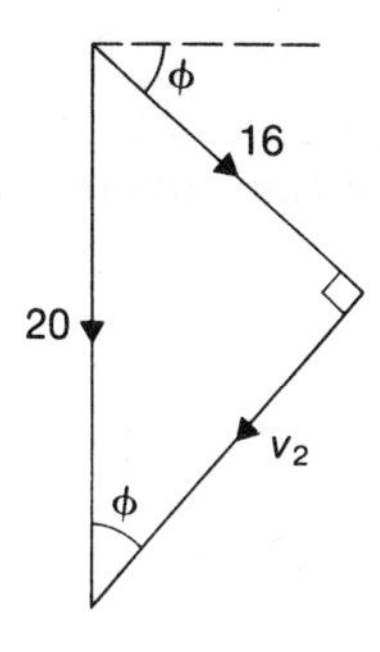

Hence $\sin\phi = \frac{4}{5}$, that is $\phi = 53°$ to the nearest degree and $v_2 = \sqrt{20^2 - 16^2} = 12$.
So the ship must steer a course E53° S and the velocity of the wind relative to the ship has magnitude 12 knots. *24m* ■

When the ship steams at 16 knots

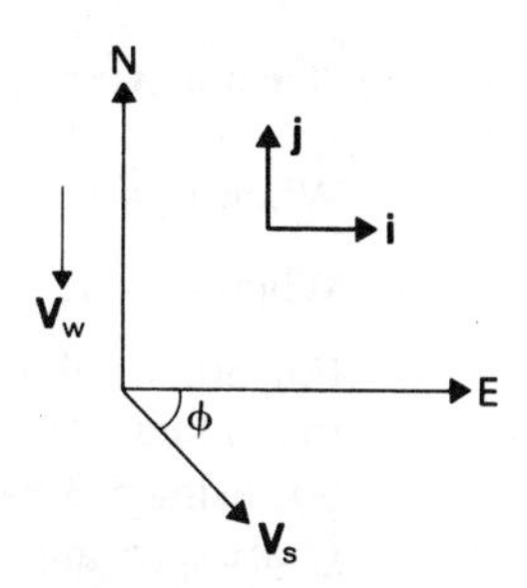

Let $\mathbf{V}_S$ be in the direction Eϕ S, that is, in the direction of the vector $\cos\phi\,\mathbf{i} - \sin\phi\,\mathbf{j}$.

Then $\mathbf{V}_S = 16(\cos\phi\,\mathbf{i} - \sin\phi\,\mathbf{j})$.

As $\mathbf{V}_S$ is perpendicular to $\mathbf{V}$, it follows that $\mathbf{V} = \lambda\,(\sin\phi\,\mathbf{i} + \cos\phi\,\mathbf{j})$, where λ is a constant.

Then, using $\mathbf{V} = \mathbf{V}_W - \mathbf{V}_S$,

$$\lambda\sin\phi\,\mathbf{i} + \lambda\cos\phi\,\mathbf{j} = -16\cos\phi\,\mathbf{i} + (16\sin\phi - 20)\mathbf{j}.$$

Equating coefficients of $\mathbf{i}$ and $\mathbf{j}$:

$$\lambda\sin\phi = -16\cos\phi$$

and $$\lambda\cos\phi = 16\sin\phi - 20.$$

Eliminating λ:
$16\sin^2\phi - 20\sin\phi = -16\cos^2\phi = 16\sin^2\phi - 16$.
$\therefore \sin\phi = \frac{4}{5}$.
It follows that $\lambda = -16\cot\phi = -12$, and as $|\mathbf{V}| = |\lambda| = 12$ the answers are as obtained in the first solution.

S6.

$$\mathbf{p} = 2a\mathbf{i} + (a\cos\omega t)\mathbf{j} + (a\sin\omega t)\mathbf{k}$$
and $$\mathbf{q} = (a\sin\omega t)\mathbf{i} - (a\cos\omega t)\mathbf{j} + 3a\mathbf{k}.$$

$$\therefore \mathbf{r} = \mathbf{p} - \mathbf{q} = (2a - a\sin\omega t)\mathbf{i} + (2a\cos\omega t)\mathbf{j} + (a\sin\omega t - 3a)\mathbf{k}.$$

$$\mathbf{v} = \dot{\mathbf{r}} = (-a\omega\cos\omega t)\mathbf{i} + (-2a\omega\sin\omega t)\mathbf{j} + (a\omega\cos\omega t)\mathbf{k}.$$

When $\mathbf{r}.\mathbf{v} = 0$,

$$(2a - a\sin\omega t)(-a\omega\cos\omega t) + (2a\cos\omega t)(-2a\omega\sin\omega t) + (a\sin\omega t - 3a)(a\omega\cos\omega t) = 0.$$

$$\therefore \cos\omega t(2\sin\omega t + 5) = 0.$$

As $-1 \leqslant \sin \omega t \leqslant 1$, we deduce that $\mathbf{r.v} = 0$ when $\cos \omega t = 0$.

$\therefore$ **r** and **v** are perpendicular when $t = \dfrac{(2n+1)\pi}{2\omega}$, $n = 0, 1, 2, \ldots$

When $n = 0$, $\quad \omega t = \pi/2$ and $\mathbf{r} = a\mathbf{i} - 2a\mathbf{k}$. $\quad \therefore |\mathbf{r}| = a\sqrt{5}$.

When $n = 1$, $\quad \omega t = 3\pi/2$ and $\mathbf{r} = 3a\mathbf{i} - 4a\mathbf{k}$. $\quad \therefore |\mathbf{r}| = 5a$.

For other values of n the periodicity of $\sin \omega t$ and $\cos \omega t$ tells us that the values $a\sqrt{5}$ and $5a$ of $|\mathbf{r}|$ will be repeated.
So, using 8.3 F4, the smallest and greatest distances between P and Q are $a\sqrt{5}$ and $5a$. *25m* ■

S7.

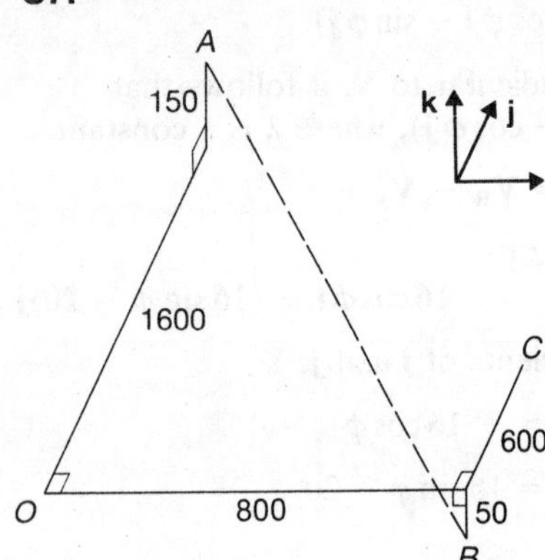

The position vectors of A, B, C are $\mathbf{r}_A$ km, $\mathbf{r}_B$ km, $\mathbf{r}_C$ km, where

$$\mathbf{r}_A = 1{\cdot}6\mathbf{j} + 0{\cdot}15\mathbf{k}$$
$$\mathbf{r}_B = 0{\cdot}8\mathbf{i} - 0{\cdot}05\mathbf{k}$$
$$\mathbf{r}_C = 0{\cdot}8\mathbf{i} + 0{\cdot}6\mathbf{j}.$$
$$\overrightarrow{AB} = \mathbf{r}_B - \mathbf{r}_A = 0{\cdot}2(4\mathbf{i} - 8\mathbf{j} - \mathbf{k}).$$

Let **m**, **n** be unit vectors in the directions of $\overrightarrow{AB}$ and $\overrightarrow{OC}$ respectively.

Then $\mathbf{m} = \dfrac{\overrightarrow{AB}}{|AB|} = \dfrac{1}{9}(4\mathbf{i} - 8\mathbf{j} - \mathbf{k})$ and $\mathbf{n} = \dfrac{\overrightarrow{OC}}{|OC|} = \dfrac{1}{5}(4\mathbf{i} + 3\mathbf{j})$.

Let $\mathbf{v}_S$ km h^{-1}, $\mathbf{v}_M$ km h^{-1} be the velocities of the ski-bucket and the man .

Then $\mathbf{v}_S = 9\mathbf{m} = (4\mathbf{i} - 8\mathbf{j} - \mathbf{k})$ and $\mathbf{v}_M = 5\mathbf{n} = (4\mathbf{i} + 3\mathbf{j})$.
So $\mathbf{r}_S$ km and $\mathbf{r}_M$ km, the position vectors of the ski-bucket and the man at time t hours are given by

$$\mathbf{r}_S = \mathbf{r}_A + \mathbf{v}_S t = 1{\cdot}6\mathbf{j} + 0{\cdot}15\mathbf{k} + (4\mathbf{i} - 8\mathbf{j} - \mathbf{k})t$$

and $\mathbf{r}_M = \mathbf{v}_M t = (4\mathbf{i} + 3\mathbf{j})t$.

The position vector of the ski-bucket relative to the man is **r** km, where

$$\mathbf{r} = \mathbf{r}_S - \mathbf{r}_M = (1{\cdot}6 - 11t)\mathbf{j} + (0{\cdot}15 - t)\mathbf{k}.$$

When $t = \frac{1}{7}$,

$$\mathbf{r} = \frac{1}{7}(0{\cdot}2\mathbf{j} + 0{\cdot}05\mathbf{k})$$

$$\therefore |\mathbf{r}| = \frac{0{\cdot}2062}{7} = 0{\cdot}029.$$

So at time $t = \frac{1}{7}$ the ski-bucket and the man are $0{\cdot}029$ km $= 29$ m apart. Hence regulations are broken. *25m* ■

Chapter 9 Vector dynamics of a particle

9.1 GETTING STARTED

In the study of one-dimensional motion, which is treated in Chapter 4, all physical quantities can be represented by real numbers. The advantage of using vector notation for the study of motion in two and three dimensions is that most of the basic formulae appear similar to their one-dimensional analogues, so that equations of motion can be expressed succinctly. The Essential Facts in this chapter will be seen to be following much the same pattern as those in Chapter 4. But now many of the physical quantities are represented by vectors, and although the equations are simple, in applications it is necessary to work with the components of the vectors. Thus, for instance, a particle moving in a vertical plane has two scalar equations governing its horizontal and vertical acceleration components. These equations are called the horizontal and vertical components of the vector equation of motion of the particle, and the process of obtaining them is called resolving forces horizontally and vertically. Much of the art of solving problems in two-dimensional motion is to decide in which direction it is best to resolve forces, as will be demonstrated in the examples worked out in Sections 9.4 and 9.5.

9.2 THE LAWS OF PARTICLE MOTION

ESSENTIAL FACTS

All the following facts relate to a particle P of mass m with position vector $\mathbf{r}$, velocity $\mathbf{v}$ and acceleration $\mathbf{f}$ with respect to a fixed origin O.

F1. Momentum The momentum of P is $m\mathbf{v}$.

F2. Newton's First Law of Motion If no force acts on P then $\mathbf{v}$ is constant.

F3. Newton's Second Law of Motion When P is acted on by a single force $\mathbf{F}$ the acceleration of P is given by

$$\mathbf{F} = m\mathbf{f}.$$

This relation between $\mathbf{F}$ and $\mathbf{f}$ is called the **equation of motion** of P. Equivalently, $\mathbf{F}$ is equal to the rate of change of momentum.

That is,
$$\mathbf{F} = \frac{\mathrm{d}}{\mathrm{d}t}(m\mathbf{v}).$$

F4. Composition of forces When P is acted on by two forces $\mathbf{F}_1$ and $\mathbf{F}_2$ the acceleration of P is identical to that which would be produced by a single force $\mathbf{F}_1 + \mathbf{F}_2$.

That is,
$$\mathbf{F}_1 + \mathbf{F}_2 = m\mathbf{f}.$$

Another way of stating this fact is to say

The **resultant** of two forces acting on a particle is a force equal to the vector sum of the two forces.

F5. The resultant force $\mathbf{R}$ of a set of n forces $\mathbf{F}_1, \mathbf{F}_2, \ldots, \mathbf{F}_n$ acting on a particle is given by

$$\mathbf{R} = \mathbf{F}_1 + \mathbf{F}_2 + \cdots + \mathbf{F}_n.$$

Since the n forces are equivalent to the single force $\mathbf{R}$,

$$\mathbf{F}_1 + \mathbf{F}_2 + \ldots \mathbf{F}_n = m\mathbf{f}.$$

F6. Impulse The impulse of a force $\mathbf{F}$ in the time interval $t_1 \leqslant t \leqslant t_2$ is

$$\mathbf{I} = \int_{t_1}^{t_2} \mathbf{F}\,\mathrm{d}t.$$

F7. Impulse and momentum The increase in momentum of P in a time interval is equal to the impulse of the resultant force acting on P during that interval. This follows from F3 by integration:

$$\mathbf{I} = \int_{t_1}^{t_2} \mathbf{F}\,\mathrm{d}t = \int_{t_1}^{t_2} \frac{\mathrm{d}}{\mathrm{d}t}(m\mathbf{v})\,\mathrm{d}t = \Big[m\mathbf{v}\Big]_{t=t_1}^{t=t_2}.$$

Denote the velocities of P at times t_1 and t_2 by $\mathbf{v}_1$ and $\mathbf{v}_2$ respectively.

Then
$$\mathbf{I} = m\mathbf{v}_2 - m\mathbf{v}_1.$$

ILLUSTRATIONS

I1. At time t s a particle of mass 3 kg has position vector $\mathbf{r}$ m, where

$$\mathbf{r} = 3t\mathbf{i} - 4\cos t\mathbf{j}.$$

Find (a) the force acting on the particle at time t s.

(b) the impulse of the force during the time interval $0 \leqslant t \leqslant \frac{\pi}{2}$.

(c) the momentum of the particle at time t s. □

(a) Let the force be $\mathbf{F}$ N.

Then $\mathbf{F} = m\ddot{\mathbf{r}} = 3(4\cos t\mathbf{j})$.

$\therefore$ the force acting on the particle is $12\cos t\mathbf{j}$ N. ■

(b) The impulse $\mathbf{I}$ N s is given by

$$\mathbf{I} = \int_0^{\pi/2} 12\cos t\,\mathbf{j}\,\mathrm{d}t = \Big[12\sin t\,\mathbf{j}\Big]_0^{\pi/2} = 12\mathbf{j}.$$

$\therefore$ the impulse of the force during the interval $0 \leqslant t \leqslant \frac{\pi}{2}$ is $12\mathbf{j}$ N s. ■

(c) $\dot{\mathbf{r}} = 3\mathbf{i} + 4\sin t\,\mathbf{j}$.
So the momentum of the particle is

$$3(3\mathbf{i} + 4\sin t\,\mathbf{j}) = 9\mathbf{i} + 12\sin t\,\mathbf{j} \text{ kg m s}^{-1}.$$ ■

I2. A particle P of mass 2 kg moves under the action of two forces $\mathbf{F}_1$ N and $\mathbf{F}_2$ N, where $\mathbf{F}_1 = 8\mathbf{i} - \mathbf{j} + 4\mathbf{k}$. At time t s the velocity of P is given by

$$(4 + t^2)\mathbf{i} + (5t - t^2)\mathbf{j} + (2 + 3t^2)\mathbf{k} \text{ m s}^{-1}.$$

(**a**) Find $\mathbf{F}_2$. □

P has acceleration $2t\mathbf{i} + (5 - 2t)\mathbf{j} + 6t\mathbf{k}$ m s^{-1}.

Then, by F4,

$$(8\mathbf{i} - \mathbf{j} + 4\mathbf{k}) + \mathbf{F}_2 = 2[2t\mathbf{i} + (5 - 2t)\mathbf{j} + 6t\mathbf{k}].$$

$$\therefore \mathbf{F}_2 = (4t - 8)\mathbf{i} + (11 - 4t)\mathbf{j} + (12t - 4)\mathbf{k}.$$ ■

(**b**) Find the impulse of each force in the time interval $1 \leqslant t \leqslant 3$. □
The impulse of $\mathbf{F}_1$ is

$$\mathbf{I}_1 = \int_1^3 (8\mathbf{i} - \mathbf{j} + 4\mathbf{k})\,\mathrm{d}t$$

$$= \Big[8t\mathbf{i} - t\mathbf{j} + 4t\mathbf{k}\Big]_1^3 = 16\mathbf{j} - 2\mathbf{j} + 8\mathbf{k} \text{ N s}.$$

The impulse of $\mathbf{F}_2$ is

$$\mathbf{I}_2 = \int_1^3 [(4t - 8)\mathbf{i} + (11 - 4t)\mathbf{j} + (12t - 4)\mathbf{k}]\,dt$$

$$= \Big[(2t^2 - 8t)\mathbf{i} + (11t - 2t^2)\mathbf{j} + (6t^2 - 4t)\mathbf{k}\Big]_1^3$$

$$= 6\mathbf{j} + 40\mathbf{k} \text{ N s.}$$

(c) Verify that, in the time interval $1 \leqslant t \leqslant 3$, the increase in the momentum of P is equal to the sum of the impulses of the two forces. □

When $t = 3$: P has velocity $13\mathbf{i} + 6\mathbf{j} + 29\mathbf{k}$ m s^{-1}

and momentum $26\mathbf{i} + 12\mathbf{j} + 58\mathbf{k}$ N s.

When $t = 1$: P has velocity $5\mathbf{i} + 4\mathbf{j} + 5\mathbf{k}$ m s^{-1}

and momentum $10\mathbf{i} + 8\mathbf{j} + 10\mathbf{k}$ N s.

So the increase in momentum is

$(26\mathbf{i} + 12\mathbf{j} + 58\mathbf{k}) - (10\mathbf{i} + 8\mathbf{j} + 10\mathbf{k}) = 16\mathbf{i} + 4\mathbf{j} + 48\mathbf{k}$ N s.

Also, $\mathbf{I}_1 + \mathbf{I}_2 = (16\mathbf{i} - 2\mathbf{j} + 8\mathbf{k}) + (6\mathbf{j} + 40\mathbf{k})$
$= 16\mathbf{i} + 4\mathbf{j} + 48\mathbf{k}$ N s. ■

I3.

A particle P of mass 4 kg is moving with velocity $(-\mathbf{i} - 3\mathbf{j} + 16\mathbf{k})$ m s^{-1} when it is acted on by a force for a fixed period of time. When the force ceases to act the velocity of the particle is u m s^{-1} in the direction of the vector $-7\mathbf{j} + 24\mathbf{k}$. Given that the impulse of the force is parallel to the vector $\mathbf{i} - 4\mathbf{j} + 8\mathbf{k}$ find u and the magnitude of the impulse. □

As $|-7\mathbf{j} + 24\mathbf{k}| = 25$, the final velocity of the particle is $\frac{u}{25}(-7\mathbf{j} + 24\mathbf{k})$ m s^{-1}.

Then the initial momentum of P is $4(-\mathbf{i} - 3\mathbf{j} + 16\mathbf{k})$ N s and the final momentum of P is $\frac{4u}{25}(-7\mathbf{j} + 24\mathbf{k})$ N s.

Let the impulse of the force be $\lambda(\mathbf{i} - 4\mathbf{j} + 8\mathbf{k})$ N s, where λ is a constant. By F7:

$$\lambda(\mathbf{i} - 4\mathbf{j} + 8\mathbf{k}) = \frac{4u}{25}(-7\mathbf{j} + 24\mathbf{k}) - 4(-\mathbf{i} - 3\mathbf{j} + 16\mathbf{k}).$$

Equating coefficients of $\mathbf{i}$, $\mathbf{j}$ and $\mathbf{k}$:

$$\lambda = 4, \qquad -4\lambda = -\frac{28u}{25} + 12, \qquad 8\lambda = \frac{96u}{25} - 64.$$

Hence $\lambda = 4$ and $u = 25$. ■

The magnitude of the impulse is

$$|\lambda(\mathbf{i} - 4\mathbf{j} + 8\mathbf{k}| = |4\mathbf{i} - 16\mathbf{j} + 32\mathbf{k}| = 36 \text{ N s. } ■$$

9.3 ENERGY, POWER AND WORK

ESSENTIAL FACTS

At time t a particle P of mass m has position vector $\mathbf{r}$ and velocity $\mathbf{v}$.

F1. Kinetic energy

The kinetic energy of P is $\frac{1}{2}m\mathbf{v}.\mathbf{v} = \frac{1}{2}mv^2$, where $v = |\mathbf{v}|$ is the speed of P.

F2. Power

The power exerted by a force $\mathbf{F}$ acting on P is $S = \mathbf{F}.\mathbf{v}$.

F3. Work

The work done in the interval $t_1 \leqslant t \leqslant t_2$ by a force which exerts power S is

$$W = \int_{t_1}^{t_2} S\,\mathrm{d}t.$$

F4. Rate of working

Let the work done by a force $\mathbf{F}$ from a fixed instant to time t be W. Then

the power exerted by $\mathbf{F}$ is the rate of working of $\mathbf{F}$,

that is, $$S = \frac{\mathrm{d}W}{\mathrm{d}t}$$

F5. Principle of energy

The increase in kinetic energy of P during an interval is equal to the total work done in that interval by the resultant of the forces acting on P.

Proof: Let $\mathbf{v} = \mathbf{v}_1$ when $t = t_1$ and $\mathbf{v} = \mathbf{v}_2$ when $t = t_2$.

The rate of change of kinetic energy is $\frac{\mathrm{d}}{\mathrm{d}t}(\frac{1}{2}m\mathbf{v}.\mathbf{v}) = m\mathbf{v}.\frac{\mathrm{d}\mathbf{v}}{\mathrm{d}t}$.

Let the resultant force $\mathbf{R}$ acting on P exert power S and do work W in the interval.

By 9.2 F5, $\quad m\frac{\mathrm{d}\mathbf{v}}{\mathrm{d}t} = \mathbf{R}.$

Taking the scalar product with $\mathbf{v}$ and using F2,

$$\frac{\mathrm{d}}{\mathrm{d}t}(\tfrac{1}{2}m\mathbf{v}.\mathbf{v}) = m\mathbf{v}.\frac{\mathrm{d}\mathbf{v}}{\mathrm{d}t} = \mathbf{R}.\mathbf{v} = S.$$

Integrating with respect to t from t_1 to t_2 and using F3,

$$\tfrac{1}{2}m\mathbf{v}_2.\mathbf{v}_2 - \tfrac{1}{2}m\mathbf{v}_1.\mathbf{v}_1 = W.$$

F6. Potential energy

Let A and B, with position vectors $\mathbf{r}_A$, $\mathbf{r}_B$, be two arbitrary points on the path of P as it moves in space. When a force $\mathbf{F}$ acting on P

depends only on **r** it sometimes happens that as P moves from A to B the work done by **F** depends only on $\mathbf{r}_A$ and $\mathbf{r}_B$. So the work done by **F** does not depend on the way in which P moves from A to B. That is, it depends neither on the time interval taken nor on the particular path taken. In such a case, the potential energy V of the force **F** is defined as

minus the work done by **F** in a displacement to the point P from some arbitrarily chosen fixed point in space.

F7. Gravitational potential energy

The gravitational force $m\mathbf{g}$ has potential energy

$$V = -m\mathbf{g}.\mathbf{r} + \text{constant}.$$

The constant is normally chosen so that $V = 0$ at some convenient point. As **g** is directed vertically downwards (towards the centre of the earth) it follows that $V = mg\mathbf{k}.\mathbf{r} + \text{constant}$, where **k** is a unit vector directed vertically upwards and $g = |\mathbf{g}|$.

Thus $\qquad V = mgh,$

where h is the vertical distance of P above some fixed point.

F8. Conservation of energy

In the case when V is the potential energy of the resultant force acting on a particle, V is called the **potential energy of the particle**. It then follows from F5 that

$$\tfrac{1}{2}mv^2 \quad + \quad V \quad = E \quad \text{(a constant)}$$

Kinetic energy of P + potential energy of P= total energy of P.

This is the **principle of conservation of energy**.

The total energy of P is constant throughout the motion, but its value E depends on the constant chosen in the potential energy V.

ILLUSTRATIONS

I1.

A **constant** force **F** acts on a particle P. Show that the work, W, done by **F** as P moves from a point P_1 to a point P_2 is given by

$$W = \mathbf{F}.\overrightarrow{P_1P_2}. \ \square$$

Let the position vector of P at time t be **r**.

Let $\mathbf{r} = \mathbf{r}_1$, $t = t_1$ at P_1 and $\mathbf{r} = \mathbf{r}_2$, $t = t_2$ at P_2.

Let $\mathbf{r} = x\mathbf{i} + y\mathbf{j} + z\mathbf{k}$ and $\mathbf{F} = X\mathbf{i} + Y\mathbf{j} + Z\mathbf{k}$.

Then, by 9.3 F2 and F3,

$$W = \int_{t_1}^{t_2} \mathbf{F}.\frac{\mathrm{d}\mathbf{r}}{\mathrm{d}t}\mathrm{d}t = \int_{t_1}^{t_2} \left(X\frac{\mathrm{d}x}{\mathrm{d}t} + Y\frac{\mathrm{d}y}{\mathrm{d}t} + Z\frac{\mathrm{d}z}{\mathrm{d}t}\right)\mathrm{d}t.$$

Since X, Y and Z are constants.

$$W = \Big[Xx + Yy + Zz\Big]_{t=t_1}^{t=t_2}$$

$$= \Big[\mathbf{F}.\mathbf{r}\Big]_{t=t_1}^{t=t_2}$$

$$= \mathbf{F}.\mathbf{r}_2 - \mathbf{F}.\mathbf{r}_1.$$

$$= \mathbf{F}.(\mathbf{r}_2 - \mathbf{r}_1)$$

$$= \mathbf{F}.\overrightarrow{P_1P_2}. \blacksquare$$

12.

A particle P of mass 3 kg moves so that its position vector $\mathbf{r}$ m at time t s is given by

$$\mathbf{r} = (2t^2 - t)\mathbf{i} + (t\cos 2t)\mathbf{j} + 3\sin t\mathbf{k}.$$

Find

(a) the momentum of P when $t = 0$. □

$$\dot{\mathbf{r}} = (4t - 1)\mathbf{i} + (\cos 2t - 2t\sin 2t)\mathbf{j} + 3\cos t\mathbf{k}.$$

When $t = 0$; $\quad \dot{\mathbf{r}} = -\mathbf{i} + \mathbf{j} + 3\mathbf{k}$.

$\therefore$ the momentum of P is $-3\mathbf{i} + 3\mathbf{j} + 9\mathbf{k}$ kg m s^{-1}. ■

(b) the kinetic energy of P when $t = \pi$. □
When $t = \pi$; $\quad \dot{\mathbf{r}} = (4\pi - 1)\mathbf{i} + \mathbf{j} - 3\mathbf{k}$.

$$\therefore\ \dot{\mathbf{r}}.\dot{\mathbf{r}} = (4\pi - 1)^2 + 1 + 9 = 16\pi^2 - 8\pi + 11.$$

Hence the kinetic energy of P
is $\quad \frac{1}{2} \times 3 \times \dot{\mathbf{r}}.\dot{\mathbf{r}} = (24\pi^2 - 12\pi + 16{\cdot}5)$ J. ■

(c) the force acting on P when $t = \dfrac{\pi}{2}$. □

$$\ddot{\mathbf{r}} = 4\mathbf{i} + (-4\sin 2t - 4t\cos 2t)\mathbf{j} - 3\sin t\mathbf{k}.$$

When $t = \dfrac{\pi}{2}$; $\quad \ddot{\mathbf{r}} = 4\mathbf{i} + 2\pi\mathbf{j} - 3\mathbf{k}$.

Hence the force acting on P is $3\ddot{\mathbf{r}} = 12\mathbf{i} + 6\pi\mathbf{j} - 9\mathbf{k}$ N. ■

(d) the power exerted by the force acting on P when $t = \dfrac{\pi}{2}$. □

When $t = \dfrac{\pi}{2}$; $\quad \dot{\mathbf{r}} = (2\pi - 1)\mathbf{i} - \mathbf{j}$.

Hence, using the force obtained in part **(c)**, the power exerted by the force is $12(2\pi - 1) + 6\pi(-1) - 9 \times 0 = 6\,(3\pi - 2)$ W. ■

(e) the work done by the force during the interval $\dfrac{\pi}{2} \leqslant t \leqslant \pi$. □

When $t = \dfrac{\pi}{2}$; $\quad \dot{\mathbf{r}}.\dot{\mathbf{r}} = (2\pi - 1)^2 + 1 = 4\pi^2 - 4\pi + 2.$

$\therefore$ the kinetic energy at $t = \dfrac{\pi}{2}$ is

$$\frac{3}{2}(4\pi^2 - 4\pi + 2) = (6\pi^2 - 6\pi + 3)\text{ J}.$$

Using the result in part (b), the increase in kinetic energy in the interval is

$$(24\pi^2 - 12\pi + 16{\cdot}5) - (6\pi^2 - 6\pi + 3) = (18\pi^2 - 6\pi + 13{\cdot}5)\text{ J}.$$

Hence, by 9.3 F5, the work done by the force is $(18\pi^2 - 6\pi + 13{\cdot}5)$ J. ■

9.4 REACTION OF A SURFACE

ESSENTIAL FACTS

F1. Reaction

When a particle A is in contact with the surface of a rigid body B, the force $\mathbf{R}$ exerted by the body on the particle is called a **reaction**. The force exerted by A on B is, by Newton's Third Law, equal to $-\mathbf{R}$. While contact continues, the reaction exerted by B on A must be directed outwards from the surface.
The reaction of a **smooth** surface is always along the **normal** (perpendicular) to the surface.

F2. Normal reaction and friction

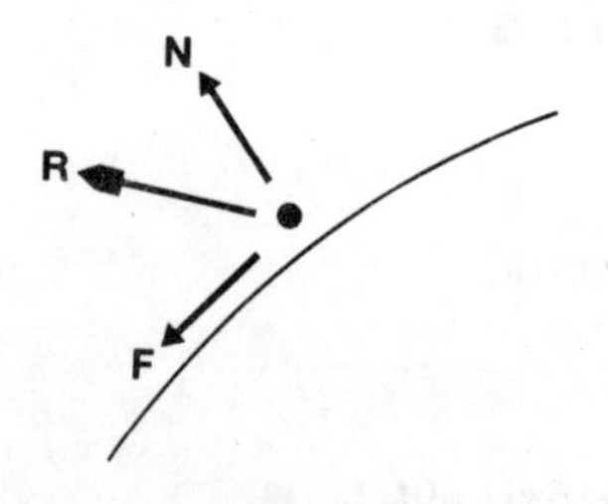

In general, the reaction $\mathbf{R}$ exerted on a particle by a surface may be decomposed into two forces: $\mathbf{R} = \mathbf{N} + \mathbf{F}$, where

$\mathbf{N}$ is perpendicular to the surface,
$\mathbf{F}$ is tangential to the surface.

$\mathbf{N}$ is called the **normal reaction** and $\mathbf{F}$ is called the force of **friction**.

F3. Rough surface

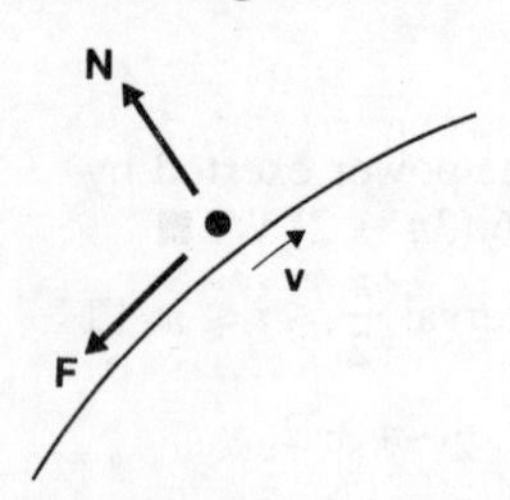

When a particle moves on a fixed rough surface, the friction $\mathbf{F} \neq \mathbf{0}$, and $\mathbf{F}$ acts in the direction opposite to the velocity. When the surface is dry,

$$|\mathbf{F}| = \mu\,|\mathbf{N}|,$$

where $\mathbf{N}$ is the normal reaction and μ is a constant called the **coefficient of dynamic friction** between the particle and the surface. Although this experimental relation does not hold for all types of surface, it is to be assumed whenever 'coefficient of friction' is mentioned.

F4.

When a particle moves on the surface of a fixed rigid body the normal reaction of $\mathbf{N}$ is, by definition, perpendicular to $\mathbf{v}$, the velocity of the particle, that is $\mathbf{N}.\mathbf{v} = 0$.
Hence, from 9.3 F2 and 9.3 F3, $\mathbf{N}$ exerts no power and does no work during the motion.
It follows that when the surface is smooth, the reaction does no work during the motion.

ILLUSTRATIONS

I1.

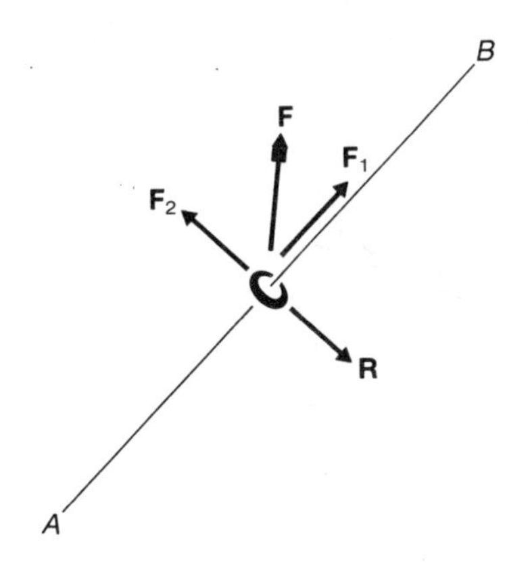

A force $\mathbf{F}$, of magnitude 117 N, acts in the direction of the vector $12\mathbf{i} - 3\mathbf{j} + 4\mathbf{k}$. A bead of mass 2 kg moves along a smooth straight wire from the point A, with position vector $(\mathbf{i} - 2\mathbf{j} + 3\mathbf{k})$ m, to the point B, with position vector $(17\mathbf{i} + 6\mathbf{j} + 5\mathbf{k})$ m under the influence of $\mathbf{F}$ and the reaction of the wire only.

(a) Find $\mathbf{F}$. □

Let $\mathbf{F} = \lambda(12\mathbf{i} - 3\mathbf{j} + 4\mathbf{k})$ N, where $\lambda > 0$.
As $|\mathbf{F}| = 117$ N, we have $117 = |\lambda(12\mathbf{i} - 3\mathbf{j} + 4\mathbf{k})| = 13\lambda$.
$\therefore \lambda = 9$ and $\mathbf{F} = 108\mathbf{i} - 27\mathbf{j} + 36\mathbf{k}$ N. ■

(b) Find $\mathbf{R}$, the reaction of the wire on the bead. □

As the wire is smooth $\mathbf{R}$ is perpendicular to the wire, that is, $\mathbf{R}.\overrightarrow{AB} = 0$.
The resultant force acting on the bead is $\mathbf{R} + \mathbf{F} = \mathbf{R} + \mathbf{F}_1 + \mathbf{F}_2$, where $\mathbf{F} = \mathbf{F}_1 + \mathbf{F}_2$ with $\mathbf{F}_1$ parallel to $\overrightarrow{AB}$ and $\mathbf{F}_2$ perpendicular to $\overrightarrow{AB}$.
As the bead moves along AB its acceleration must be parallel to $\overrightarrow{AB}$.
Hence, by 9.2 F4, $\mathbf{R} + \mathbf{F}_1 + \mathbf{F}_2$ is parallel to $\overrightarrow{AB}$,

$$\therefore \mathbf{R} + \mathbf{F}_2 = \mathbf{0}.$$

To find $\mathbf{F}_2$, and hence $\mathbf{R}$, we first find $\mathbf{F}_1$.
If $\mathbf{n}$ is a unit vector in the direction $\overrightarrow{AB}$, then $\mathbf{F}_1 = (\mathbf{F}.\mathbf{n})\mathbf{n}$.

Now $\overrightarrow{AB} = 16\mathbf{i} + 8\mathbf{j} + 2\mathbf{k}$. $\quad \therefore \mathbf{n} = \dfrac{\overrightarrow{AB}}{|\overrightarrow{AB}|} = \dfrac{1}{9}(8\mathbf{i} + 4\mathbf{j} + \mathbf{k})$.

$$\therefore \mathbf{F}_1 = 88\mathbf{n}.$$

Hence $\mathbf{R} = -\mathbf{F}_2 = \mathbf{F}_1 - \mathbf{F} = \dfrac{1}{9}(-268\mathbf{i} + 595\mathbf{j} - 236\mathbf{k})$ N. ■

(c) Find the work done by $\mathbf{F}$ in the motion from A to B. □

As $\mathbf{F}$ is a constant force we may use the result established in 9.3 I1.
Thus the work done $= \mathbf{F}.\overrightarrow{AB}$

$$= (108\mathbf{i} - 27\mathbf{j} + 36\mathbf{k}).(16\mathbf{j} + 8\mathbf{j} + 2\mathbf{k})$$

$$= 1584 \text{ J.} \blacksquare$$

(d) Given that at A the bead has a velocity of $21\ \mathrm{m\,s^{-1}}$ in the direction AB, find the speed of the bead when it reaches B. □

As **R** is perpendicular to AB it does no work in the motion. Hence, if u m s^{-1} is the speed of the bead at B, we have (using 9.3 F5)

$$1584 = \tfrac{1}{2}(2)u^2 - \tfrac{1}{2}(2)(21)^2.$$
$$\therefore u^2 = 2025.$$

So the bead has a speed of 45 m s^{-1} when it reaches B. ■

I2. A particle of mass m slides down a line of greatest slope of a rough plane of inclination α. Initially the particle is moving with speed u and after travelling a distance d its speed is v. Find an expression for the coefficient of friction between the particle and the plane. □

During the motion the particle is subject to two forces; its weight $m\mathbf{g}$ and the reaction **R** of the plane.
So the acceleration **f** of the particle is given by the equation of motion

$$m\mathbf{f} = \mathbf{R} + m\mathbf{g} \quad (1)$$
$$= \mathbf{N} + \mathbf{F} + m\mathbf{g},$$

where **N** is the normal reaction and **F** is the friction.
Let **i** be the unit vector down the line of greatest slope, and **j** the downward unit vector normal to the plane.

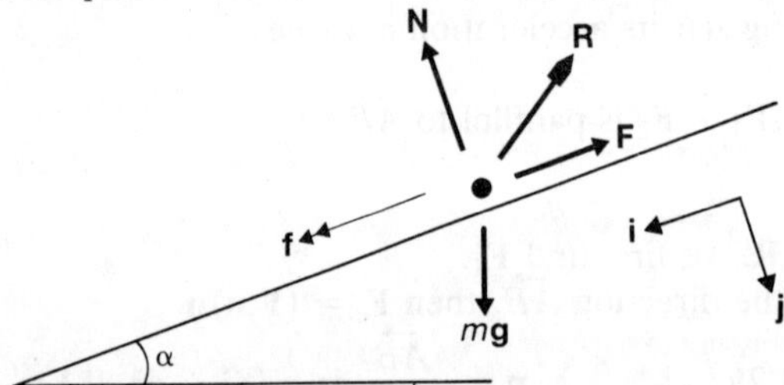

Then $\mathbf{g} = g\sin\alpha\,\mathbf{i} + g\cos\alpha\,\mathbf{j}$.

Let $\mathbf{f} = f\mathbf{i}$, $\mathbf{N} = -N\mathbf{j}$, $\mathbf{F} = -F\mathbf{i}$.

Then

$$mf\mathbf{i} = (mg\sin\alpha - F)\mathbf{i} + (mg\cos\alpha - N)\mathbf{j}.$$

$$\therefore mf = mg\sin\alpha - F \quad (2)$$

and $0 = mg\cos\alpha - N$ (3) [A]

N is positive because the reaction must act outwards, and F is positive because **F** acts up the slope.

$\therefore F = \mu N$ [B], using F3.

[A] In questions of this type we do not normally write down the vector equation of motion (equation (1)), nor do we introduce the unit vectors **i** and **j**.
The normal procedure would be to draw a diagram in which the scalar components of the acceleration and the forces are indicated in the relevant directions.

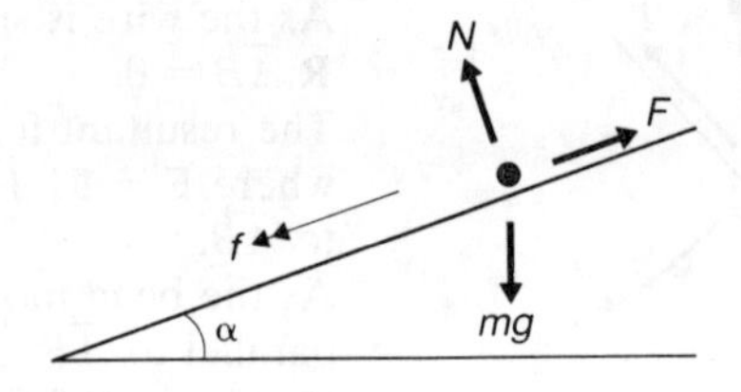

We then say:
Resolving parallel to the plane:

$$mf = mg\sin\alpha - F \quad (2)$$

Resolving perpendicular to the plane:

$$N - mg\cos\alpha = 0. \quad (3)$$

The phrase 'resolving in a given direction' is an accepted abbreviation for 'taking the components of the equation of motion in a given direction' or, equivalently, 'taking the scalar product of the equation of motion with a unit vector in a given direction'.
An even shorter abbreviation is to write

$\swarrow$: $mf = mg\sin\alpha - F$ (2)

$\nwarrow$: $N - mg\cos\alpha = 0$ (3)

Here the symbol $\swarrow$: means 'taking the components of the equation of motion (that is,

Substitute for F and N in (2) and (3):

$f = g\sin\alpha - \mu g\cos\alpha.$

Since f is constant we may use 3.3 F7D, and obtain

$v^2 = u^2 + 2g(\sin\alpha - \mu\cos\alpha)d.$

$\therefore \mu = \tan\alpha + \dfrac{u^2 - v^2}{2gd}\sec\alpha.$ ■

resolving) in the direction of the arrow.'
Both abbreviations are used in this book and no marks would be lost if they were used in an examination.

[B] The question refers to the 'coefficient of friction' when it should really refer to the 'coefficient of dynamic friction'. This is a commonly used abbreviation. However, if the question requires the use of a coefficient of dynamic friction **and** a coefficient of static friction (see Chapter 13) then care must be taken to avoid any confusion.

13.

A smooth narrow tube in the form of an arc AB of a circle of centre O and radius a is fixed so that A is vertically above O and OB is horizontal. Particles P of mass m and Q of mass $2m$, with a light inextensible string of length $\dfrac{\pi a}{2}$ connecting them, are placed inside the tube, with P at A and Q at B, and released from rest. Assuming that the string remains taut during the motion, find the speeds of the particles when P reaches B. □

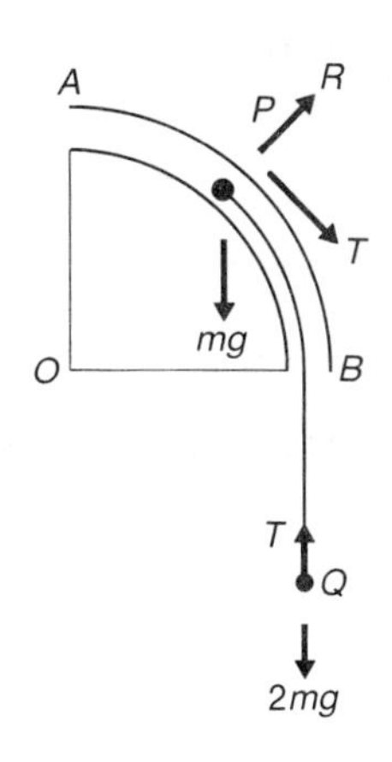

The forces acting on P are its weight mg, the tension T and the reaction R.
The forces acting on Q are its weight $2mg$ and the tension in the string, which is also T because the tube is smooth (6.2 F2).
The total work done by the tension is zero. (6.2 F3).
Since the tube is smooth, R acts in a direction perpendicular to the motion of P, and therefore the only forces doing work are the weights of the particles.
Take the total potential energy to be zero at the start of the motion. Then, when P reaches B,

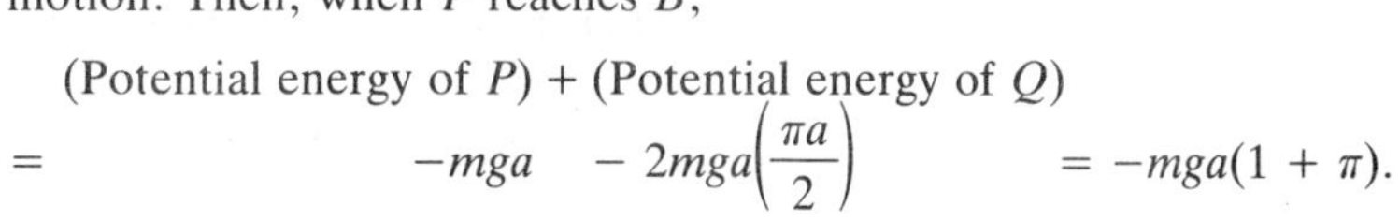

$$\text{(Potential energy of } P) + \text{(Potential energy of } Q)$$
$$= -mga - 2mga\left(\frac{\pi a}{2}\right) = -mga(1 + \pi).$$

Let the speed of P and Q then be v.
Initially the speed of P and Q is zero.
Therefore, by conservation of energy;

$$\text{(Final total energy)} = \text{(Initial total energy)}$$

$$\therefore -mga(1 + \pi) + \tfrac{1}{2}mv^2 + \tfrac{1}{2}(2m)v^2 = 0 + 0$$

$$\therefore v^2 = \frac{2}{3}(1 + \pi)ga.$$

Hence the speed is $\sqrt{\frac{2}{3}(1 + \pi)ga}$. ■

14. A smooth wire in the form of a circle of radius $5a$ is fixed in a vertical plane. A small bead of mass m is threaded on the wire and a light elastic string of natural length $6a$ and modulus λ joins the bead to the highest point of the wire. The bead is released from rest in a position where the string is just taut. Given that the bead reaches the lowest point of the wire, show that $\lambda \leqslant 4{\cdot}8mg$. □

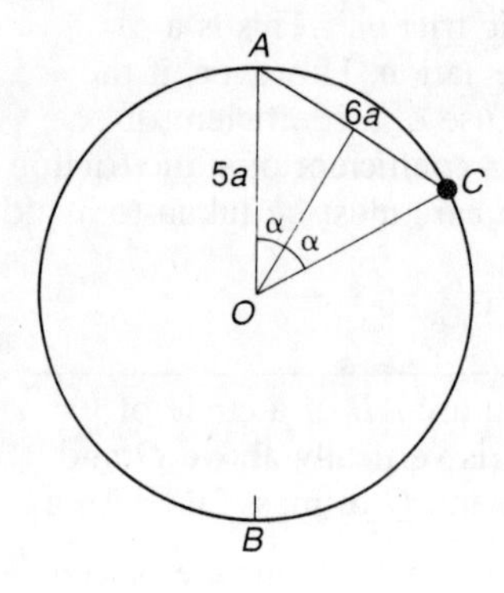

Let A and B be the highest and lowest points of the wire, O the centre of the circle and C the initial position of the bead P.
Initially the string is just taut, which means that $AC = 6a$, the natural length of the string. Thus angle $AOC = 2\alpha$, where $\sin\alpha = \frac{3}{5}$ and $\cos\alpha = \frac{4}{5}$.
In a general position P is subjected to three forces; a tension T in the string, a reaction R with the wire and its weight mg.
As the wire is smooth, R is a normal reaction which does no work in the motion. The other two forces both have potential energies and so, by 9.3 F8, the total energy of the system is constant throughout the motion. Take the zero position of gravitational potential energy to be when OP is horizontal.

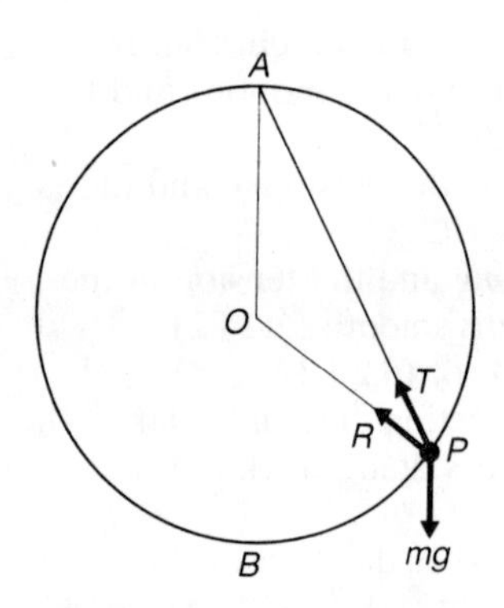

When the bead is at C:
The gravitational potential energy is

$$mg.5a\cos 2\alpha = 5mga(\cos^2\alpha - \sin^2\alpha) = \frac{7}{5}mga.$$

The string has its natural length, so the potential energy of the string is zero.
The bead is at rest, so the kinetic energy is zero.
If the bead reaches B:
The gravitational potential energy is $mg(-5a) = -5mga$.
The string has length $10a$, so the potential energy is $\frac{1}{2}\lambda\frac{(4a)^2}{6a} = \frac{4}{3}\lambda a$.

Let the speed of P when it reaches B be u. Then its kinetic energy at B is $\frac{1}{2}mu^2$.
As energy is conserved.

$$\frac{1}{2}mu^2 - 5mga + \frac{4}{3}\lambda a = 0 + \frac{7}{5}mga + 0.$$

$$\therefore\ 15u^2 = 192mga - 40\lambda a = 4a(48mg - 10\lambda).$$

$$\therefore\ 4{\cdot}8mg - \lambda \geqslant 0.$$

So, if P reaches B, $\lambda \leqslant 4{\cdot}8mg$. ■

9.5 EXAMINATION QUESTIONS AND SOLUTIONS

Q1.

In the interval $0 \leqslant t \leqslant 6$, where t denotes time in seconds, a particle of mass 2 kg is acted upon by two forces. The first force acts in the direction of the vector $8\mathbf{i} - \mathbf{j} + 4\mathbf{k}$ and has a constant magnitude of 18 N. The second force acts in the direction of the vector $-2\mathbf{i} + 2\mathbf{j} - \mathbf{k}$ and its magnitude decreases uniformly with time from 18 N when $t = 0$ to zero when $t = 6$. Show that the acceleration of the particle during the interval $0 \leqslant t \leqslant 6$ is

$$(2 + t)\mathbf{i} + (5 - t)\mathbf{j} + (1 + \tfrac{1}{2}t)\mathbf{k}\ \mathrm{m\,s^{-2}}.$$

Given that the particle has velocity $2\mathbf{i}\ \mathrm{m\,s^{-1}}$ when $t = 0$, find its velocity when $t = 6$. (JMB 1985)

Q2.

A particle of mass 3 units moves under the action of a force $\mathbf{F}$. At time t the velocity $\mathbf{v}$ of the particle is given by

$$\mathbf{v} = 3\mathbf{i} - \mathbf{j} + 2t\mathbf{k}.$$

(i) Find the force $\mathbf{F}$.
(ii) Find the kinetic energy of the particle at time t.
(iii) Given that the position of the particle when $t = 0$ is $\mathbf{i} + \mathbf{j}$, find its position vector when $t = 2$. (JMB 1984)

Q3.

A body of mass 2 kg moves under the action of a variable force whose value at time t seconds is

$$(3t - 2)\mathbf{i} - 8\mathbf{k}\ \text{newtons}.$$

No other force acts on the body. When $t = 0$ the velocity of the body is $(4\mathbf{i} - 2\mathbf{j} + 5\mathbf{k})\ \mathrm{m\,s^{-1}}$. Find
(i) the velocity of the body at time t,
(ii) the values of t for which the velocity of the body is perpendicular to its initial direction of motion,
(iii) the power of the force acting on the body when $t = 4$.
(JMB 1986)

Q4.

A particle of mass 3 units moves under the action of a force $\mathbf{F}$ whose value at time t is given by

$$\mathbf{F} = \mathbf{i} - 2t\mathbf{j} + 3t^2\mathbf{k}.$$

When $t = 1$ the velocity of the particle is $-(\mathbf{i} + \mathbf{k})$. Find
(i) the impulse of $\mathbf{F}$ in the interval from $t = 1$ to $t = 2$,
(ii) the work done by $\mathbf{F}$ in the same interval. (JMB 1984)

Q5. A particle of unit mass moves under the action of a variable force so that its acceleration $\mathbf{a}$ at time t is given by

$$\mathbf{a}(t) = \cos t\,\mathbf{i} + \sin t\,\mathbf{j} + 2\sin 2t\,\mathbf{k}$$

where $\mathbf{i}$, $\mathbf{j}$, $\mathbf{k}$ are mutually perpendicular unit vectors.

(i) What is the momentum of the particle at time $t = \dfrac{\pi}{2}$ if the particle is at rest at time $t = 0$?

(ii) What is the impulse of the force acting on the particle from $t = 0$ to $t = \pi$? (NI 1983)

Q6. (*Take the acceleration due to gravity to be* $10\ \mathrm{m\,s^{-2}}$.)

A small curtain ring of mass 0·5 kg is moving with constant velocity along a rough horizontal rail, and the coefficient of friction between the ring and the rail is 0·3. The motion is caused by a girl pulling on a light inextensible cord which is attached to the ring and which lies in the same vertical plane as the rail. Given that the cord makes an acute angle θ with the downward vertical and that the girl pulls the cord with tension T N, show that

(a) $T = \dfrac{1{\cdot}5}{\sin\theta - 0{\cdot}3\cos\theta}$, (b) $\tan\theta > 0{\cdot}3$.

The girl suddenly doubles her pull to $2T$ N without changing θ. Show that the ring then moves with acceleration $3\ \mathrm{m\,s^{-2}}$. Given that the ring's initial velocity was $1\ \mathrm{m\,s^{-1}}$ and assuming that it continues to slide along the rail with acceleration $3\ \mathrm{m\,s^{-2}}$ with the tension unaltered at $2T$ N, find, in terms of θ, the work W J done by the girl during the first second of accelerated motion, and show that $W > 7{\cdot}5$. (AEB 1986)

Q7. A particle of mass 0·05 kg moving along a straight line AB at $1{\cdot}5\ \mathrm{m\,s^{-1}}$ is given an impulse which doubles its speed and causes the particle to move along a line at an angle of 60° to its original direction.

If a particle of mass 0·2 kg at rest were given the same impulse, in what direction would it move with respect to the line AB and what would be its speed? (NI 1983)

SOLUTIONS

S1. The first force of $\mathbf{F}_1$ N acts in the direction of the vector $8\mathbf{i} - \mathbf{j} + 4\mathbf{k}$ and has magnitude 18 N.

Hence, $\mathbf{F}_1 = 18\,\dfrac{8\mathbf{i} - \mathbf{j} + 4\mathbf{k}}{|8\mathbf{i} - \mathbf{j} + 4\mathbf{k}|} = 16\mathbf{i} - 2\mathbf{j} + 8\mathbf{k}$.

The second force $\mathbf{F}_2$ N acts in the direction of the vector $-2\mathbf{i} + 2\mathbf{j} - \mathbf{k}$ and its magnitude decreases uniformly from 18 N at $t = 0$ to zero at $t = 6$.

$$\therefore \mathbf{F}_2 = \lambda(-2\mathbf{i} + 2\mathbf{j} - \mathbf{k})(6 - t), \text{ where } \lambda(> 0) \text{ is a constant.}$$

At $t = 0$: $|\mathbf{F}_2| = \lambda\sqrt{4 + 4 + 1} \times 6 = 18\lambda$.

But $|\mathbf{F}_2| = 18$ (given).

$\therefore \lambda = 1$ and $\mathbf{F}_2 = (-2\mathbf{i} + 2\mathbf{j} - \mathbf{k})(6 - t)$.

Let the acceleration of the particle be $\mathbf{f}$ m s^{-2}.

By 9.2 F4,
$$2\mathbf{f} = \mathbf{F}_1 + \mathbf{F}_2$$
$$= (4 + 2t)\mathbf{i} + (10 - 2t)\mathbf{j} + (2 + t)\mathbf{k}.$$
$$\therefore \mathbf{f} = (2 + t)\mathbf{i} + (5 - t)\mathbf{j} + (1 + \tfrac{1}{2}t)\mathbf{k}.$$

Hence, on integrating with respect to t, the velocity $\mathbf{v}$ m s^{-1} is given by

$$\mathbf{v} = (2t + \tfrac{1}{2}t^2)\mathbf{i} + (5t - \tfrac{1}{2}t^2)\mathbf{j} + (t + \tfrac{1}{4}t^2)\mathbf{k} + \mathbf{A}.$$

When $t = 0$, $\mathbf{v} = 2\mathbf{i}$. $\therefore 2\mathbf{i} = \mathbf{A}$.
So when $t = 6$, the velocity is

$$(12 + 18)\mathbf{i} + (30 - 18)\mathbf{j} + (6 + 9)\mathbf{k} + 2\mathbf{i}$$
$$= 32\mathbf{i} + 12\mathbf{j} + 15\mathbf{k} \text{ m s}^{-1}.$$ *15m* ■

S2.

(i) $\mathbf{v} = 3\mathbf{i} - \mathbf{j} + 2t\mathbf{k}$.

$\therefore$ the acceleration $\mathbf{f} = \dfrac{d\mathbf{v}}{dt} = 2\mathbf{k}$.

By Newton's Law (9.2 F3): $\mathbf{F} = 3\mathbf{f} = 6\mathbf{k}$. ■

(ii) Kinetic energy $= \tfrac{1}{2} \times 3 \times \mathbf{v}.\mathbf{v}$
$= \tfrac{1}{2} \times 3(9 + 1 + 4t^2)$
$= 15 + 6t^2$. ■

(iii) The position vector at time t is $\mathbf{r}$, where

$$\mathbf{r} = \int \mathbf{v}\,dt = 3t\mathbf{i} - t\mathbf{j} + t^2\mathbf{k} + \mathbf{A}.$$

At time $t = 0$, $\mathbf{r} = \mathbf{i} + \mathbf{j}$. $\therefore \mathbf{i} + \mathbf{j} = \mathbf{A}$. [A]

Hence at $t = 2$; $\mathbf{r} = 7\mathbf{i} - \mathbf{j} + 4\mathbf{k}$. *11m* ■

[A] Do not forget that the position vector when $t = 0$ is $\mathbf{i} + \mathbf{j}$, not $\mathbf{0}$. This is a very common error, which is more likely to occur if you use the definite integral method,

$$\dot{\mathbf{r}} = 3\mathbf{i} - \mathbf{j} + 2t\mathbf{k}.$$

$$\Big[\mathbf{r}\Big]_{t=0}^{t=2} = \Big[3t\mathbf{i} - t\mathbf{j} + t^2\mathbf{k}\Big]_0^2.$$

Correct so far, but

× $\therefore [\mathbf{r}]_0^2 = 6\mathbf{i} - 2\mathbf{j} + 4\mathbf{k}$ ×

is notationally incorrect on the L.H.S. because t has disappeared. It leads straight to the pitfall:

× At $t = 2$, $\mathbf{r} - \mathbf{0} = 6\mathbf{i} - 2\mathbf{j} + 4\mathbf{k}$ ×

instead of

✓ $\mathbf{r} - (\mathbf{i} + \mathbf{j}) = 6\mathbf{i} - 2\mathbf{j} + 4\mathbf{k}$ ✓.

It is generally safer to use the indefinite integral and the constant $\mathbf{A}$, and it is just as quick.

S3.

(i) Let $\mathbf{f}$ m s^{-2} be the acceleration of the body.

Then $2\mathbf{f} = (3t - 2)\mathbf{i} - 8\mathbf{k}$.

$\therefore$ the velocity $\mathbf{v}$ m s^{-1} is given by

$2\mathbf{v} = (\frac{3}{2}t^2 - 2t)\mathbf{i} - 8t\mathbf{k} + \mathbf{A}$.

At $t = 0$, $\mathbf{v} = 4\mathbf{i} - 2\mathbf{j} + 5\mathbf{k}$.

$\therefore 8\mathbf{i} - 4\mathbf{j} + 10\mathbf{k} = \mathbf{A}$.

$\therefore \mathbf{v} = (\frac{3}{4}t^2 - t + 4)\mathbf{i} - 2\mathbf{j} + (5 - 4t)\mathbf{k}$. ■

(ii) $\mathbf{v}.(4\mathbf{i} - 2\mathbf{j} + 5\mathbf{k}) = 0$ [A] when

$4(\frac{3}{4}t^2 - t + 4) + 4 + 5(5 - 4t) = 0$.

$\therefore t^2 - 8t + 15 = 0$.

$\therefore (t - 3)(t - 5) = 0$.

So the velocity is perpendicular to its initial direction of motion when $t = 3$ and when $t = 5$. ■

(iii) When $t = 4$; $\mathbf{v} = 12\mathbf{i} - 2\mathbf{j} - 11\mathbf{k}$

and the force acting is $10\mathbf{i} - 8\mathbf{k}$. [B]

$$(12\mathbf{i} - 2\mathbf{j} - 11\mathbf{k}).(10\mathbf{i} - 8\mathbf{k}) = 120 + 88 = 208. \quad \text{[C]}$$

So at $t = 4$ the force exerts a power 208 W. [D]

17m ■

[A] The initial direction of motion of the body is the direction of its velocity when $t = 0$.

[B] We could find the power at an arbitrary time t seconds and then substitute $t = 4$. However the calculation is easier, and less prone to error, if we find the force and velocity at $t = 4$ before taking the scalar product.

[C] Using 9.3 F2.

[D] Be careful not to confuse this **vector** problem with the rectilinear case and write

✗ Power = force × velocity,

and then use the magnitudes.

At $t = 4$; force $= \sqrt{100 + 64} = \sqrt{164}$,

velocity $= \sqrt{144 + 4 + 121} = \sqrt{269}$.

$\therefore$ Power $= \sqrt{164} \times \sqrt{269}$. ✗

S4.

(i) Impulse $= \displaystyle\int_1^2 \mathbf{F}\,dt$ [A]

$$= \int_1^2 (\mathbf{i} - 2t\mathbf{j} + 3t^2\mathbf{k})\,dt$$

$$= \Big[t\mathbf{i} - t^2\mathbf{j} + t^3\mathbf{k}\Big]_1^2$$

$$= \mathbf{i} - 3\mathbf{j} + 7\mathbf{k}.$$

[A] Using 9.2 F6. Not

✗ Impulse = (force) × (time interval) ✗

because $\mathbf{F}$ is not constant

(ii) The acceleration $\mathbf{f}$ is given by

$3\mathbf{f} = \mathbf{i} - 2t\mathbf{j} + 3t^2\mathbf{k}$.

$\therefore$ the velocity $\mathbf{v}$ is given by

$3\mathbf{v} = t\mathbf{i} - t^2\mathbf{j} + t^3\mathbf{k} + \mathbf{A}$.

When $t = 1$, $\mathbf{v} = -\mathbf{i} - \mathbf{k}$. [B]

[B] Beware the very common error of failing to notice that $\mathbf{v} = -\mathbf{i} - \mathbf{k}$ when $t = 1$, *not*

✗ when $t = 0$ ✗.

$\therefore\ -3\mathbf{i} - 3\mathbf{k} = \mathbf{i} - \mathbf{j} + \mathbf{k} + \mathbf{A}.$

$\therefore\ \mathbf{A} = -4\mathbf{i} + \mathbf{j} - 4\mathbf{k}.$

When $t = 2$,
$3\mathbf{v} = 2\mathbf{i} - 4\mathbf{j} + 8\mathbf{k} + \mathbf{A} = -2\mathbf{i} - 3\mathbf{j} + 4\mathbf{k}.$

Kinetic energy is $\frac{1}{2} \times 3\ \mathbf{v.v}.$ [C]
$\therefore$ at $t = 1$, kinetic energy $= \frac{1}{2} \times 3 \times 2 = 3$, and
when $t = 2$, kinetic energy $= \frac{1}{2} \times 3 \times \frac{29}{9} = \frac{29}{6}.$

$\therefore$ the force $\mathbf{F}$ does $\frac{29}{6} - 3 = \frac{11}{6}$ units of work.

[D] [E] [F] *14m* ■

[C] Using 9.3 F1.
[D] Using 9.3 F5.
[E] Note that we could alternatively find the power $S = \mathbf{F.v}$, and then the work done is given by $\int_1^2 S\,dt$. However this is more complicated and prone to error.
[F] A common error is to say that
✘ in the interval $1 \leqslant t \leqslant 2$, the particle undergoes
a displacement $\mathbf{d}$, where $\mathbf{d} = \int_1^2 \mathbf{v}\,dt$.

Then the work done is $\mathbf{F.d}$. ✘
This is wrong, because the result of 9.3 I1 is only true for a **constant** force.

S5.

(i) The acceleration $\mathbf{a}(t) = \cos t\,\mathbf{i} + \sin t\,\mathbf{j} + 2\sin 2t\,\mathbf{k}.$

$\therefore$ the velocity $\mathbf{v}(t) = \sin t\,\mathbf{i} - \cos t\,\mathbf{j} - \cos 2t\,\mathbf{k} + \mathbf{A}.$

When $t = 0$, $\mathbf{v} = 0$, so $\mathbf{0} = -\mathbf{j} - \mathbf{k} + \mathbf{A}.$

$\therefore\ \mathbf{v}(t) = \sin t\,\mathbf{i} + (1 - \cos t)\mathbf{j} + (1 - \cos 2t)\mathbf{k}.$

Momentum $= 1 \times \mathbf{v}(t).$

$\therefore$ at $t = \frac{\pi}{2}$, momentum $= \mathbf{v}(\frac{\pi}{2}) = \mathbf{i} + \mathbf{j} + 2\mathbf{k}.$

(ii) Let $\mathbf{F}$ be the force acting.
Then $\mathbf{F} = 1 \times \mathbf{a}(t)$, (using 9.2 F3).
The impulse

$$\mathbf{I} = \int_0^{\pi} \mathbf{F}\,dt$$

$$= \int_0^{\pi} (\cos t\,\mathbf{i} + \sin t\,\mathbf{j} + 2\sin 2t\,\mathbf{k})\,dt$$

$$= \Big[\sin t\,\mathbf{i} - \cos t\,\mathbf{j} - \cos 2t\,\mathbf{k}\Big]_0^{\pi}$$

$= 2\mathbf{j}.$ *11m* ■

S6.

In addition to the tension the ring is acted on by its weight $0{\cdot}5g = 5$ N, a normal reaction R N and a frictional force F N.

Let the resulting acceleration of the ring be f m s^{-2}.

(a) When the velocity is constant $f = 0$. So the equations of motion are

$\uparrow$: $\quad R - 5 - T\cos\theta = 0.$

$\rightarrow$: $\quad T\sin\theta - F = 0.$

$F = 0{\cdot}3R$ (using 9.4 F3). $\quad \therefore T\sin\theta = 0{\cdot}3R.$

Eliminating R: $\quad T\sin\theta = 0{\cdot}3(5 + T\cos\theta).$

$$\therefore T = \frac{1{\cdot}5}{\sin\theta - 0{\cdot}3\cos\theta}. \qquad (1)$$

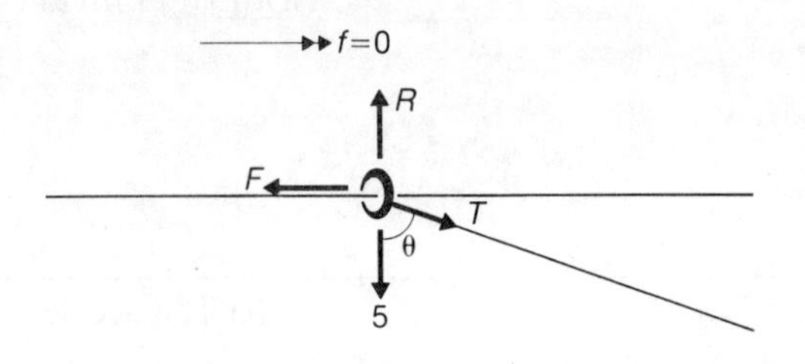

(b) The tension in the string must be positive, that is, $T > 0$.

$\therefore \sin\theta - 0{\cdot}3\cos\theta > 0. \qquad \therefore \tan\theta > 0{\cdot}3.$

When the tension in the string is $2T$, we have

$\uparrow$: $\quad R - 5 - 2T\cos\theta = 0.$

$\rightarrow$: $\quad 0{\cdot}5f = 2T\sin\theta - F$
$\quad = 2T\sin\theta - 0{\cdot}3R.$

Eliminating R:

$$\begin{aligned} 0{\cdot}5f &= 2T\sin\theta - 1{\cdot}5 - 0{\cdot}6T\cos\theta \\ &= 2T(\sin\theta - 0{\cdot}3\cos\theta) - 1{\cdot}5 \\ &= 2 \times 1{\cdot}5 - 1{\cdot}5, \text{ using the result in (a).} \end{aligned}$$

$\therefore f = 3.$ ■

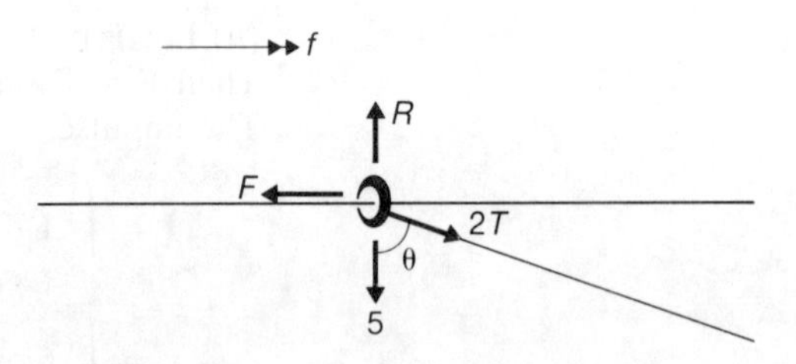

The bead starts with velocity 1 m s^{-1} and moves with acceleration 3 m s^{-2} for 1 second. During this time it travels a distance

$S = 1 \times 1 + \frac{1}{2} \times 3 \times 1^2 = 2{\cdot}5$ m (using 3.3 F7C).

Using 9.3 I1, the work done by the force $2T$ in the displacement 2·5 m is given by

$$W = 2T\sin\theta \times \frac{5}{2} = 5T\sin\theta \quad \boxed{A}$$

$$= \frac{7{\cdot}5\sin\theta}{\sin\theta - 0{\cdot}3\cos\theta} \quad \text{from (1)}$$

$$= \frac{7{\cdot}5}{1 - 0{\cdot}3\cot\theta}.$$

$\boxed{A}$ Read the question carefully again before continuing. You could make the error

✗ $W =$ (total force) $\times$ 2·5
$= (2T\sin\theta - 0{\cdot}3 \times 2T\cos\theta - 0{\cdot}3 \times 5) \times 2{\cdot}5.$ ✗

It is the force exerted by the *girl* that must be used.

Hence the work done by the girl is

$$\frac{7 \cdot 5}{1 - 0 \cdot 3 \cot \theta} \text{ J.}$$

Since $0 \cdot 3 \cot \theta > 0$, $\quad 1 - 0 \cdot 3 \cot \theta < 1$
and so $W > 7 \cdot 5$ $\quad$ [B] *25m* ■

[B] Not

× Because $\tan \theta > 0 \cdot 3$. ×

This simply tells us that $0 \cdot 3 \cot \theta < 1$, and so we can only deduce that $W > 0$.

S7.

Before impulse

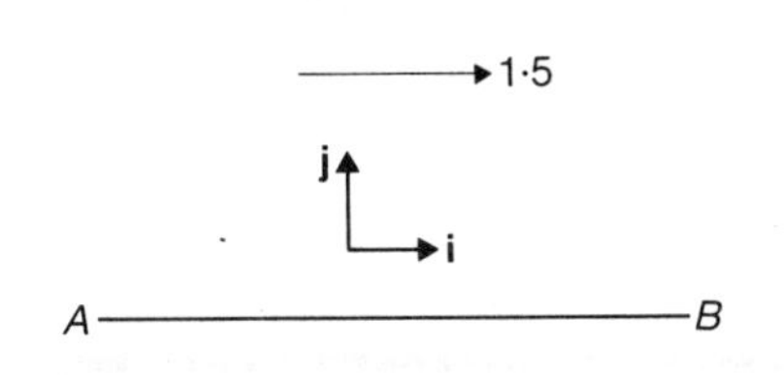

After impulse

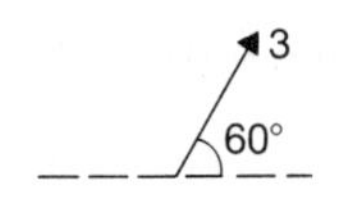

Let **i** and **j** be unit vectors parallel and perpendicular to the direction AB. Then if $\mathbf{u}\ \text{m s}^{-1}$ and $\mathbf{v}\ \text{m s}^{-1}$ are the velocities of the particle before and after the impulse $\mathbf{I}$ N s is applied,

$$\mathbf{u} = 1 \cdot 5\mathbf{i} \text{ and } \mathbf{v} = 3\cos 60°\mathbf{i} + 3\sin 60°\mathbf{j} = 1 \cdot 5\mathbf{i} + \frac{3\sqrt{3}}{2}\mathbf{j}.$$

Using 9.2 F7, $\quad \mathbf{I} = 0 \cdot 05\left(1 \cdot 5\mathbf{i} + \frac{3\sqrt{3}}{2}\mathbf{j}\right) - 0 \cdot 05 \times 1 \cdot 5\mathbf{i}$

$$= \frac{3\sqrt{3}}{40}\mathbf{j}.$$

Let the particle of mass 0·2 kg acquire a velocity $\mathbf{w}\ \text{m s}^{-1}$.

Then $\frac{3\sqrt{3}}{40}\mathbf{j} = 0 \cdot 2\mathbf{w}$.

$$\therefore \mathbf{w} = \frac{3\sqrt{3}}{8}\mathbf{j}. \ ■$$

Hence the particle moves in a direction perpendicular to AB.

The speed of the particle is $\frac{3\sqrt{3}}{8}\ \text{m s}^{-1}$. $\quad$ *11m* ■

Chapter 10 Projectiles

10.1 GETTING STARTED

Any object projected into the air is called a projectile. In reality a projectile experiences two principle forces; these are a frictional force due to air resistance and the gravitational force due to the presence of the earth.

In all the work in this Chapter resistance is neglected and, as the motion takes place close to the surface of the Earth, the gravitational force on the projectile is taken to be constant and equal to mg vertically downwards for a body of mass m. Consequently the motion will be confined to a fixed vertical plane containing the velocity of projection. So we are concerned with motion in two dimensions. Finally, all projectiles, whether they be shells fired from a gun or balls kicked by footballers, are taken to be particles. That is, the physical dimensions of the projectile are neglected.

10.2 THE MOTION OF A PROJECTILE

ESSENTIAL FACTS

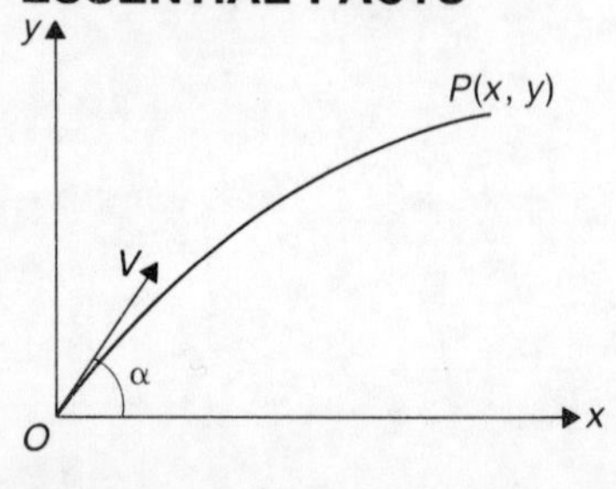

In the following facts a particle moves in a cartesian Oxy plane with Ox horizontal and Oy vertically upwards. Unit vectors **i** and **j** are parallel to Ox and Oy respectively, so that when the particle is at the point with coordinates (x, y) the position vector of the particle is given by **r**, where

$$\mathbf{r} = x\mathbf{i} + y\mathbf{j} = \begin{pmatrix} x \\ y \end{pmatrix}.$$

The particle is projected at time $t = 0$ with speed V at an inclination α to the horizontal and, unless stated otherwise, the initial point of projection is taken to be O, the origin of the coordinate system.

F1. The initial velocity of the particle is $\mathbf{V}$, where

$$\mathbf{V} = V\cos\alpha\,\mathbf{i} + V\sin\alpha\,\mathbf{j} = \begin{pmatrix} V\cos\alpha \\ V\sin\alpha \end{pmatrix}.$$

F2. The acceleration of the particle is $\ddot{\mathbf{r}} = \mathbf{g}$, so that

$$\ddot{x}\mathbf{i} + \ddot{y}\mathbf{j} = -g\mathbf{j}, \quad \text{that is,} \quad \begin{pmatrix} \ddot{x} \\ \ddot{y} \end{pmatrix} = \begin{pmatrix} 0 \\ -g \end{pmatrix}.$$

$$\therefore\ \ddot{x} = 0 \quad \text{and} \quad \ddot{y} = -g.$$

Thus the horizontal component of the acceleration is zero and the vertical component is $-g$.

F3. At time t after projection the velocity of the particle is $\dot{\mathbf{r}}$, where

$$\dot{\mathbf{r}} = \mathbf{V} + \mathbf{g}t.$$

So $$\dot{x}\mathbf{i} + \dot{y}\mathbf{j} = V\cos\alpha\,\mathbf{i} + (V\sin\alpha - gt)\mathbf{j};$$

that is, $$\begin{pmatrix} \dot{x} \\ \dot{y} \end{pmatrix} = \begin{pmatrix} V\cos\alpha \\ V\sin\alpha - gt \end{pmatrix}.$$

$$\therefore\ \dot{x} = V\cos\alpha \quad \text{and} \quad \dot{y} = V\sin\alpha - gt.$$

Thus the horizontal component of the velocity is constant and equal to $V\cos\alpha$, and the vertical component is $V\sin\alpha - gt$.

F4. The position vector of the particle at time t is $\mathbf{r}$, where

$$\mathbf{r} = \mathbf{V}t + \tfrac{1}{2}\mathbf{g}t^2.$$

So $$x\mathbf{i} + y\mathbf{j} = Vt\cos\alpha\,\mathbf{i} + (Vt\sin\alpha - \tfrac{1}{2}gt^2)\mathbf{j},$$

that is $$\begin{pmatrix} x \\ y \end{pmatrix} = \begin{pmatrix} Vt\cos\alpha \\ Vt\sin\alpha - \tfrac{1}{2}gt^2 \end{pmatrix}.$$

$$\therefore\ x = Vt\cos\alpha \quad \text{and} \quad y = Vt\sin\alpha - \tfrac{1}{2}gt^2.$$

F5. From F3 and F4 we derive the formula connecting the height with the vertical component of the velocity.

$$\dot{y}^2 = V^2 \sin^2 \alpha - 2gy.$$

F6. When the particle is initially projected from a point with position vector $\mathbf{a} = (a_1\mathbf{i} + a_2\mathbf{j})$ its position vector at time t is $\mathbf{r}$, where

$$\mathbf{r} = \mathbf{a} + \mathbf{V}t + \tfrac{1}{2}\mathbf{g}t^2.$$

So $\qquad x\mathbf{i} + y\mathbf{j} = (a_1 + Vt\cos\alpha)\mathbf{i} + (a_2 + Vt\sin\alpha - \tfrac{1}{2}gt^2)\mathbf{j},$

that is, $\begin{pmatrix} x \\ y \end{pmatrix} = \begin{pmatrix} a_1 + Vt\cos\alpha \\ a_2 + Vt\sin\alpha - \frac{1}{2}gt^2 \end{pmatrix}.$

F7.

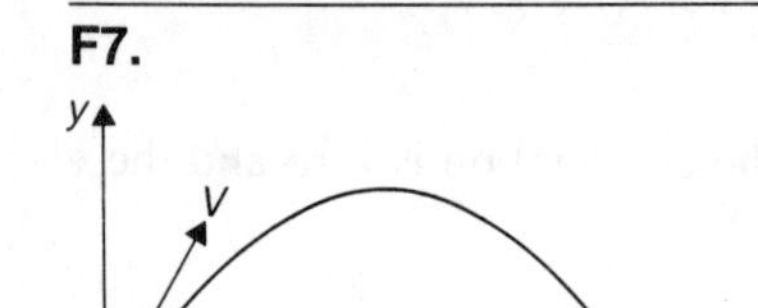

The particle moves in a parabolic path. The equation of the path is found by eliminating t from the equations of F4 and is

$$y = x\tan\alpha - \frac{gx^2}{2V^2}(1 + \tan^2\alpha).$$

F8.

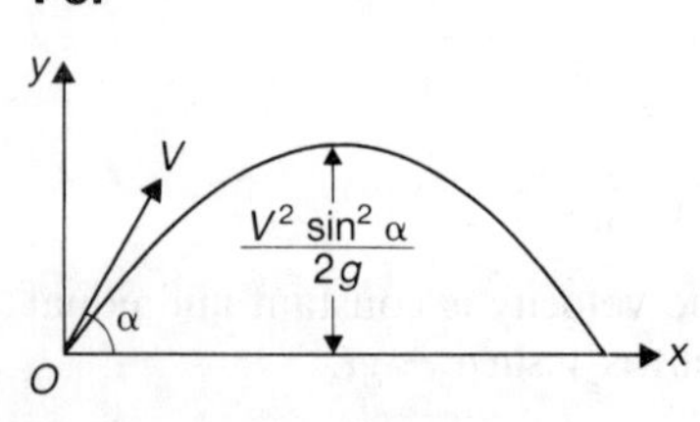

The maximum height of the particle above the horizontal plane through O is attained when the particle is moving horizontally. This occurs when $\dot{\mathbf{r}}.\mathbf{j} = 0$, that is when $\dot{y} = 0$.

The maximum height is $\dfrac{V^2\sin^2\alpha}{2g}$.

The time taken to reach the maximum height is $\dfrac{V}{g}\sin\alpha$.

F9.

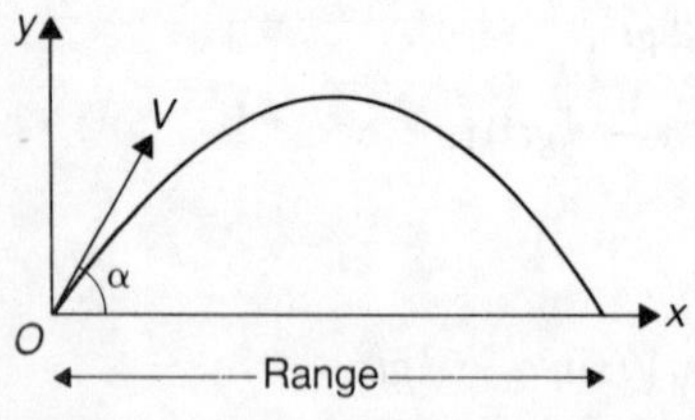

The **horizontal range** (often simply called the **range**) of a projectile is the distance from O of the point where the projectile returns to the horizontal plane through O. It returns to this plane when $\mathbf{r}.\mathbf{j} = 0$, that is, when $y = 0$. This occurs at time $\dfrac{2V\sin\alpha}{g}$ from the instant of projection, and the range is $\dfrac{V^2\sin 2\alpha}{g}$.

F10. For a fixed speed V of projection the maximum possible value of the range is $\dfrac{V^2}{g}$, and it occurs when $\alpha = 45°$.

ILLUSTRATIONS

I1.

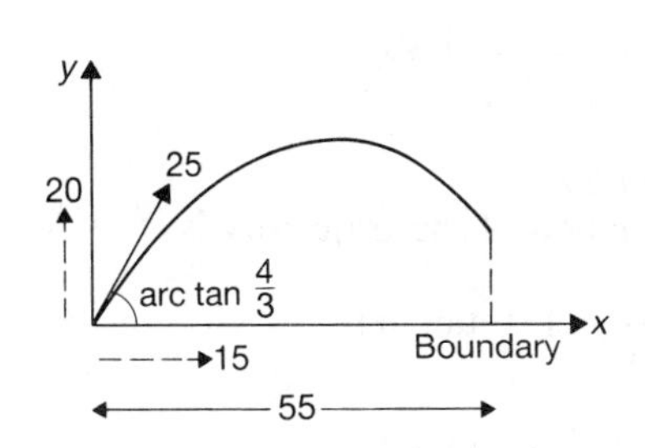

A batsman hits a cricket ball, giving it a velocity of 25 m s^{-1} at an elevation of $\arctan\left(\frac{4}{3}\right)$. Find the greatest height attained, the range on level ground and the time of flight.
Find the height of the ball when it reaches the boundary which is 55 m from the batsman. □
Initially the velocity of the ball has a horizontal component of 15 m s^{-1} and a vertically upward component of 20 m s^{-1}.
Then denoting the horizontal and vertically upward components of the displacement of the ball by x m and y m respectively, we see, from F4 and F3, that t seconds after the ball has been hit

$$x = 15t, \qquad y = 20t - 5t^2$$

and

$$\dot{x} = 15, \qquad \dot{y} = 20 - 10t.$$

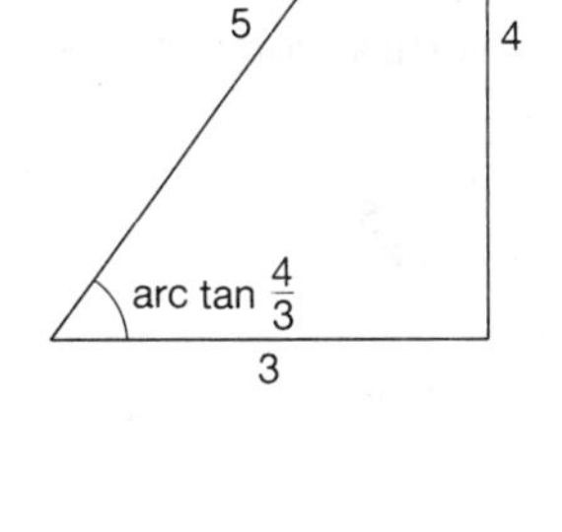

At the greatest height $\dot{y} = 0$. That is, $t = 2$, and so

$$y = 20 \times 2 - 5 \times 4 = 20.$$

Thus the greatest height is 20 m. ■
The ball meets the ground when $y = 0$.

$$\therefore 20t - 5t^2 = 0.$$

$$\therefore t = 0 \quad \text{or} \quad t = 4.$$

The value $t = 0$ corresponds to the initial instant when the ball is hit and so the required value is $t = 4$.
When $t = 4$, $\quad x = 15 \times 4 = 60.$
So the range of the ball on level ground is 60 m and the time of flight is 4 s.
When $x = 55$, $\quad t = \frac{55}{15} = \frac{11}{3}$ and $y = 20 \times \frac{11}{3} - 5 \times \left(\frac{11}{3}\right)^2 = \frac{55}{9}$.
Thus the ball is $6\frac{1}{9}$ m above the ground when it reaches the boundary. ■

I2.

A particle is projected with speed $\sqrt{10ag}$ from a point on level ground at a horizontal distance $5a$ from the foot of a vertical pole of height $\frac{5a}{2}$, so as just to clear the pole. Find the possible angles of elevation at the instant of projection and the speed of the particle as it passes over the pole. □

Let the particle be projected from O with elevation α and let A be the top of the pole. If the particle passes through A at time t after projection we have, from F4,

$$5a = \sqrt{10ag}\,t\cos\alpha$$

and $$\frac{5a}{2} = \sqrt{10ag}\,t\sin\alpha - \frac{1}{2}gt^2.$$

Eliminating t gives

$$\frac{5a}{2} = 5a\tan\alpha - \frac{1}{2}g\frac{25a^2}{10ag}\sec^2\alpha$$

$$= 5a\tan\alpha - \frac{5a}{4}(1+\tan^2\alpha).$$

$$\therefore \tan^2\alpha - 4\tan\alpha + 3 = 0.$$

$$\therefore (\tan\alpha - 2)^2 = 1.$$

So $\tan\alpha = 2 \pm 1$. That is $\tan\alpha = 3$ or 1.
Hence the possible angles of projection

are 71·6° and 45°. 

Let v be the speed of the particle at A.
As gravity is the only force acting on the particle, energy is conserved.
Let the mass of the particle be m.

Then $$\frac{1}{2}mv^2 + mg\frac{5a}{2} = \frac{1}{2}m(10ag).$$

$$\therefore v^2 = 5ag.$$

So the speed of the particle as it passes over the pole is $\sqrt{5ag}$. B ■

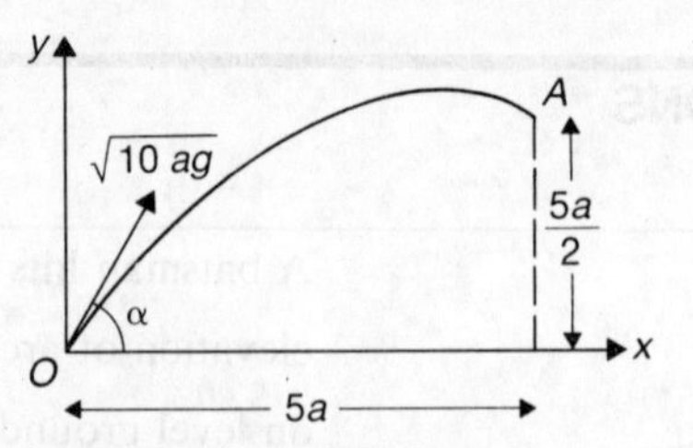

A *Alternative solution*
From F7, the equation of the trajectory is

$$y = x\tan\alpha - \frac{gx^2}{2V^2}(1+\tan^2\alpha)$$

$$= x\tan\alpha - \frac{x^2}{20a}(1+\tan^2\alpha).$$

The particle passes through $A\left(5a, \frac{5a}{2}\right)$.

$$\therefore \frac{5a}{2} = 5a\tan\alpha - \frac{5a}{4}\left(1+\tan^2\alpha\right).$$

The solution then proceeds as before.

B *Alternative solution*
Consider the horizontal and vertical components of the velocity at A.

By F3, $\dot{x} = \sqrt{10ag}\cos\alpha$.

By F5, $\dot{y}^2 = 10ag\sin^2\alpha - 2g\frac{5a}{2}$.

Then
$$v^2 = \dot{x}^2 + \dot{y}^2$$
$$= 10ag\,(\cos^2\alpha + \sin^2\alpha) - 5ag$$
$$= 5ag.$$

I3. Points A and B are on horizontal ground and a distance 800 m apart. A shell is projected from A and towards B with a velocity of $80\sqrt{3}$ m s^{-1} at an elevation 30°. At the same instant a target is projected from B with velocity 80 m s^{-1} at an elevation α in the plane of the shell and directed towards A. Given that the shell hits the target, find α. Find also the time which elapses between the initial projection and the collision. □

At time t s after projection the shell and target are at heights y_1 m and y_2 m repectively above the horizontal level AB.

Using F4: $$y_1 = (80\sqrt{3}\sin 30°)t - \tfrac{1}{2}gt^2$$

and $$y_2 = (80\sin\alpha)t - \tfrac{1}{2}gt^2.$$

When the shell and target collide, $y_1 = y_2$.

$$\therefore \quad 80\sqrt{3}\sin 30^\circ = 80\sin\alpha. \qquad (1)$$

$$\therefore \quad \sin\alpha = \frac{\sqrt{3}}{2}. \qquad \therefore \quad \alpha = 60^\circ.$$

Let the collision occur after T s.
Let the horizontal distances covered by the shell and the target be x_1 m and x_2 m respectively.

By F4; $\quad x_1 = (80\sqrt{3}\cos 30^\circ)T = 120T$

and $\quad x_2 = (80\cos\alpha)T = 40T$.

The sum of the horizontal distances covered is equal to AB.

$$\therefore x_1 + x_2 = 800. \qquad \therefore 160T = 800. \qquad \therefore T = 5.$$

So the time which elapses between the projection and the collision is 5 seconds. ■

14.

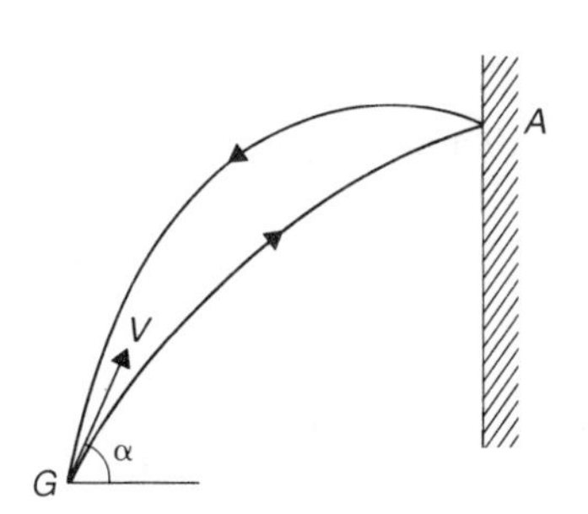

A girl throws a ball with initial velocity V at an inclination α. The ball strikes a smooth vertical wall at a horizontal distance d from the girl and, after rebounding, returns to her hand. Find the total time of flight of the ball and the coefficient of restitution between the ball and the wall. □

When the ball hits the wall at A, the vertical component of the impulse it receives is zero, because the wall is smooth. Hence the vertical component of its velocity is unaltered.
Consequently, throughout both parts of the flight, the height of the ball is the same at any time as it would be in the absence of the wall. So the total time of flight T is given by

$$0 = (V\sin\alpha)T - \tfrac{1}{2}gT^2 \quad \text{(using F4).} \qquad \therefore T = \frac{2V}{g}\sin\alpha. \ ■$$

eV cos α
V cos α

When the ball is moving from G to A, the horizontal component of the velocity is constant at $V\cos\alpha$.

So the time of flight from G to A is $t_1 = \dfrac{d}{V\cos\alpha}$.

Let the coefficient of restitution be e.
From the law of restitution (6.4F7), the horizontal component of the velocity of the ball immediately after impact has magnitude $eV\cos\alpha$. This component remains constant whilst the ball is moving from A back to G and so the time from A to G is

$$t_2 = \frac{d}{eV\cos\alpha}.$$

Now $T = t_1 + t_2$.

$$\therefore \frac{2V}{g}\sin\alpha = \frac{d}{V\cos\alpha} + \frac{d}{eV\cos\alpha}.$$

$$\therefore 1 + \frac{1}{e} = \frac{V^2}{gd}\sin 2\alpha.$$

Hence the coefficient of restitution is $\dfrac{gd}{V^2\sin 2\alpha - gd}$. ■

---

15. A stone is thrown from the top of a tower 20 m high, which stands on level ground, with a velocity of 25 m s^{-1} at elevation arc tan $\frac{3}{4}$. Find the horizontal distance from the foot of the tower of the point where it hits the ground. □

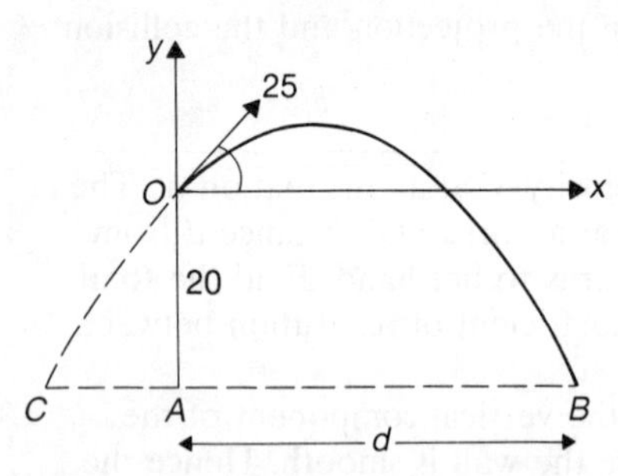

Take axes with O at the top of the tower, Ox horizontal and Oy vertically upwards. Let T s be the time of flight of the stone from O to the point B where it hits the ground.

Vertical component of displacement

As the point B is 20 m below O and the initial vertical component of the velocity is $25 \times \frac{3}{5} = 15$ m s^{-1}, we have

$$-20 = 15T - \tfrac{1}{2} \times 10T^2, \text{ taking } g = 10 \text{ and using F4.}$$

$$\therefore T^2 - 3T - 4 = 0.$$

$$\therefore (T-4)(T+1) = 0.$$

As $T > 0$ at B we have $T = 4$.

Note that the value $T = -1$ corresponds to the point C which is where the stone would land if it were projected in the opposite direction at an angle of *depression* arc tan $\frac{3}{4}$.

Horizontal component of displacement

The initial horizontal component of the velocity

is $25 \times \dfrac{4}{5} = 20$ m s^{-1}.

This remains constant.

Hence $\quad d = 20T = 80. \qquad$ So $AB = 80$ m. ■

10.3 EXAMINATION QUESTIONS AND SOLUTIONS

Q1. A particle P of mass m moves freely under gravity in the plane of a horizontal axis Ox and an upward vertical axis Oy. The particle is projected from O, at time $t = 0$, with speed u and at an angle α

above Ox. Write down expressions for the coordinates (x, y) of P at time t. Given that u is fixed and that α may vary between 0 and $\frac{1}{2}\pi$ radians, derive, in terms of u and g, the maximum range R along Ox.
In the case when α is such that the range along Ox is a maximum, show that

(i) the greatest height reached is $\frac{1}{4}R$.
(ii) the kinetic energy of the particle at any instant is proportional to the difference between $\frac{1}{2}R$ and y.

(JMB 1986)

Q2.

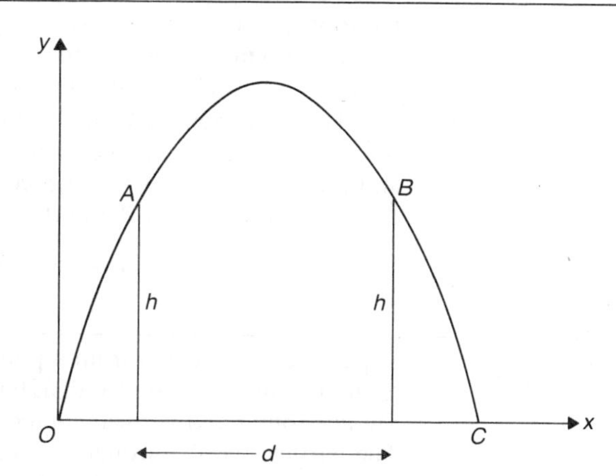

A particle P moves freely under gravity in the plane of a fixed horizontal axis Ox, which lies on flat ground, and a fixed upward vertical axis Oy. The particle is projected from O with a velocity whose components along Ox and Oy are u and v, respectively. Prove that the particle returns to the ground at a point C distant $2uv/g$ from O.

The particle passes, in succession, through two points A and B, at a height h above Ox and at a distance d apart, as shown in the diagram. Write down the horizontal and vertical components of the velocity of the particle at A.

Hence show that

$$d = 2u(v^2 - 2gh)^{\frac{1}{2}}/g.$$

Given that the direction of the motion of the particle as it passes through A is inclined to the horizontal at an angle of $\tan^{-1}\frac{1}{2}$, show that

$$u = \sqrt{gd}$$

and obtain an expression for v in terms of g, d and h. (JMB 1984)

Q3. A projectile is fired with speed V from a point on an inclined plane, the direction of projection lying in a vertical plane containing a line of greatest slope of the plane and making an angle θ with this line of greatest slope. The plane makes an angle α with the horizontal where $\theta < \frac{\pi}{2} - \alpha$. Show that the time before the projectile strikes the plane is

$$(2V \sin\theta)/(g \cos\alpha).$$

(LON 1982)

Q4. A particle is projected from a point O on horizontal ground with speed v at an angle of elevation α. The point O is 20 m away from the foot, F, of a vertical post. The particle subsequently strikes the post at a point 15 m above the ground, and at the moment of impact the particle is travelling horizontally. Find α and v.

The coefficient of restitution between the particle and the post is $\frac{1}{2}$. The particle next meets the ground at a point B where O, B and F are in a straight line. Calculate OB and the angle which the direction of motion of the particle makes at B with the horizontal.

[*Take g to be 9·8* $m\,s^{-2}$] (AEB 1986)

Q5. A particle P projected from a point O on level ground strikes the ground again at a distance of 120 m from O after a time 6 s. Find the horizontal and vertical components of the initial velocity of P. The particle passes through a point Q whose horizontal displacement from O is 30 m. Find

(a) the height of Q above the ground,
(b) the tangent of the angle between the horizontal and the direction of motion of P when it is at Q and also the speed of P at this instant,
(c) the horizontal displacement from O of the point at which P is next at the level of Q.

[*Take g to be 10* $m\,s^{-2}$] (AEB 1985)

Q6. A particle P, of mass m, is projected from a point O with speed u at an angle of inclination α to the horizontal. Find the time taken for P to reach the highest point A of its path.

When P is at A it collides directly, and coalesces, with a particle Q, of mass $3m$. Just before impact Q was moving horizontally with the same speed as P but in the opposite direction. Find the speed of the composite particle immediately after impact and show that the loss of kinetic energy is $\frac{3}{2}mu^2\cos^2\alpha$.

Given that the composite particle meets the horizontal plane through O at a point B, express OB in terms of u, α and g.

(LON 1984)

Q7.

A particle P is projected from a point A which is at a height $6h$ above horizontal ground. The particle moves freely under gravity, attains a greatest height $8h$ above the ground and strikes the ground at the point B. Find, in terms of h and g, the vertical component of the velocity of P at the instant when P leaves A and at the instant when P arrives at B. Hence, or otherwise, show that the time taken for P to reach B from A is $6\sqrt{(h/g)}$.

Given that the horizontal distance between A and B is $8h$, find, in terms of h and g, the horizontal component of the velocity of P.

Find also the tangent of the acute angle between the horizontal and the velocity of P at time $3\sqrt{(h/g)}$ after P leaves A. (LON 1986)

SOLUTIONS

S1.

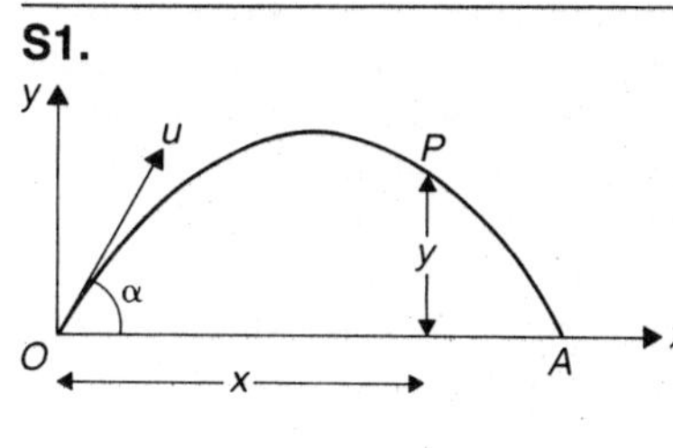

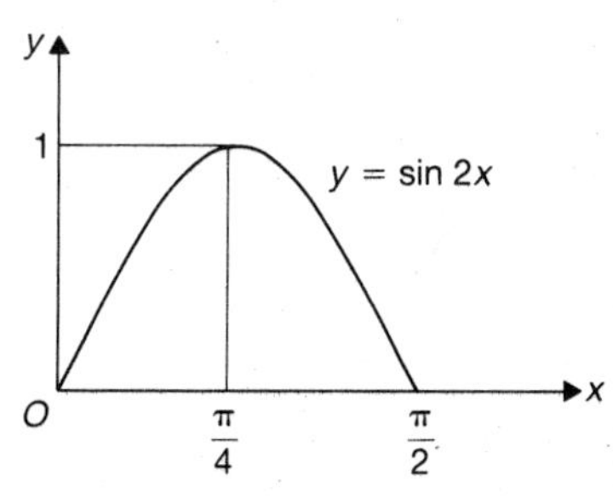

$x = (u\cos\alpha)t, \qquad y = (u\sin\alpha)\,t - \frac{1}{2}gt^2.$ ■

At A, $y = 0$. $\qquad \therefore t = 0$ or $t = 2\dfrac{u}{g}\sin\alpha$.

As $t = 0$ at O, we deduce that $t = \dfrac{2u}{g}\sin\alpha$ at A.

Hence, at A, $\qquad x = u\cos\alpha\,\dfrac{2u}{g}\sin\alpha = \dfrac{u^2}{g}\sin 2\alpha.$

The maximum value of $\sin 2\alpha$ is 1.

$\therefore$ the maximum range is $R = \dfrac{u^2}{g}$. ■

This is achieved when $2\alpha = \dfrac{\pi}{2}$, that is, $\alpha = \dfrac{\pi}{4}$.

(i) $\dot{y} = u\sin\alpha - gt.$

Maximum height is attained when $\dot{y} = 0$, that is, when $t = \dfrac{u\sin\alpha}{g}$.

At this time, $y = u\sin\alpha\,\dfrac{u\sin\alpha}{g} - \dfrac{g}{2}\,\dfrac{u^2\sin^2\alpha}{g^2} = \dfrac{u^2\sin^2\alpha}{2g}.$

So, when $\alpha = \dfrac{\pi}{4}$, the maximum height

is $\dfrac{u^2}{2g}\sin^2\dfrac{\pi}{4} = \dfrac{u^2}{4g} = \dfrac{R}{4}$. ■

(ii) Let v be the velocity of P at the point of its path with coordinates (x, y).
As gravity is the only force acting on P energy is conserved.

$$\therefore \tfrac{1}{2}mu^2 = \tfrac{1}{2}mv^2 + mgy.$$

$$\therefore \tfrac{1}{2}mv^2 = \tfrac{1}{2}mu^2 - mgy = mg(\tfrac{1}{2}R - y).$$

Thus the kinetic energy is proportional to the difference between $\tfrac{1}{2}R$ and y. *17m* ■

S2.

$x = ut, \quad y = vt - \tfrac{1}{2}gt^2$ [A]

$y = 0$ when $t = 0$ or $t = \dfrac{2v}{g}$.

As $t = 0$ at O, $\quad t = \dfrac{2v}{g}$ at C.

Hence, at C, $\quad x = 2\dfrac{uv}{g}$.

$\therefore OC = \dfrac{2uv}{g}$. ■

At A, $\quad \dot{x} = u, \quad \dot{y}^2 = v^2 - 2gh$,
so the horizontal and vertical velocity components are u and $\sqrt{v^2 - 2gh}$. [B] ■
Hence, from the first part of the question, it follows that $d = \dfrac{2u\sqrt{v^2 - 2gh}}{g}$. (1) [C]

At A, $\quad \dfrac{\dot{y}}{\dot{x}} = \dfrac{\sqrt{v^2 - 2gh}}{u} = \dfrac{1}{2}$. (2) [D]

Eliminating $\sqrt{v^2 - 2gh}$ from (1) and (2),

$\dfrac{2u}{g}\dfrac{u}{2} = d. \quad \therefore u = \sqrt{gd}$. ■

From (2), $\quad v^2 - 2gh = \dfrac{u^2}{4}$.

$\therefore v^2 = 2gh + \dfrac{gd}{4}$.

$\therefore v = \tfrac{1}{2}\sqrt{8gh + gd}$. *20m* ■

[A] These results may be quoted immediately by using F4.
[B] $\dot{x} = u$ as the horizontal velocity component is constant.
The equation $\dot{y} = v - gt$ would not directly help us here, because we are given the height h of A, *not* the time when A is reached. Therefore we use F5 to obtain y (at A) in terms of the data.
[C] This part of the question may be done by starting anew and considering the equations of motion of a particle projected from A. However, we should always remember that the examiner has a reason for demanding preliminary results. Here, it is that the first results are directly applicable to the motion from A to B.
[D] At any general position of P the direction of motion makes angle θ with the horizontal, where $\tan\theta = \dfrac{\dot{y}}{\dot{x}}$. So here, $\dfrac{\dot{y}}{\dot{x}} = \dfrac{1}{2}$.

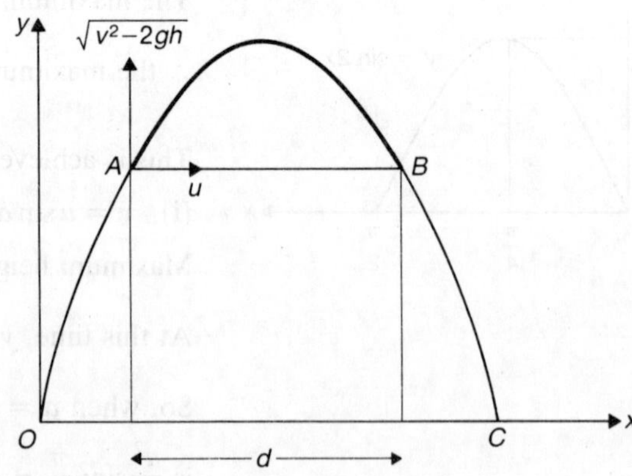

S3.

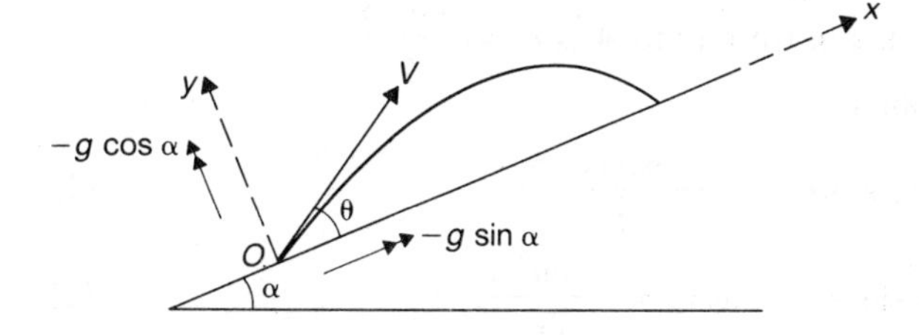

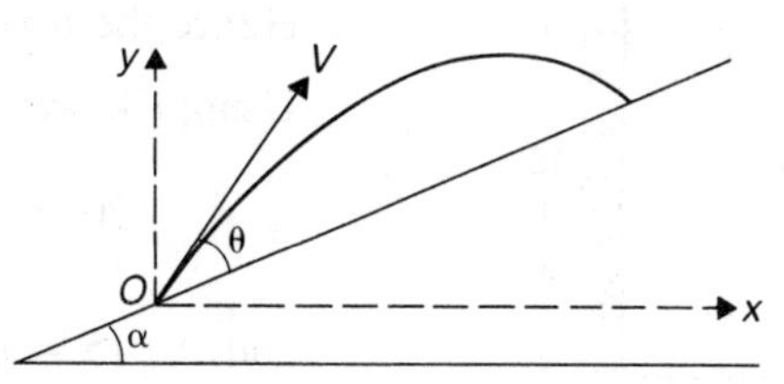

Take O as the point of projection, and axes Ox along a line of greatest slope and Oy perpendicular to the plane and directed away from the plane.

As the acceleration g is vertically downwards, it has components $-g\sin\alpha$ and $-g\cos\alpha$ parallel to the Ox and Oy axes respectively.

Thus $\quad \ddot{y} = -g\cos\alpha.$

$\therefore \quad \dot{y} = -gt\cos\alpha + V\sin\theta$

and $\quad y = -\tfrac{1}{2}gt^2\cos\alpha + Vt\sin\theta.$

The projectile is on the plane when $y = 0$, that is, when $t = 0$ (at O) and when

$V\sin\theta - \tfrac{1}{2}gt\cos\alpha = 0.$

So the time of flight is $\dfrac{2V\sin\theta}{g\cos\alpha}$. *9m* ■

Alternative solution.

Take O as the point of projection and Ox, Oy in the horizontal and vertically upwards directions respectively. The line of greatest slope has equation $y = x\tan\alpha$.

As the initial velocity components along Ox and Oy are $V\cos(\theta+\alpha)$ are $V\sin(\theta+\alpha)$ respectively, we have

$$x = V\cos(\theta+\alpha)t, \qquad y = V\sin(\theta+\alpha)t - \tfrac{1}{2}gt^2.$$

The projectile is on the plane when $y = x\tan\alpha$, that is, when

$V\sin(\theta+\alpha)t - \tfrac{1}{2}gt^2 = V\cos(\theta+\alpha)t\tan\alpha.$

So $t = 0$ (at O) or

$$t = \frac{V\sin(\theta+\alpha) - V\cos(\theta+\alpha)\tan\alpha}{\tfrac{1}{2}g}.$$

So the time of flight is

$$\frac{2V}{g}[\sin(\theta+\alpha) - \cos(\theta+\alpha)\tan\alpha]$$

$$= \frac{2V}{g\cos\alpha}[\sin(\theta+\alpha)\cos\alpha - \cos(\theta+\alpha)\sin\alpha]$$

$$= \frac{2V}{g\cos\alpha}[\sin(\theta+\alpha-\alpha)]$$

$$= \frac{2V\sin\theta}{g\cos\alpha}.$$ ■

S4.

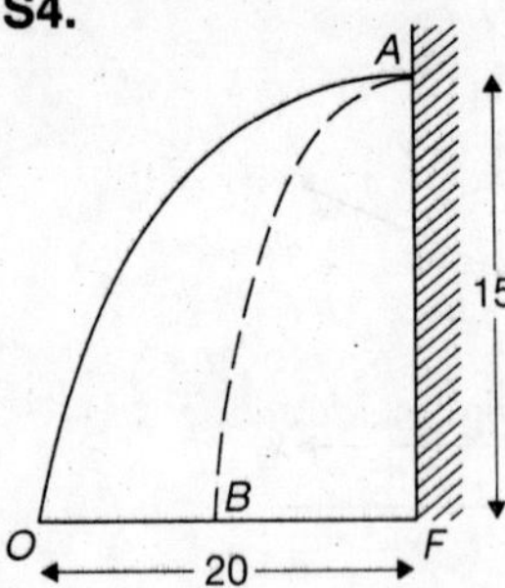

At O the vertical component of the velocity is $v \sin\alpha$, and at A it is zero.

Hence the time taken from O to A is $t_1 = \dfrac{v\sin\alpha}{g}$.

Using F4, we obtain

$$20 = v\cos\alpha\, t_1 = \frac{v^2 \sin\alpha\cos\alpha}{g} \tag{1}$$

and $$15 = v\sin\alpha t_1 - \tfrac{1}{2}gt_1^2 = \frac{v^2\sin^2\alpha}{2g}. \tag{2}$$

Hence, eliminating v^2,

$$\frac{15}{20} = \frac{\tan\alpha}{2}. \quad \text{That is} \quad \tan\alpha = \tfrac{3}{2}.$$

Therefore $\alpha = 56{\cdot}3°$ to 1dp.

Substitute $\sin\alpha = \dfrac{3}{\sqrt{13}}$ and $g = 9{\cdot}8$. in (2).

$$\frac{v^2}{2 \times 9{\cdot}8}\,\frac{9}{13} = 15. \qquad \therefore v = \sqrt{\frac{1274}{3}}.$$

So the speed of projection is $20{\cdot}6\ \mathrm{m\,s^{-1}}$ (to 1dp). ■

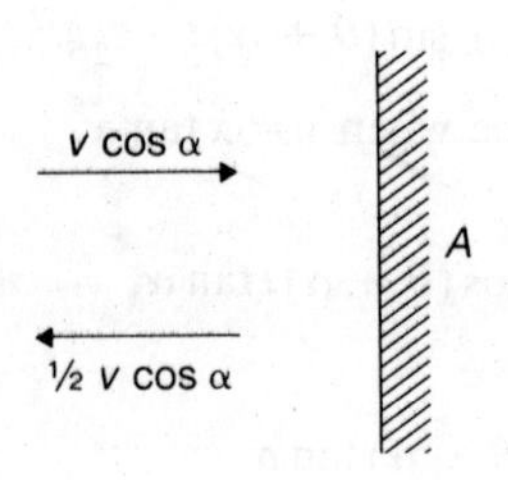

The impact

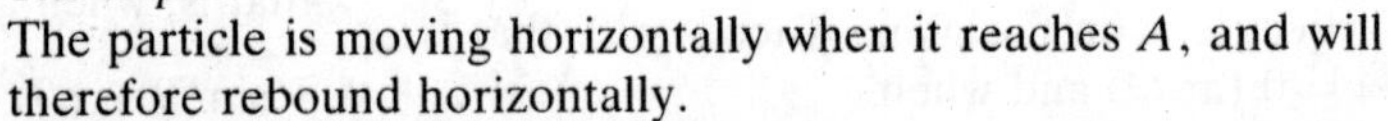

The particle is moving horizontally when it reaches A, and will therefore rebound horizontally.

From the law of restitution (6.4F7), its speed on rebounding is $\frac{1}{2}v\cos\alpha$.

The descent

Using F5, the magnitude of the vertical component of the velocity at B is

$$\sqrt{2g \times 15} = v\sin\alpha \qquad \text{[from (2)]}.$$

The time for the descent is the time taken, with acceleration g, to acquire a vertical velocity component $v\sin\alpha$, which is $\dfrac{v\sin\alpha}{g}$.

The horizontal component of the velocity has constant magnitude $\frac{1}{2}v\cos\alpha$.

$$\therefore FB = \tfrac{1}{2}v\cos\alpha\left(\frac{v\sin\alpha}{g}\right)$$

$$= \tfrac{1}{2}(20), \qquad \text{from (1)}.$$

$$\therefore OB = 10\ \text{m}.$$

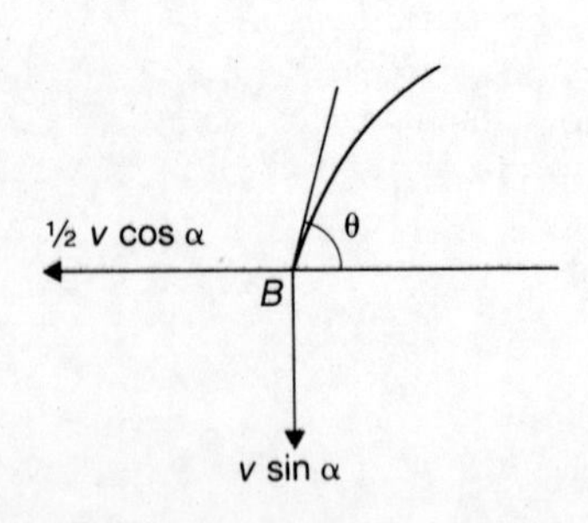

Let the angle the direction of motion makes at B with the horizontal be θ.

$$\text{Then } \tan\theta = \frac{v\sin\alpha}{\frac{1}{2}v\cos\alpha} = 2\tan\alpha = 3.$$

$\therefore \theta = 71{\cdot}6°$ (to 1dp) *27m* ■

S5.

Let $U\ \text{m s}^{-1}$ and $V\ \text{m s}^{-1}$ be the horizontal and vertical components of the velocity at O.
As the time of flight is 6 s, we have

$$U = \frac{120}{6} = 20 \quad \boxed{A}$$

and $\quad 0 = V \times 6 - \frac{1}{2}(10)(6)^2. \quad \boxed{B}\ \therefore\ V = 30.$

So the initial horizontal and vertical components of the velocity are $20\ \text{m s}^{-1}$ and $30\ \text{m s}^{-1}$.
(a) Let h m be the height of Q above the ground. The time of flight from O to Q is t_1 s, where

$$t_1 = \frac{30}{20} = 1{\cdot}5. \quad \boxed{C}$$

Hence $h = 30 \times 1{\cdot}5 - \frac{1}{2} \times 10 \times (1{\cdot}5)^2 = 33{\cdot}75.$ $\boxed{D}$

So Q is 33·75 m above the ground.
(b) The horizontal component of the velocity at Q is $20\ \text{m s}^{-1}$ and the vertical component is $v\ \text{m s}^{-1}$, where

$$v = 30 - 10 \times 1{\cdot}5 = 15. \quad \boxed{E}$$

Let θ be the angle between the direction of motion at Q and the horizontal.
Then $\tan\theta = \frac{15}{20} = \frac{3}{4}$. $\boxed{F}$ ■

The speed at $Q = \sqrt{20^2 + 15^2} = 25\ \text{m s}^{-1}$. $\boxed{G}$ ■
(c) If P is a height 33·75 m above the ground at time t_2 s,
then $33{\cdot}75 = 30t_2 - \frac{1}{2} \times 10 \times t_2^2$. $\boxed{D}$

$\therefore\ t_2^2 - 6t_2 + 6{\cdot}75 = 0. \quad \boxed{H}$

$\therefore\ (t_2 - 1{\cdot}5)(t_2 - 4{\cdot}5) = 0.$

As $t_2 = 1{\cdot}5$ corresponds to Q, the particle is next at the height of Q at time $t_2 = 4{\cdot}5$.
At this time the horizontal displacement from O is $20 \times 4{\cdot}5 = 90$ m. *27m* ■

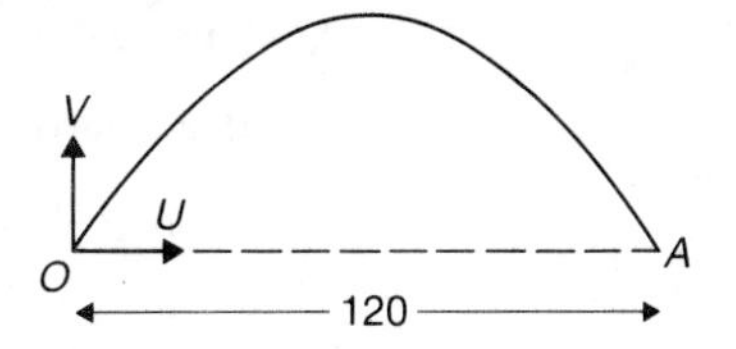

$\boxed{A}$ The horizontal component of the velocity is constant.
So $U = \dfrac{\text{distance}}{\text{time}}$.
$\boxed{B}$ Using F4.

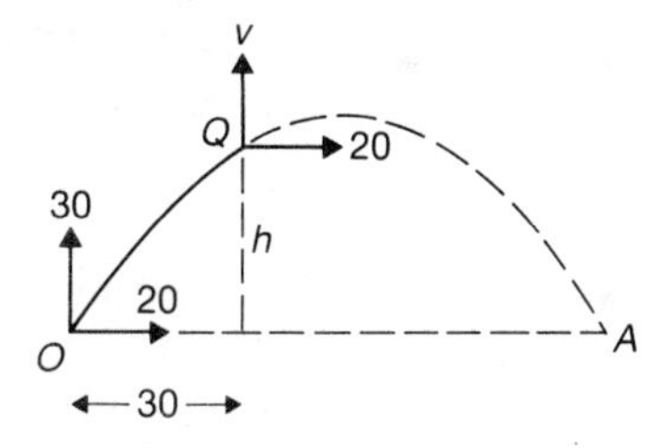

$\boxed{C}$ Again, as the horizontal component of the velocity is constant $t_1 = \dfrac{\text{distance}}{\text{speed}}$.
$\boxed{D}$ Using F4.
$\boxed{E}$ Using F3.
$\boxed{F}$ $\tan\theta = \dfrac{\text{vertical component of the velocity}}{\text{horizontal component of the velocity}}$.

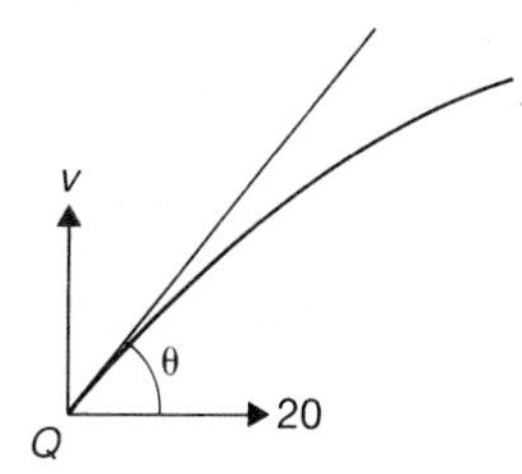

$\boxed{G}$ We could use the conservation of energy to find the speed at Q. However, as we have already found the horizontal and vertical velocity components, this method is easier.
$\boxed{H}$ As we already know that $t_2 = 1{\cdot}5$ is a root of this equation the factorisation is eased, and we do not need to complete the square or to use the 'formula' for the roots.

S6.

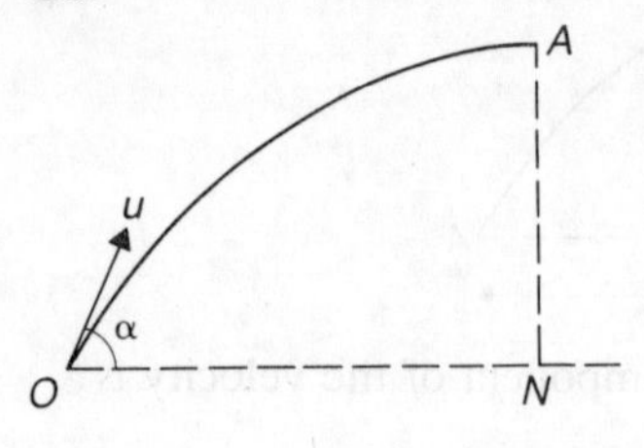

Let T be the time of flight from O to A.
At A the velocity is horizontal. So, considering the vertical motion,

$$0 = u \sin\alpha - gT.$$

$$\therefore T = \frac{u \sin\alpha}{g}. \blacksquare$$

During the motion from O to A, P has a constant horizontal speed $u \cos\alpha$.
So immediately before the impact both particles have speed $u \cos\alpha$.
Let the speed of the composite particle, of mass $4m$, be v immediately after the impact, the directions of the motion of the particles being as shown in the diagram.

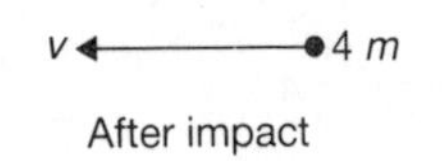

Using the principle of conservation of momentum

$$3mu\cos\alpha - mu\cos\alpha = 4mv.$$

$$\therefore v = \tfrac{1}{2}u\cos\alpha.$$

So, after impact, the composite particle is moving horizontally, towards O, with speed $\frac{1}{2}u\cos\alpha$. $\blacksquare$
The kinetic energy before impact is

$$\tfrac{1}{2}m\,(u\cos\alpha)^2 + \tfrac{1}{2}.3m\,(u\cos\alpha)^2 = 2mu^2\cos^2\alpha.$$

The kinetic energy after impact is

$$\tfrac{1}{2}.4mv^2 = \tfrac{1}{2}.mu^2\cos^2\alpha.$$

Hence the loss of kinetic energy is

$$2mu^2\cos^2\alpha - \tfrac{1}{2}.mu^2\cos^2\alpha = \tfrac{3}{2}mu^2\cos^2\alpha. \blacksquare$$

The horizontal distance covered whilst moving from O to A is

$$ON = u\cos\alpha.T = \frac{u^2}{g}\sin\alpha\cos\alpha.$$

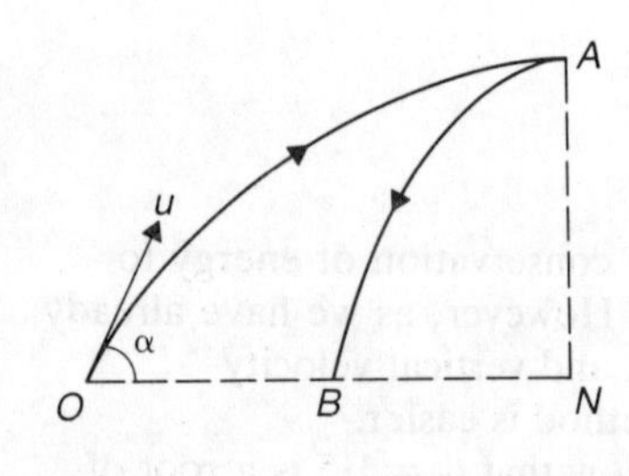

The vertical distance covered whilst moving from O to A is AN, where

$$0 = (u\sin\alpha)^2 - 2g \times AN. \qquad \therefore AN = \frac{u^2\sin^2\alpha}{2g}.$$

The composite particle has no vertical velocity component at A, so (using F4) the time of flight from A to B is

$$\sqrt{\frac{2AN}{g}} = \frac{u\sin\alpha}{g}.$$

The horizontal distance covered in this time is

$$BN = v\left(\frac{u\sin\alpha}{g}\right) = \frac{u^2\sin\alpha\cos\alpha}{2g}.$$

Hence $OB = ON - BN = \dfrac{u^2}{2g}\sin\alpha\cos\alpha.$ **24m** $\blacksquare$

S7.

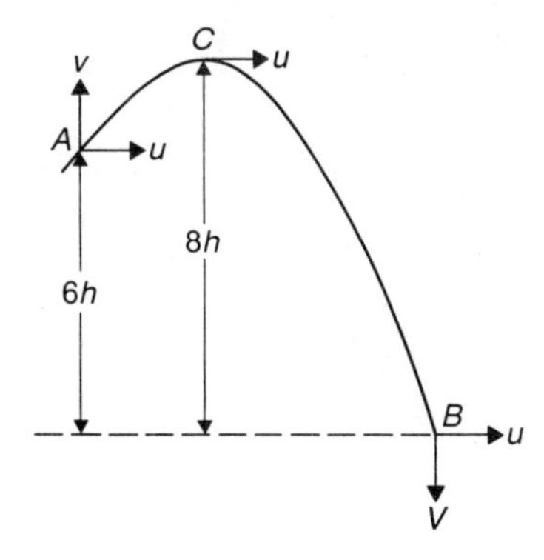

Let the vertically upward components of the velocity of P at A and B be v and $-V$ respectively. We choose $-V$ as the particle is clearly moving downwards at B. At the highest point, C, there is no vertical component of the velocity, and throughout the motion let u be the constant horizontal component of the velocity.

Vertical components

From A to C: P has risen a distance $2h$.

$$\therefore 0 = v^2 - 2g \times 2h.$$

$$\therefore v = 2\sqrt{gh}.$$

From A to B: P has fallen a distance $6h$.

$$\therefore (-V)^2 = v^2 - 2g(-6h) = 4gh + 12gh.$$

$$\therefore V = 4\sqrt{gh}.$$

So the vertical component of the velocity at A is $2\sqrt{gh}$ upwards and at B it is $4\sqrt{gh}$ downwards. ■

Let T be the time of flight from A to B.

Then $\qquad -V = v - gT.$

$\therefore \qquad -4\sqrt{gh} = 2\sqrt{gh} - gT.$

Hence $\qquad T = 6\sqrt{\dfrac{h}{g}}.$ ■

Horizontal component

The horizontal distance between A and B, $8h$, is covered in time T.

$$\therefore u = \frac{8h}{T} = \frac{4}{3}\sqrt{gh}. \text{ ■}$$

At time $3\sqrt{\dfrac{h}{g}}$:

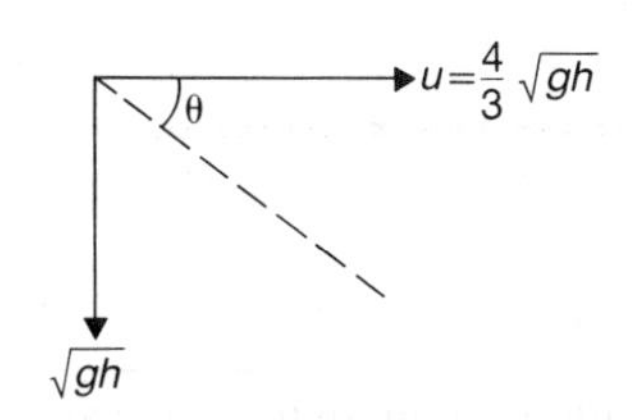

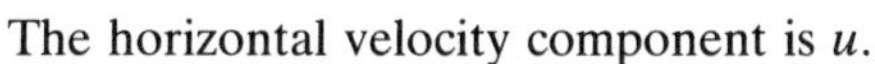

The horizontal velocity component is u.

The vertically upward velocity component is W, where

$$W = v - g \times 3\sqrt{\frac{h}{g}} = 2\sqrt{gh} - 3\sqrt{gh} = -\sqrt{gh}.$$

That is, the vertical component of the velocity is $\sqrt{gh}$ downwards.

Let θ be the angle of *depression* of the velocity with the horizontal.

Then $\tan\theta = \dfrac{\sqrt{gh}}{\frac{4}{3}\sqrt{gh}} = \dfrac{3}{4}$. $\qquad$ *24m* ■

Chapter 11 Circular motion

11.1 GETTING STARTED

In studying the motion of a particle P which moves in a circular path it is normally convenient to take the fixed centre of the circle as the origin O of coordinates. However, the components of acceleration in fixed directions are often not the most convenient to use. Many problems are more easily solved by considering the components of the acceleration in directions parallel and perpendicular to OP. Newton's Second Law is then applied by resolving forces in each of these directions.

11.2 VELOCITIES AND ACCELERATIONS

ESSENTIAL FACTS

In the following facts a particle P is moving in a circular path with centre O and radius a.

F1.

The position of P in its path may be specified by an angle θ which is the angle measured in a fixed chosen direction (anticlockwise in our diagram) from a fixed radius OA to the radius OP.

F2. Angular velocity

The angular velocity of P at time t is denoted by ω, where

$$\omega = \frac{\mathrm{d}\theta}{\mathrm{d}t}.$$

So ω is measured in radians per second.

F3.

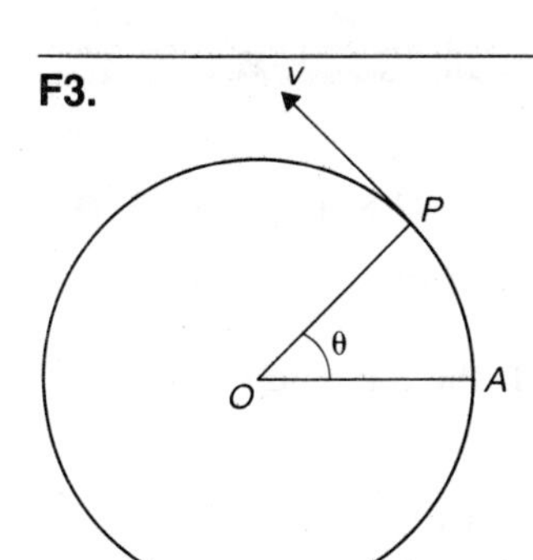

The velocity of P is directed along the tangent to the circle at P. Its measure, v, in the sense of increasing θ is called the arcual velocity, and

$$v = a\omega = a\frac{d\theta}{dt}.$$

In our diagram, when $v > 0$, P is rotating about O in the anticlockwise sense, and when $v < 0$ the motion is clockwise.

The arcual velocity v is often called just 'the velocity of P' in cases where confusion with the true (vector) velocity is unlikely.

F4.

The angular acceleration of P is the rate of change of the angular velocity, denoted by $\frac{d\omega}{dt}$ or $\frac{d^2\theta}{dt^2}$.

F5. Components of acceleration

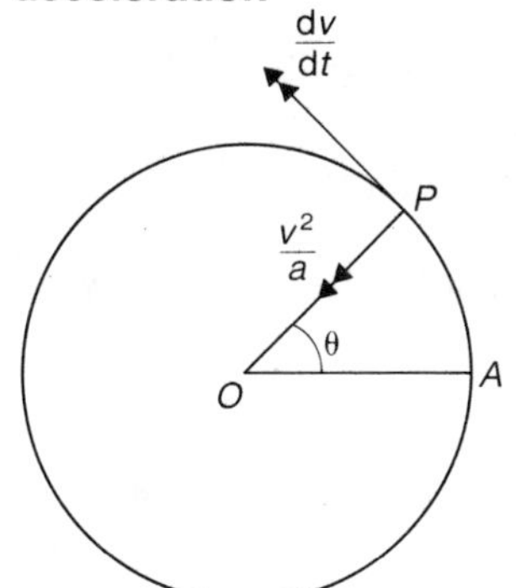

The acceleration of P may be resolved in two perpendicular directions, one along the radius OP and the second, called the tangential direction, along the tangent at P in the direction of increasing θ.
The component of the acceleration along PO is

$$a\omega^2 = \frac{v^2}{a} = a\left(\frac{d\theta}{dt}\right)^2.$$

The component of the acceleration in the tangential direction is

$$\frac{dv}{dt} = a\frac{d\omega}{dt} = a\frac{d^2\theta}{dt^2}.$$

ILLUSTRATIONS

I1.

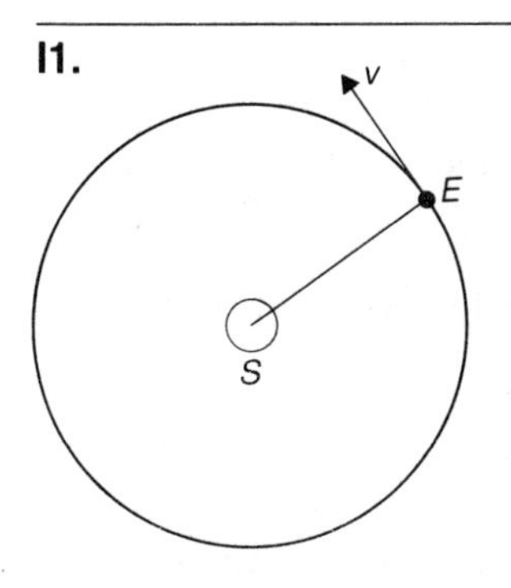

Assuming the Earth to move with constant speed, completing a circular orbit of radius $1{\cdot}49 \times 10^8$ km around the Sun as centre once every 365 days, find the angular velocity, the speed and the acceleration of the Earth.□
The speed of the Earth, $v = a\omega$, is given as constant.
Hence the angular velocity is also constant.
Therefore T, the period of revolution (or the time for one complete orbit), is given by $T = \frac{2\pi}{\omega}$.
Here $T = 365$ days or $365 \times 24 \times 60 \times 60$ s.

Hence $\omega = \dfrac{2\pi}{365 \times 24 \times 60 \times 60} = 1{\cdot}99 \times 10^{-7}\ .\ .\text{s}^{-1}$. ■
The speed of the Earth is

$$v = (1{\cdot}49 \times 10^8 \times 10^3) \times (1{\cdot}99 \times 10^{-7}) = 2{\cdot}97 \times 10^4.\ .\ \text{m s}^{-1}. \ ■$$

As ω is constant the tangential component of the acceleration is zero.
Hence the acceleration is directed towards the Sun and it has magnitude

$$\omega^2 a = \left(\frac{2\pi}{365 \times 24 \times 60 \times 60}\right)^2 (1{\cdot}49 \times 10^{11})$$

$$= 5{\cdot}91 \times 10^{-3}.\ .\text{m s}^{-2}. \ ■$$

I2. A particle P moves on a circle with centre O and radius 2 m. At time t s the radius OP has rotated from a fixed position OA through an angle of θ radians, where $\theta = 3\sin 2t$. Find the velocity in the sense of increasing θ, the speed and the magnitude of the acceleration of P when $t = \pi/3$.□

$$\omega = \frac{d\theta}{dt} = 6\cos 2t \text{ s}^{-1}, \qquad v = a\omega = 12\cos 2t \text{ m s}^{-1},$$

$$\frac{v^2}{a} = 72\cos^2 2t \text{ m s}^{-2} \quad \text{and} \quad \frac{dv}{dt} = -24\sin 2t \text{ m s}^{-2}.$$

Hence, when $t = \pi/3$: the velocity $v = 12\cos 2\pi/3 = -6 \text{ m s}^{-1}$.■
The speed of P is $|v| = 6 \text{ m s}^{-1}$.■
The magnitude of the acceleration is

$$\sqrt{\left(\frac{v^2}{a}\right)^2 + \left(\frac{dv}{dt}\right)^2} = \sqrt{(18)^2 + (-12\sqrt{3})^2} = 6\sqrt{21} \text{ m s}^{-2}. \ ■$$

I3. A particle P moves on a semi-circular path in such a way that the acceleration of P is always perpendicular to the bounding diameter AOB, where O is the mid-point of the diameter. Given that at time t the angle $POB = \theta$, show that $\dfrac{d\theta}{dt}\sin\theta$ is constant.

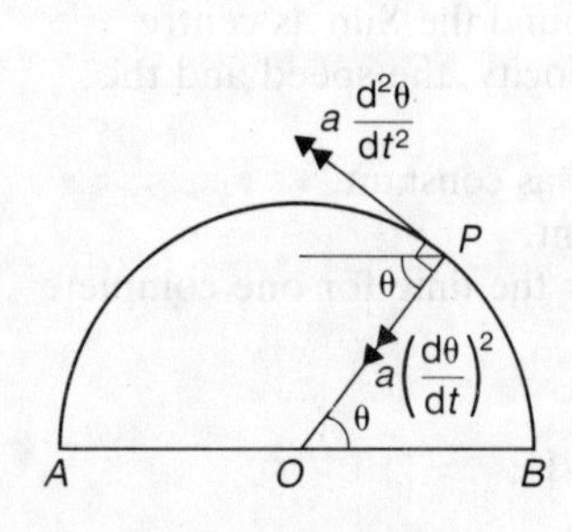

Hence, or otherwise, show that the magnitude of the acceleration of P is proportional to d^{-3}, where d is the distance of P from the bounding diameter.□
The component of the acceleration parallel to AB must be zero.
Hence

$$a\left(\frac{d\theta}{dt}\right)^2 \cos\theta + a\frac{d^2\theta}{dt^2}\sin\theta = 0.$$

That is $\dfrac{d}{dt}\left(\dfrac{d\theta}{dt}\sin\theta\right) = 0$ and hence $\dfrac{d\theta}{dt}\sin\theta = \lambda$, a constant.

It follows that $\dfrac{d^2\theta}{dt^2} = -\lambda\dfrac{\cos\theta}{\sin^3\theta}$.

The component of the acceleration perpendicular to AB is f, where

$$f = a\left(\frac{d\theta}{dt}\right)^2 \sin\theta - a\frac{d^2\theta}{dt^2}\cos\theta$$

$$= a\lambda^2\frac{\sin\theta}{\sin^2\theta} + a\lambda^2\frac{\cos^2\theta}{\sin^3\theta}$$

$$= \frac{a\lambda^2}{\sin^3\theta}(\sin^2\theta + \cos^2\theta) = \frac{a\lambda^2}{\sin^3\theta}.$$

Now $d = a\sin\theta$, so that $f = a^4\lambda^2/d^3$. ■

11.3 UNIFORM MOTION IN A HORIZONTAL CIRCLE

ESSENTIAL FACTS

In the following facts a particle P of mass m is moving in a horizontal circular path with centre O and radius a.

F1.

When P moves with constant arcual velocity v the angular velocity ω is constant and

the acceleration of P is $a\omega^2 = \dfrac{v^2}{a}$ directed towards O.

The tangential component of the acceleration is zero.

F2.

The sum of the vertical components of the forces acting on P is zero.

F3.

The sum of the components along PO of the forces acting on P is equal to $ma\omega^2 = \dfrac{mv^2}{a}$.

ILLUSTRATIONS

I1.

A particle is placed on a rough horizontal table at a distance 0·75 m from a point O on the table. The table rotates about a vertical axis through O. Given that the coefficient of friction between the particle and the table is $\frac{1}{2}$, show that the particle will not move relative to the table if the number of revolutions per minute does not exceed 24. □

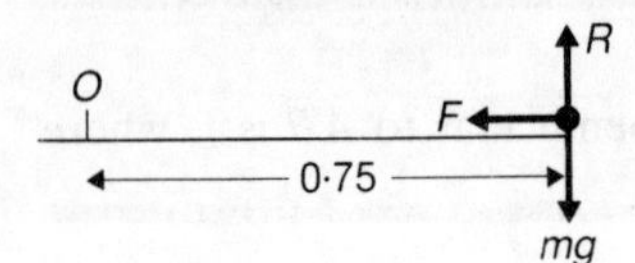

The forces acting on the particle are its weight mg N, the normal reaction R N and the frictional force F N.

Then, from F2, $\quad R = mg$

and, from F3, $\quad F = m(0{\cdot}75)\omega^2$,

where $\omega\ \text{s}^{-1}$ is the angular speed and m kg the mass of the particle. Since, for no motion relative to the table, $F \leqslant \frac{1}{2}R$, it follows that

$$0{\cdot}75\frac{\omega^2}{g} \leqslant \frac{1}{2}.$$

$$\therefore\ \omega^2 \leqslant \frac{2g}{3}, \quad \text{that is} \quad \omega \leqslant \sqrt{\frac{20}{3}}.$$

$$\text{Now}\ \sqrt{\frac{20}{3}}\ \text{rad}\,\text{s}^{-1} = \sqrt{\frac{20}{3}} \times \frac{60}{2\pi}\ \text{rev/min} = 24{\cdot}7\ldots\ \text{rev/min}.$$

So the particle will not move relative to the table if the number of revolutions per minute does not exceed 24. ■

---

12. A light inextensible string of length 3 m passes through a small fixed smooth ring O and carries a particle A, of mass 5 kg, at one end and a particle B, of mass 3 kg, at the other end. The particle A is at rest vertically below O. The particle B moves in a horizontal circle, the centre N of which is vertically below O, with constant angular speed $\frac{7}{\sqrt{3}}\ \text{rad}\,\text{s}^{-1}$.

Given that the string remains taut, find the distance NA. □

Let θ be the angle AOB and T N the tension in the string.
As A is in equilibrium, $T = 5g$.

Then, from F2, $\quad 3g = T\cos\theta = 5g\cos\theta$.

$$\therefore\ \cos\theta = \tfrac{3}{5}.$$

Now, let $OB = x$ m. Then, as $NB = OB\sin\theta$, we have from F3 that

$$T\sin\theta = 3(x\sin\theta)\left(\frac{7}{\sqrt{3}}\right)^2.$$

$$\therefore\ 3x\left(\frac{49}{3}\right) = 5g = 49 \qquad (\text{taking } g = 9{\cdot}8).$$

Hence $x = 1$.
Thus $OA = 2$ m and $ON = OB\cos\theta = 1 \times \frac{3}{5} = 0{\cdot}6$ m.
So, finally, $AN = 2 - 0{\cdot}6 = 1{\cdot}4$ m. ■

11.4 MOTION IN A VERTICAL CIRCLE

ESSENTIAL FACTS

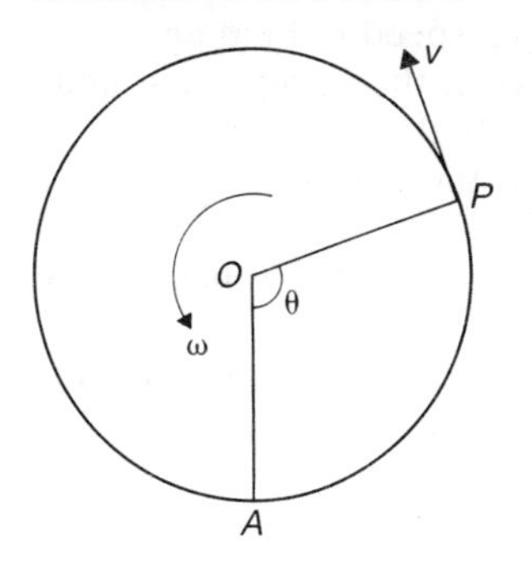

In the following facts a particle P, of mass m, is moving in a vertical circle with centre O, radius a, and lowest point A. At time t the angle POA (measured anticlockwise in our diagram) is θ, and P has angular velocity $\omega = \dfrac{d\theta}{dt}$ and arcual velocity $v = a\omega = a\dfrac{d\theta}{dt}$.

F1.

The sum of the components along PO of the forces acting on P is equal to

$$ma\omega^2 = \frac{mv^2}{a}.$$

F2.

The sum of the components in the tangential direction of the forces acting on P is equal to $m\dfrac{dv}{dt} = ma\dfrac{d\omega}{dt} = ma\dfrac{d^2\theta}{dt^2}$.

As $\dfrac{d^2\theta}{dt^2} = \dfrac{d}{d\theta}\left[\frac{1}{2}\left(\dfrac{d\theta}{dt}\right)^2\right]$, that is $\ddot{\theta} = \dfrac{d}{d\theta}(\frac{1}{2}\dot{\theta}^2)$, the sum of the tangential components of the forces acting on P is also equal to

$$ma\frac{d}{d\theta}(\tfrac{1}{2}\dot{\theta}^2).$$

F3.

The kinetic energy of P is $\frac{1}{2}mv^2 = \frac{1}{2}ma^2\omega^2 = \frac{1}{2}ma^2\dot{\theta}^2$.

F4.

When the forces acting on P consist of its weight acting vertically downwards together with other forces which all act along the line OP, the potential energy of P is

$$-mg\cos\theta + \text{constant},$$

and the total energy is conserved.

ILLUSTRATION

I1. A particle of mass m is suspended from a fixed point O by a light inextensible string of length a and hangs in equilibrium. The particle is then projected with a horizontal speed u. Find an expression for the tension in the string when the string is taut and inclined at θ to the downward vertical through O.
Given that the string becomes slack when the particle is a vertical distance $a/2$ above O, show that $2u^2 = 7ga$.□

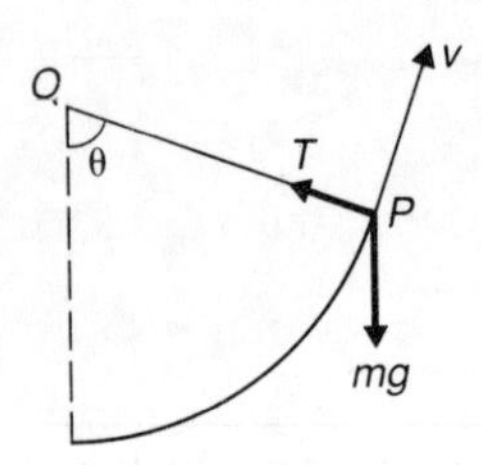

Consider a general position of the particle P, when OP has rotated through an angle θ.
The forces acting on P are its weight, of magnitude mg vertically downwards, and the tension in the string, T, directed towards O.

From F1, $\dfrac{mv^2}{a} = T - mg\cos\theta$.

The kinetic energy of P is $\frac{1}{2}mv^2$.
Take the gravitational potential energy to be zero when P is at A, the lowest point of its path.
Then the potential energy of P is $mg(AM) = mga(1 - \cos\theta)$.

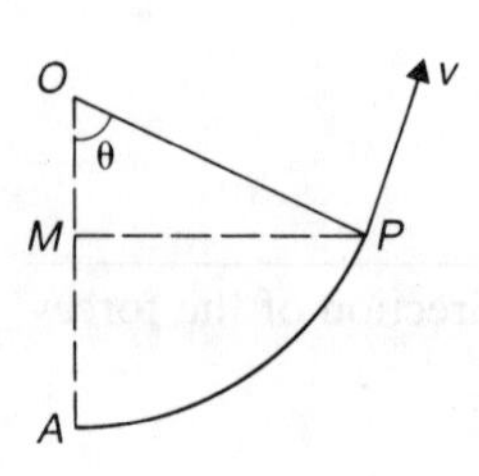

$\therefore$ Total energy of P is $\frac{1}{2}mv^2 + mga(1 - \cos\theta)$.

The total energy when P is at A is $\frac{1}{2}mu^2$.
Therefore, by F4,

$$\tfrac{1}{2}mv^2 + mga\,(1 - \cos\theta) = \tfrac{1}{2}mu^2.$$

Eliminating v^2 then gives $T = \dfrac{mu^2}{a} - mg(2 - 3\cos\theta)$. (1) ■

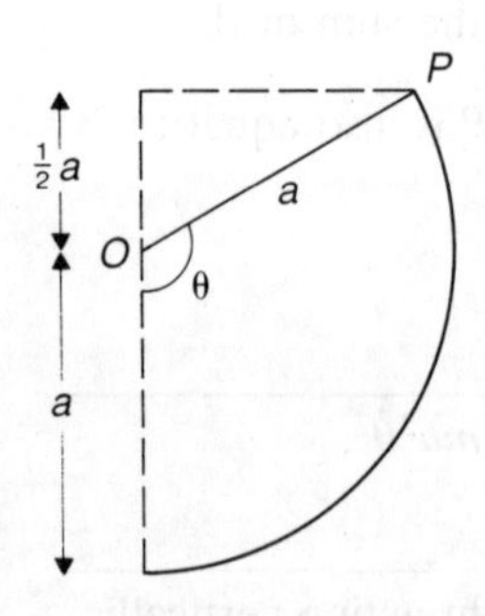

When the string becomes slack $T = 0$, and this occurs when P is a vertical distance $a/2$ above O.

In this position $\theta = \theta_1$, where $\cos\theta_1 = -\dfrac{a/2}{a} = -\frac{1}{2}$.

Hence $\quad 0 = \dfrac{mu^2}{a} - mg\,(2 + 3 \times \frac{1}{2})$. [Using equation (1)]

That is, $2u^2 = 7ga$.■
Note: The relation between v^2 and θ may alternatively be derived from the tangential component of the equation of motion.

By F2, $\quad ma\ddot{\theta} = -\,mg\sin\theta$.

$$\therefore a\frac{d}{d\theta}(\tfrac{1}{2}\dot{\theta}^2) = -g\sin\theta.$$

Integrate with respect to θ.

$$\tfrac{1}{2}a\dot{\theta}^2 = g\cos\theta + C, \text{ where } C \text{ is a constant.}$$

When $\theta = 0$, $\quad \dot{\theta} = \dfrac{u}{a}$ so $\quad C = \dfrac{u^2}{2a} - g$.

$$\therefore \tfrac{1}{2}a^2\dot{\theta}^2 = \tfrac{1}{2}u^2 - ag(1 - \cos\theta).$$

Substituting $v = a\dot{\theta}$ then gives the relation between v^2 and θ.

11.5 EXAMINATION QUESTIONS AND SOLUTIONS

Q1.

A circular race track, of radius 200 m, is banked at 45°. Find the speed at which a car may travel round the track without any frictional force acting perpendicular to the direction of motion of the car.

Given that the coefficient of friction between the wheels and the track is $\frac{1}{2}$, find, to one decimal place, the maximum speed at which a car can travel round the track without skidding.

[*Take g as 10* $m\,s^{-2}$.] (LON 1984)

Q2.

A particle of mass M is attached to an end A of a light inextensible string of length a. The other end O of the string is fixed. Initially, with A hanging vertically below O, the mass is projected at right angles to OA with speed $3\sqrt{ag}$ so that it begins to move in a vertical circle of centre O. Find the tension in the string when it makes an angle θ with the downward vertical through O. Does the string ever become slack? Justify your answer. (WJEC 1985)

Q3.

One end of a light inextensible string of length a is fastened to a fixed point and a particle is attached to the other end. The particle is held so that the string is taut and horizontal and is then projected vertically upwards with speed u. Given that the string slackens when it is inclined at an angle of 30° to the horizontal, find u in terms of a and g. (JMB 1985)

Q4.

A bead B of mass m is threaded on a smooth circular wire of radius a, centre C, fixed in a vertical plane. The bead moves freely along the wire and its speed at the lowest point is u. When CB makes an angle θ with the *upward* vertical, express its speed v and the reaction R, in the direction BC, of the wire on the bead in terms of m, g, a, u and θ. Show that the bead makes complete revolutions of the wire if $u^2 > 4ag$. If $u^2 = \dfrac{9ag}{2}$, find the value of R when the bead is at the highest point of the wire and interpret the sign of R.

The bead is set in motion with speed v_0 at the *highest* point of the wire. When it reaches the lowest point, it collides with and attaches itself to a stationary bead of half its mass also threaded on the wire and free to move on it. If after the collision the combined beads move freely along the wire, show that they will make complete revolutions of the wire if $v_0^2 > 5ag$. (WJEC 1984)

Q5. A weather satellite moves with constant speed around the earth at a constant height of 200 kilometres above the earth's surface. Given that the earth's radius is 6400 kilometres and that the acceleration due to gravity at the earth's surface is 10 m s^{-2}, how many revolutions does the satellite complete each day? (NI 1985)

Q6. A particle P moves in the x–y plane in a circle of radius a with its centre at the origin. The particle is initially at the point $(a, 0)$ and moves anti-clockwise with constant angular speed ω. Write down, in terms of a, ω, t and the unit vectors $\mathbf{i}$ and $\mathbf{j}$, the position vector $\mathbf{r}$ of P at time t. Obtain, by differentiation, an expression for $\ddot{\mathbf{r}}$ in terms of $\mathbf{r}$ and ω, and hence determine the magnitude and the direction of the acceleration of P.

Two particles are connected by a light inextensible string which passes through a small smooth hole in a smooth horizontal table. One particle has mass $3m$ and moves on the table, in a circle of radius a, with constant angular speed ω; the other particle has mass m and hangs at rest. Find an expression for ω in terms of a and g. (JMB 1986)

SOLUTIONS

S1.

The car is modelled as a particle of mass m kg. [A]
As there is no frictional force perpendicular to the motion the car experiences a normal reaction R N in addition to its weight mg N.
If u m s^{-1} is the speed of the car the equation of circular motion gives

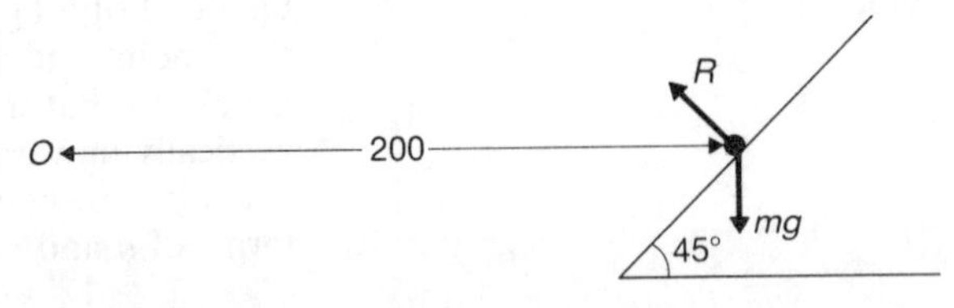

$$m\frac{u^2}{200} = R \times \frac{1}{\sqrt{2}}.$$

As the car has no vertical motion, $R \times \dfrac{1}{\sqrt{2}} = mg$.

So $\dfrac{u^2}{200} = g$, that is $u^2 = 2000$.

$\therefore u = 20\sqrt{5}$ and the car has speed $20\sqrt{5}$ m s^{-1}.■

[A] As the problem is not concerned with the possibility of the car overturning the simple particle model is sufficient. In 'overturning problems' the car must be treated as a rigid body with more than one point of contact with the bank. However, such problems are not on the A-level single mathematics syllabuses and so are beyond the scope of this book.

When frictional forces act let the car have speed v m s^{-1} and let the friction exerted by the road on the car be a force F N down the banking. [B]

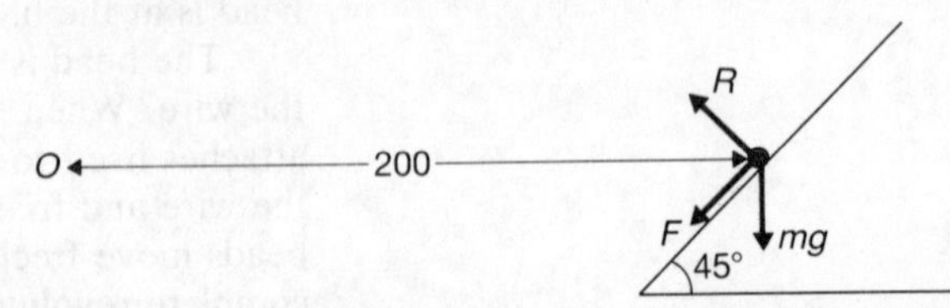

The equation of circular motion gives

$$R \times \frac{1}{\sqrt{2}} + F \times \frac{1}{\sqrt{2}} = m\frac{v^2}{200}$$

and the vertical equation of motion gives

$$R \times \frac{1}{\sqrt{2}} = F \times \frac{1}{\sqrt{2}} + mg.$$

Thus $\quad \sqrt{2}R = \dfrac{mv^2}{200} + mg$

and $\quad \sqrt{2}F = \dfrac{mv^2}{200} - mg.$

For no slipping, $F \leqslant \frac{1}{2}R$. [C]

Thus $\quad \dfrac{v^2}{200} - g \leqslant \dfrac{1}{2}\left(\dfrac{v^2}{200} + g\right).$

$\therefore v^2 \leqslant 3g(200) = 6000.$ [D]

So the maximum speed of the car is $20\sqrt{15}\ \mathrm{m\,s^{-1}}$ or $77{\cdot}5\ \mathrm{m\,s^{-1}}$ to 1dp. *25m* ■

[B] As the maximum speed is required, we wish to consider $v > u$. We suspect that, when $v > u$, the friction will need to act down the banking to prevent the car from slipping upwards, and so F will be positive.

[C] Using 13.5F1.

[D] If we had chosen the friction to be a force Q up the plane, we would have got

$$Q\sqrt{2} = mg - \frac{mv^2}{200}.$$

The relation $Q \leqslant \frac{1}{2}R$ would then have led to a minimum speed for no slipping *down* the plane. However, 13.5F1 says

$$|Q| \leqslant \mu R$$

which also implies $-Q \leqslant \frac{1}{2}R$.

So we would have got the same final result by using this inequality.

S2.

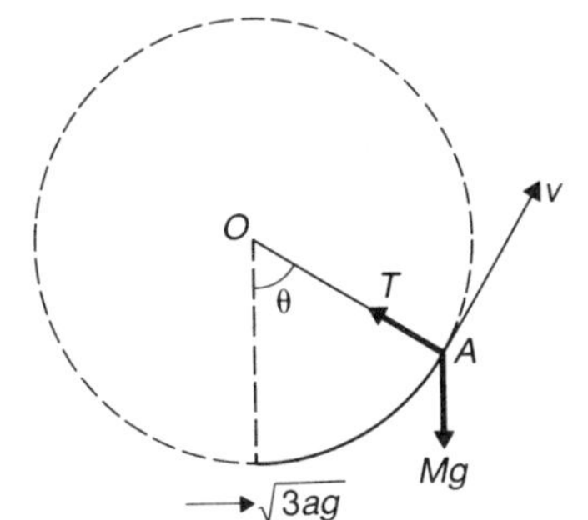

In the general position the particle has speed v as shown. The equation of motion along AO is

$$\frac{Mv^2}{a} = T - Mg\cos\theta$$

and conservation of energy gives

$$\tfrac{1}{2}Mv^2 + Mga\,(1 - \cos\theta) = \tfrac{1}{2}M \times 3ag.$$

Eliminating v gives $T = Mg + 3Mg\cos\theta$.

The string becomes slack when $T = 0$, that is, when $\cos\theta = -\frac{1}{3}$. We need to show that the particle actually reaches this position. When $\cos\theta = -\frac{1}{3}$, the energy equation gives

$$v^2 = 3ag - 2ag\,(1 + \tfrac{1}{3}) = \frac{ag}{3}.$$

Therefore the particle is still moving when $\cos\theta = -\frac{1}{3}$, so the string becomes slack. *13m* ■

S3.

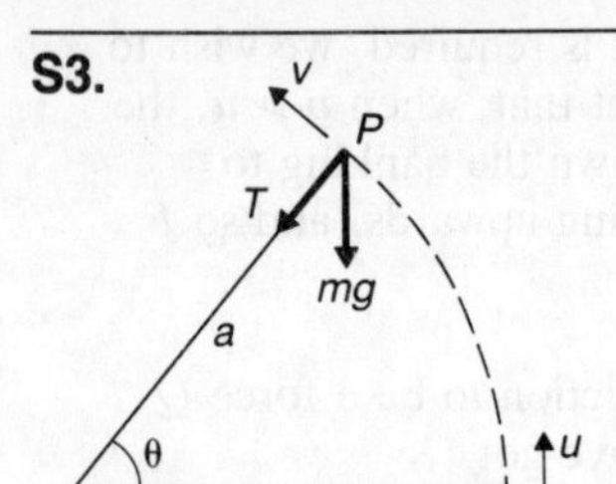

Using conservation of energy:

$\frac{1}{2}mv^2 = \frac{1}{2}mu^2 - mga\sin\theta.$

$\therefore v^2 = u^2 - 2ag\sin\theta.$ (1) [B]

The equation of motion along PO is

$$T + mg\sin\theta = \frac{mv^2}{a}$$

$$= \frac{mu^2}{a} - 2mg\sin\theta. \quad (2)$$

$$\therefore T = \frac{mu^2}{a} - 3mg\sin\theta.$$

T falls to zero when $\theta = 30°$.

$\therefore u^2 = 3ag\sin 30° = \frac{3ga}{2}.$ *9m* ■

[A] Although we have been asked to find only one thing, we have been given no lead through the question, so we have to invent symbols in order to solve the problem.
Thus we draw a diagram showing the speed of the particle, v, when the string has rotated through an angle θ, T the tension in the string and u the initial speed of the particle.

✗ $T + mg\sin\theta = mu^2/a$ ✗

is what happens if you try to do the problem without a clear diagram.

[B] ✗ $v^2 = u^2 - 2ga\sin\theta$
$= u^2 - ga$ when $\theta = 30°$.
String slackens when $v = 0$. $\therefore u^2 = ga$✗

It is important to realise that you must have *both* equations (1) and (2) in order to solve the problem.

S4.

The equations of circular motion and conservation of energy give

$$\frac{mv^2}{a} = R + mg\cos\theta$$

and $\frac{1}{2}mu^2 = \frac{1}{2}mv^2 + mga(1 + \cos\theta).$

Hence $v^2 = u^2 - 2ga(1 + \cos\theta)$

and $R = \frac{mu^2}{a} - mg(2 + 3\cos\theta).$

For complete revolutions $v^2 > 0$ when $\theta = 0$. [A]

Hence $u^2 - 2ga\ (1 + 1) > 0$, that is, $u^2 > 4ag$. ■

In the case $u^2 = \frac{9ag}{2}$, the value of R at the highest point, $\theta = 0$, is

$$\frac{9mg}{2} - 5mg = -\frac{mg}{2}.$$

The negative sign indicates that R acts vertically upwards. ■

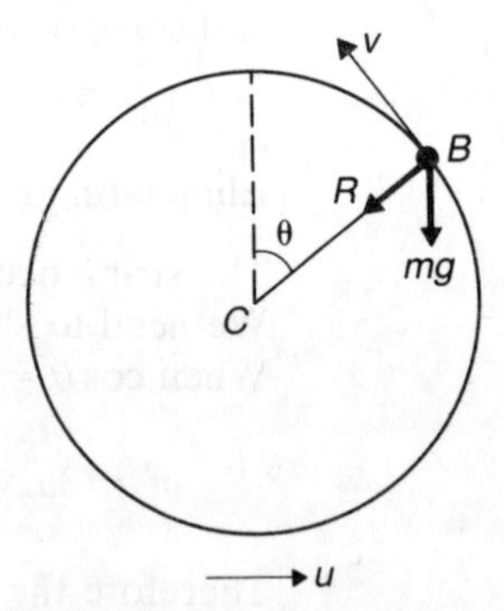

[A] As the bead is threaded on the wire it cannot fall off. Thus for complete revolutions the only condition is that the bead has a non-zero speed at the top of the wire.

When $v = v_0$ at $\theta = 0$ the equation for v gives

$v_0^2 = u^2 - 4ag$, that is, $u^2 = v_0^2 + 4ag$. [B]

The speed of the composite particle after impact is given by V, where

$mu = (m + \frac{1}{2}m)V$. [C]

Thus $V = \frac{2u}{3}$.

For complete revolutions $V^2 > 4ag$. [D]

$$V^2 - 4ag = \frac{4}{9}(v_0^2 + 4ag) - 4ag = \frac{4}{9}(v_0^2 - 5ag).$$

Hence $V^2 > 4ag$ if $v_0^2 - 5ag > 0$.

Therefore the bead will make complete revolutions if $v_0^2 > 5ag$. *25m* ■

[B] At this stage the particular case of $u^2 = \frac{9ag}{2}$ does not apply, and our first result is used to find u, the speed at the lowest point.

[C] Using the principle of conservation of momentum (6.3 F2).

[D] Using the result established in the first part of the question.

S5.

[A]
For a satellite of mass m kg moving with speed V m s^{-1} in a circle of radius 6600 km the equation of motion is

$$\frac{mV^2}{6600 \times 10^3} = F. \quad \text{[B]}$$

Now $F = \frac{\mu}{r^2}$. [C]

When $r = 6400 \times 10^3$, $F = 10m$. [D]

$\therefore \mu = (6400)^2 \times 10^6 \times 10m$.

So when $r = 6600 \times 10^3$, $F = \left(\frac{64}{66}\right)^2 \times 10m$.

Thus $V^2 = \left(\frac{64}{66}\right)^2 \times 10 \times 6600 \times 10^3 = \frac{(64\,000)^2}{66}$.

The time for one complete revolution is T s,

where $T = \frac{2\pi \times 6600 \times 10^3}{V}$. [E]

Thus the number of revolutions in one day is

$$\frac{24 \times 60 \times 60}{T} = \frac{24 \times 3600}{2\pi \times 6600 \times 10^3} \times \frac{64000}{\sqrt{66}} = 16$$

to the nearest revolution. *11m* ■

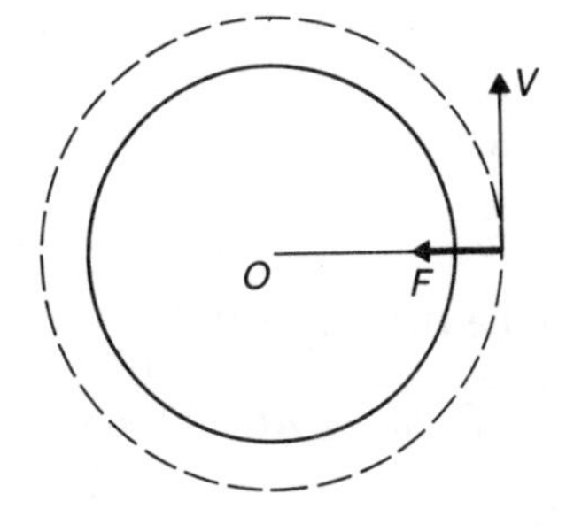

[A] The only force acting on the satellite is F N, the gravitational attraction towards Earth's centre.
[B] Remembering to change the radius into metres instead of kilometres.
[C] The gravitational attraction is an inverse square attraction, so for two particles a distance r apart, $F = \frac{\mu}{r^2}$. This is also true for a particle and a sphere, where the distance r is measured from the centre of the sphere.

[D] $F = mg = 10m$ on Earth's surface.
[E] The time for one period is the distance covered (i.e. the circumference of the circle) divided by the speed.

S6.

$\mathbf{r} = a\cos\omega t\,\mathbf{i} + a\sin\omega t\,\mathbf{j}$. [A]

$\therefore\ \dot{\mathbf{r}} = -\,\omega a\sin\omega t\,\mathbf{i} + \omega a\cos\omega t\,\mathbf{j}$.

$\therefore\ \ddot{\mathbf{r}} = -\,\omega^2 a\cos\omega t\,\mathbf{i} - \omega^2 a\sin\omega t\,\mathbf{j}$ [B]

$\quad = -\,\omega^2(a\cos\omega t\,\mathbf{i} + a\sin\omega t\,\mathbf{j})$.

$\therefore\ \ddot{\mathbf{r}} = -\,\omega^2\mathbf{r}$. ■

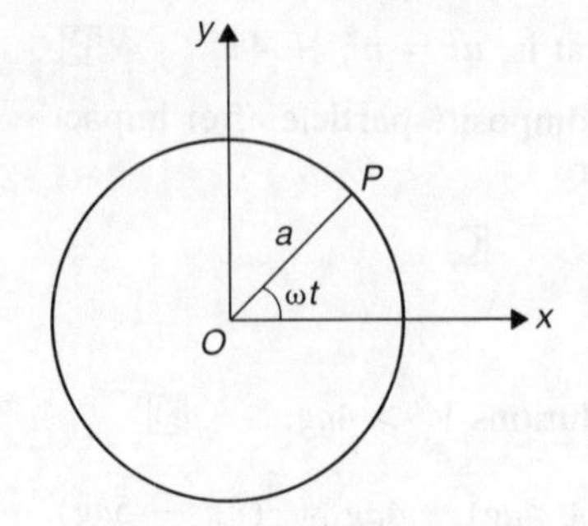

Now $\ddot{\mathbf{r}}$ is the acceleration of P.
$\therefore$ the magnitude of the acceleration is $|\ddot{\mathbf{r}}|$ and

$|\ddot{\mathbf{r}}| = |-\omega^2\mathbf{r}| = \omega^2|\mathbf{r}| = \omega^2 a$.

The direction of the acceleration is the direction of $-\mathbf{r}$.
That is, from P to O. [C]

[D]

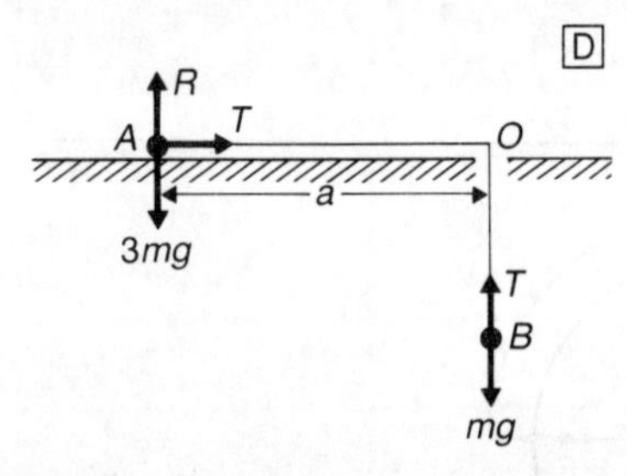

Let T be the tension in the string and R the normal reaction between A and the table. [E]
The acceleration of A is $\omega^2 a$ along AO. [F] Hence

N2 $\rightarrow$: $\quad T = 3ma\omega^2$.

B is permanently stationary, so it has no acceleration. Hence

N2 $\uparrow$: $\quad T - mg = 0$.

Eliminating T, $\quad 3ma\omega^2 = mg$.

$\therefore\ \omega^2 = \dfrac{g}{3a}$, that is $\omega = \sqrt{\dfrac{g}{3a}}$. 9m ■

[A] Although the question asks you to write down the expression for $\mathbf{r}$ it is useful to first draw a diagram showing the position of P at time t, when OP has rotated through an angle ωt. In this position the coordinates of P are $(a\cos\omega t,\ a\sin\omega t)$ and so $\mathbf{r}$ may be written down.

[B] Remembering that $\ddot{\mathbf{r}} = \mathrm{d}\dot{\mathbf{r}}/\mathrm{d}t$.

[C] $\mathbf{r} = \overrightarrow{OP}$. $\quad \therefore\ -\mathbf{r} = -\overrightarrow{OP} = \overrightarrow{PO}$.

[D] Draw a diagram, labelling the important points, and insert the forces acting on each particle.

[E] As the table is smooth the string has the same tension throughout its length and the reaction between A, the $3m$ mass, and the table will be a normal reaction.
[F] Using the first part of the question.

Chapter 12

Coplanar forces acting on a rigid body

12.1 GETTING STARTED

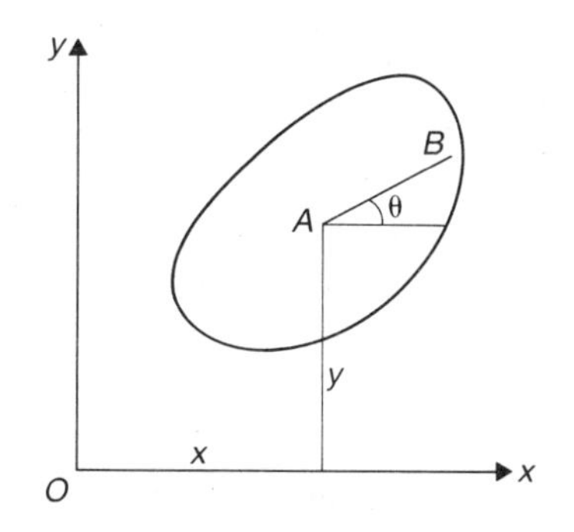

A rigid body is a collection of particles which are constrained in such a way that the distance between each pair of particles remains constant. It is held together by internal forces, which are the forces that the particles of the body exert on each other. Chapters 12–14 deal only with bodies which are restricted to move parallel to a fixed plane. In any such system the position of the body can be specified by means of three numbers. These are the coordinates x and y of a known point A of the body and the angular displacement θ of a known line AB, fixed in the body, from the x-axis. There are other ways of choosing the numbers, but they are all equivalent, and there must always be three of them.

To determine the motion of a rigid body it is not necessary, indeed not possible in practice, to obtain an equation of motion for each of its particles. Only three equations of motion are needed (corresponding to the three quantities x, y and θ), and so only three quantities connected with the forces which act on the body are required. This means that different sets of forces can produce the same effect on the motion of a rigid body; if they do, they are called **equivalent systems of forces**. The motion of the body due to a system of forces forms the subject of rigid-body dynamics, which is beyond the scope of this book. In this chapter only systems of forces are considered, without reference to motion. Indeed, it is conventional even to omit reference to a rigid body at all. For instance, when we say 'This system of forces is in equilibrium' we mean 'A rigid body could be in equilibrium under this system of forces'. Again, the phrase 'force acting at a point' means 'force acting on a particle situated at a certain point of a rigid body'.

12.2 SYSTEMS OF FORCES

ESSENTIAL FACTS

In the following Facts the forces are assumed to be acting on a rigid body at points which all lie in the same plane. The forces all act in directions which lie in this plane. This is what is meant by a **coplanar system of forces**. The statement that two systems of forces, S_1 and S_2, are **equivalent** means that S_2 would produce the **same effect** as S_1 **on the motion of any rigid body**.

F1. Transmissibility of force

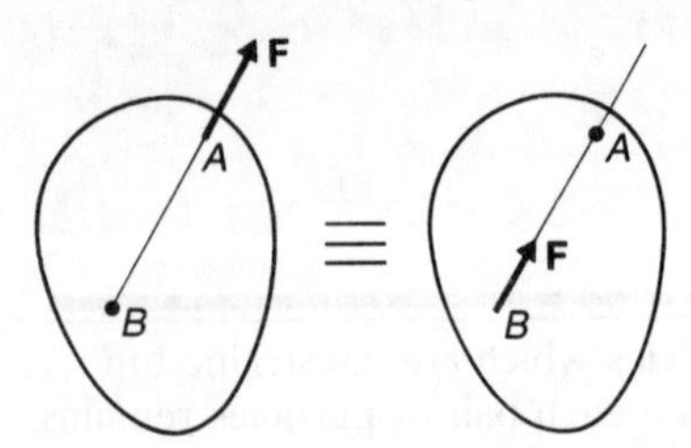

The **line of action** of a force acting at a point A is the line through A parallel to the direction of the force.
A force **F** acting at a point A of a rigid body is equivalent to a force **F** acting at any other point B on its line of action.

This is called the **principle of transmissibility of force**. It means that the effect of a force on the motion of a rigid body is determined if the magnitude and direction and line of action of the force is known. Knowledge is not needed of the particular point on the line of action at which the force is applied.

F2. Composition of forces acting at a point

Forces **P** and **Q** acting at a point A are together equivalent to a single force $\mathbf{R} = \mathbf{P} + \mathbf{Q}$ acting at A.

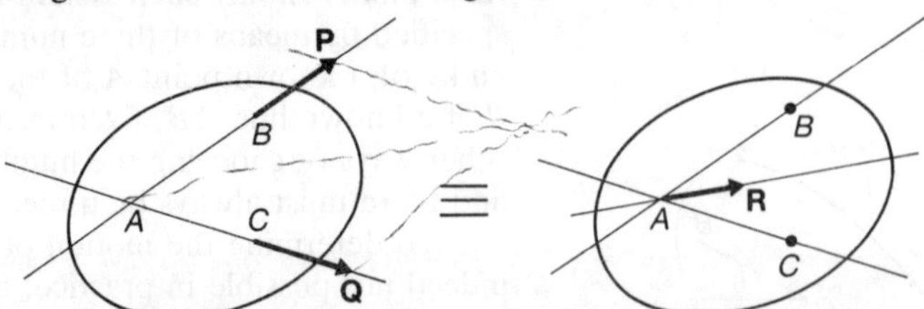

From F1 it follows that two forces with intersecting lines of action are equivalent to their vector sum acting at the point of intersection, as shown in the diagram.

F3. Moment of a force about a point

The moment about a point A of a force of magnitude F acting along a line L is a quantity which has a magnitude and a sense (or 'direction'). Let the perpendicular distance of A from L be h. Then the **magnitude of the moment** is Fh.

The **sense (direction)** of the moment is the sense of rotation about A of a point which moves along L in the direction of the force.
In the accompanying diagram the sense of the moment is clockwise.
It is acceptable to indicate the sense by means of a curved arrow: ↺ for anticlockwise and ↻ for clockwise.
A common convention is that anticlockwise moments are positive

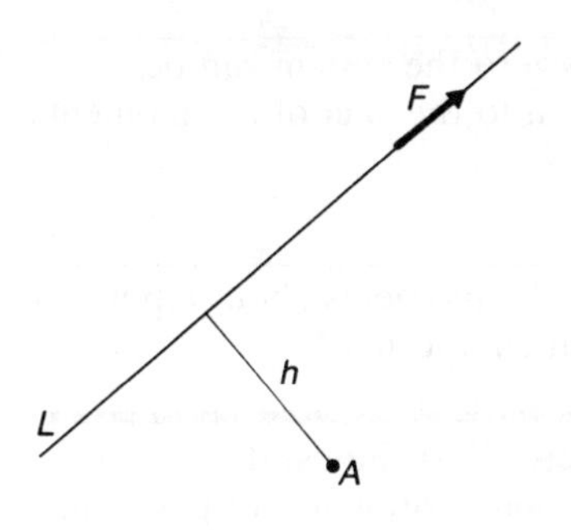

and clockwise moments negative; under this convention the moment about A of the force F acting along the line L in the direction shown is $-Fh$.

The dimension of moment of force is ML^2T^{-2} and its SI unit is the newton-metre (N m).

F4.

Two systems of forces are equivalent if and only if

both the vector sum of the forces,
and the sum of the moments of the forces about a point A,

are the same in each system.

F5. Couple

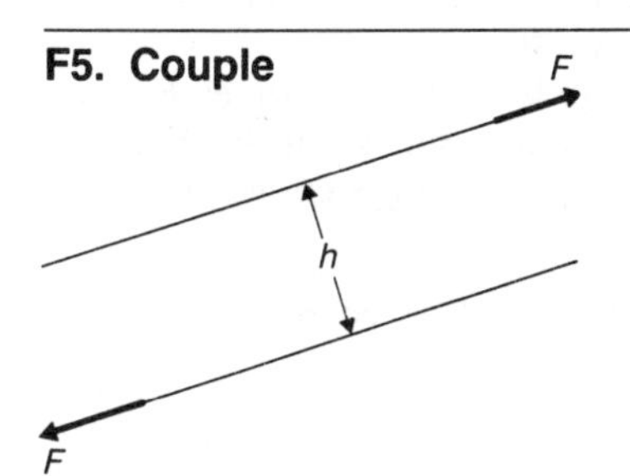

A couple is a system of two forces equal in magnitude and opposite in direction. The sum of the moments of the forces of a couple is the same about any point, and is called the **moment of the couple**.

The moment of the couple shown in the diagram is $\{Fh, \circlearrowright\}$, which may also be written $-Fh$, using the anticlockwise-positive convention.

The word 'couple' is sometimes used as an abbreviation for 'moment of couple'.

F6. Reduction to a force at a point and a couple

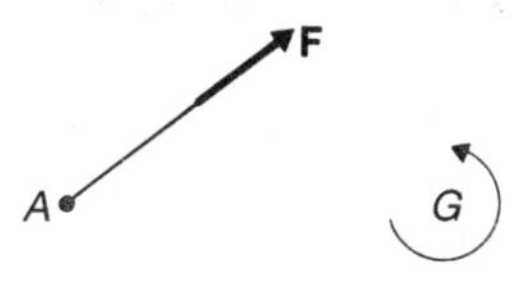

A coplanar system of forces is equivalent to a force acting at a given point together with a couple.

Let the given point be A. Then the system can be reduced to a force $\mathbf{F}$ at A equal to the **vector sum** of all the forces

together with

a couple of moment G equal to the **sum of the moments** about A of all the forces.

F7. Reduction to a resultant force
$\mathbf{F} \neq 0$

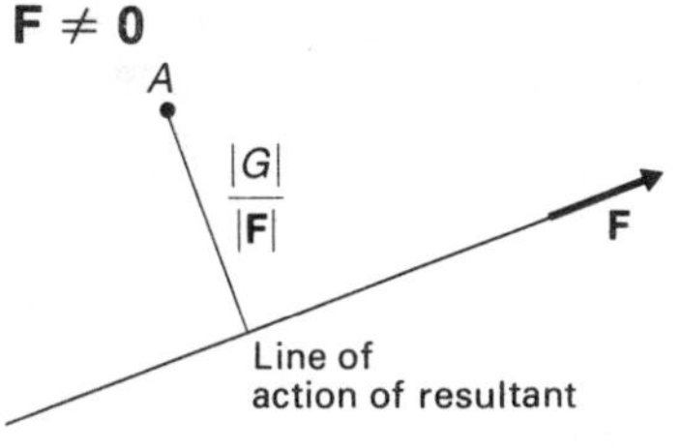

If the vector sum $\mathbf{F}$ of all the forces of a system is not zero, the system is equivalent to a single resultant force $\mathbf{F}$. The resultant acts along a line such that its moment about any point A is equal to the sum, G, of the moments about A of all the forces of the system.

The distance of the line of action from A is $\dfrac{|G|}{|\mathbf{F}|}$.

Moving along the line of action in the direction of $\mathbf{F}$ the sense of rotation about A is the same as the sense of G.

F8. Reduction to a resultant couple
F = 0

If the vector sum **F** of all the forces is zero the system can be reduced to a couple of moment G equal to the sum of the moments of all the forces about any point.

F9. Equilibrium
F = 0, G = 0

If the vector sum **F** and the sum G of the moments about a point A are both zero the system of forces is in equilibrium.

ILLUSTRATIONS

Where numerical data are given in these Illustrations, it is understood that forces are in newtons and coordinates of points in metres.

I1.

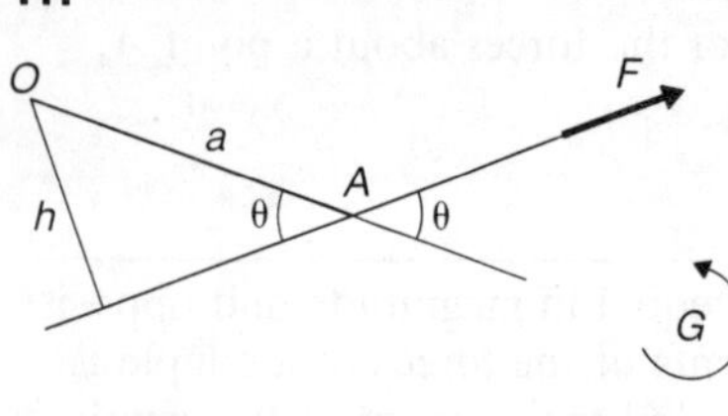

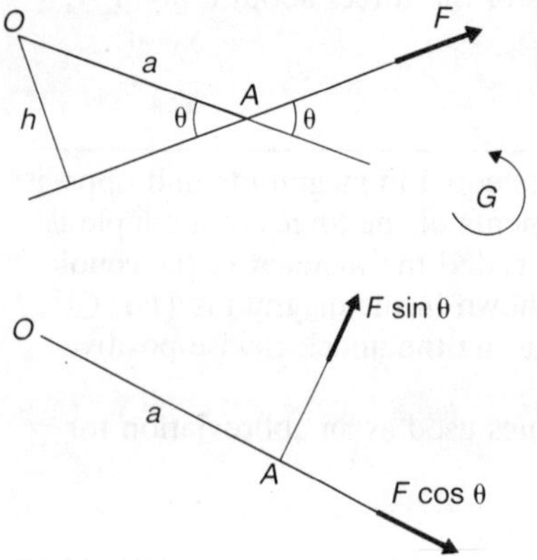

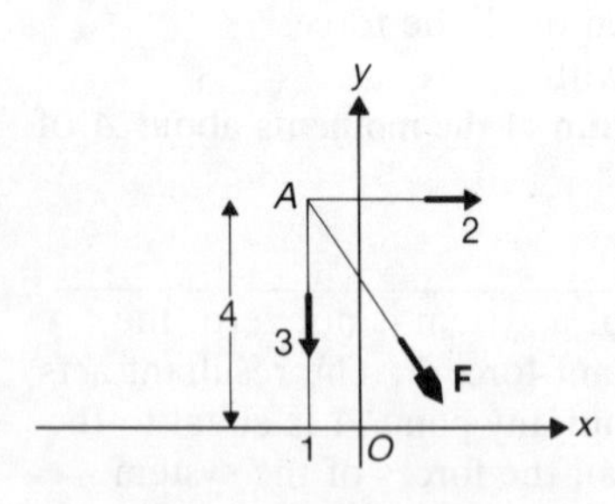

Find the moment of a force of magnitude F, acting at a point A, about a point O, where $OA = a$ and the direction of F is at an angle θ to OA as shown in the diagram. □

Method 1 The perpendicular distance from O to the line of action of the force is $h = a \sin\theta$. Therefore the required moment is $G = Fa \sin\theta$ in the direction ↺ shown. ■

Method 2 Using F2, replace the single force by two forces $F\cos\theta$ and $F\sin\theta$ acting along the lines shown in the diagram. By F4, the sum of the moments about O of this equivalent system is equal to the required moment G.

$$\therefore G = F\cos\theta\,(0) + F\sin\theta(a) = Fa\sin\theta \text{ as before.}$$

I2.

Find the moment about the origin O of a force $\mathbf{F} = 2\mathbf{i} - 3\mathbf{j}$ acting at the point $A(-1, 4)$. □

We do *not* try to use F3 directly, which would mean calculating the perpendicular distance from O to the line of action. Rather we use F2 immediately to replace **F** by a force of magnitude 2 and a force of magnitude 3, as shown in the diagram, and then use F4 as in I1, method 2.

The required moment, in the anticlockwise sense, is G N m, where

$$G = -2 \times 4 + 3 \times 1 = -5.$$

The moment is therefore clockwise and of magnitude 5 N m. ■

I3.

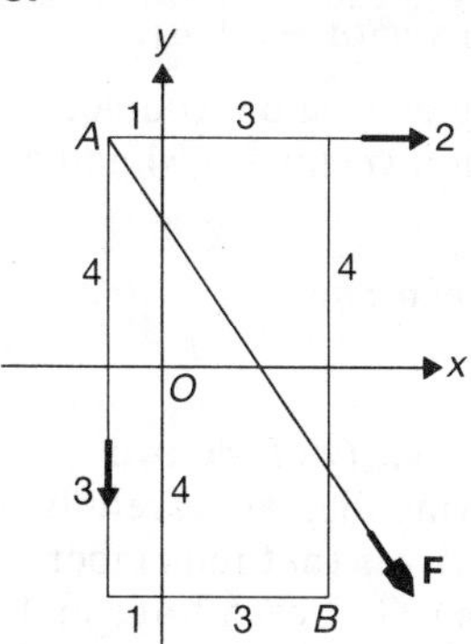

Find the moment about the point $B(3, -4)$ of a force $\mathbf{F} = 2\mathbf{i} - 3\mathbf{j}$ acting at the point $A(-1, 4)$. □

Here we must remember to use the distances from B, not from O, to the lines of action of the forces. The required moment is $-2 \times 8 + 3 \times 4 = -4$; that is, 4 N m clockwise. ■

I4.

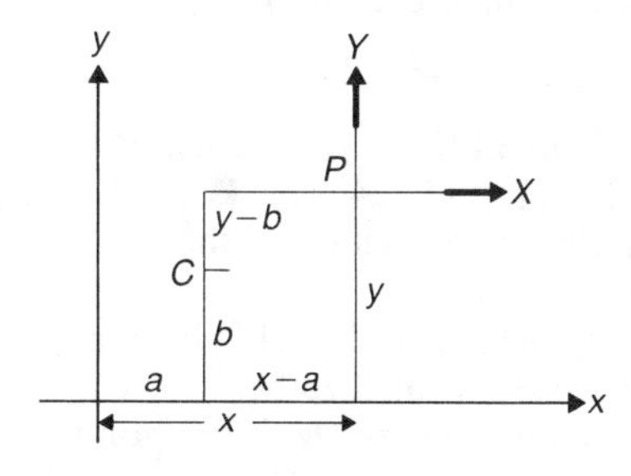

Find the moment (i) about the origin, (ii) about the point $C(a, b)$, of a force $\mathbf{F}$ of magnitude F acting at the point $P(x, y)$ in the direction inclined at an angle θ to Ox. □

Let $\mathbf{F} = X\mathbf{i} + Y\mathbf{j}$, so that $X = F\cos\theta$ and $Y = F\sin\theta$.

(i) From the diagram, in which x and y have been chosen to be positive, we see that the moment about O is

$$G_O = xY - yX \qquad (1)$$

$$= xF\sin\theta - yF\cos\theta.$$

You can verify the formula in the cases when P is in each of the other three quadrants.

(ii) From the diagram we see that the moment about C is

$$G_C = (x - a)Y - (y - b)X \qquad (2)$$

$$= (x - a)F\sin\theta - (y - b)F\cos\theta.$$ ■

I5.

The moment about the origin O of a force $X\mathbf{i} + Y\mathbf{j}$ is G, counting the anticlockwise sense as positive. Find the equation of the line of action of the force. □

Suppose the point $P(x, y)$ lies on the line of action. Then the moment of the force about O is $xY - yX$ (see I4).

Therefore $\qquad Yx - Xy = G.$

Since Y, X and G are given constants, this is the equation of a straight line which is the locus of P. Hence it is the required equation of the line of action. ■

I6. A force $\mathbf{P} = \mathbf{i} - \mathbf{j}$ acts at the point A with position vector $2\mathbf{i} + 3\mathbf{j}$ and a force $\mathbf{Q} = 3\mathbf{i} - 2\mathbf{j}$ at B with position vector $-\mathbf{i} + 4\mathbf{j}$.

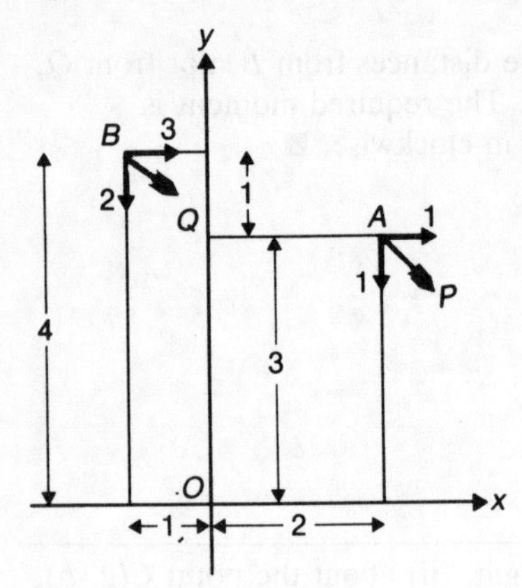

(a) Reduce this system to (i) a force acting at O and a couple, (ii) a force acting at the point C with position vector $\mathbf{i} - 3\mathbf{j}$ and a couple. □

(i) By F6, the force at O is equal to the vector sum

$\mathbf{P} + \mathbf{Q} = 4\mathbf{i} - 3\mathbf{j}$.

The couple G is the sum of the moments about O of the two forces. You can either draw the accompanying diagram carefully or draw a sketch like the one in I4 so as to recover (not remember) equation I4(1). Then find the moment about O of each force and add them up.

$$G = (\text{moment of } \mathbf{P}) + (\text{moment of } \mathbf{Q})$$
$$= -1 \times 2 - 1 \times 3 + -3 \times 4 + 2 \times 1 = -15. \qquad (1)$$

Hence the system is equivalent to a force $(4\mathbf{i} - 3\mathbf{j})$ N acting at O together with a clockwise couple of magnitude 15 N m. ■

(ii) The force acting at C is again $\mathbf{P} + \mathbf{Q}$, the vector sum of the forces of the system, which we have already found.

The couple, H, is the sum of the moments about C, which we shall derive by means of a good diagram in which C is marked.

$$H = -1 \times 6 - 1 \times 1 - 3 \times 7 + 2 \times 2 = -24.$$

Hence the system is equivalent to a force $(4\mathbf{i} - 3\mathbf{j})$ N acting at C together with a clockwise couple of magnitude 24 N m. ■

(b) Find the magnitude, direction and line of action of the resultant of the system of forces. □

By F7, the resultant is a force $\mathbf{P} + \mathbf{Q}$, which is $4\mathbf{i} - 3\mathbf{j}$. This has magnitude 5 N and direction at an angle arc tan 0·75 below Ox. Suppose the line of action of the resultant meets Ox at the point $D(h, 0)$. By F4, the moment of the resultant about O is equal to G, which is -15, from (1).

$$\therefore\ -3h = -15. \quad \therefore\ h = 5.$$

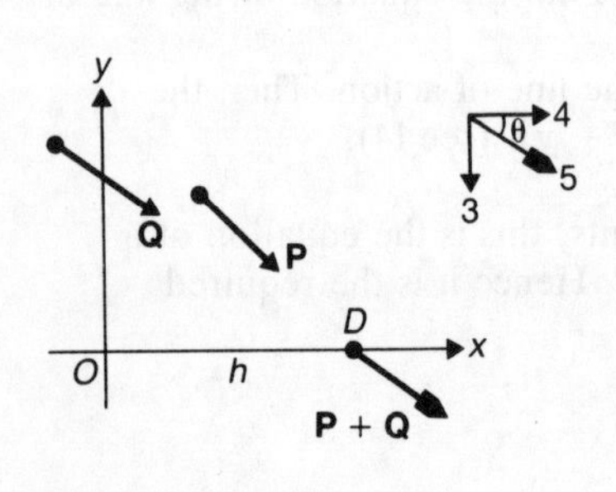

Hence the line of action passes through the point (5, 0) and makes an angle arc tan 0·75 with Ox, as shown in the diagram. ■

Note: In *parametric vector form* (see *PM*13.5F1) the equation of the line of action is $\mathbf{r} = 5\mathbf{i} + \lambda(4\mathbf{i} - 3\mathbf{j})$.

Alternatively the line of action of the resultant can be found by the method of I5. If the resultant $4\mathbf{i} - 3\mathbf{j}$ acts at the point (x, y) its moment about O is $-3x - 4y$. We know that this moment is -15.

Therefore $\qquad 3x + 4y = 15,$

and so this is the equation of the line of action.

I7.

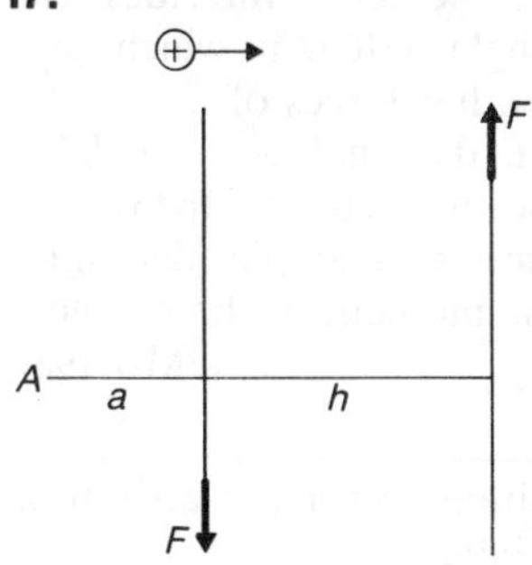

Show that the sum of the moments of the forces of a couple is the same about any point. □

This is asserted in F5; we now prove it. Let the couple consist of two forces of magnitude F acting along parallel lines a distance h apart. Let A be any point. Through A draw a line perpendicular to the lines of action, and choose a positive direction ⊕→ on this line. Let the displacements from A along this line of the intersections with the lines of action be a and $a + h$. Then the sum of the moments of the two forces about A is

$$-Fa + F(a + h) = Fh.$$

This does not depend on the position of A. We conclude that the sum of the moments is always the same, and so Fh can be called just the *moment of the couple*. ■

12.3 EXAMINATION QUESTIONS AND SOLUTIONS

Q1.

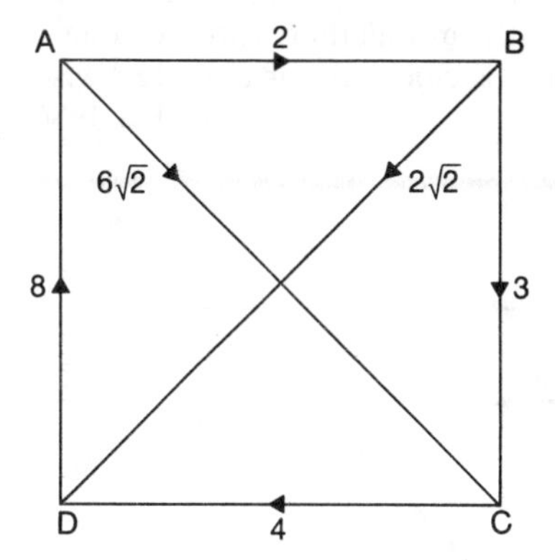

Forces of magnitude and direction given in the diagram act along the sides and diagonals of the square ABCD of side L.

(i) What is the magnitude of their resultant force?

(ii) What angle does the resultant force make with DC?

(iii) At what distance from D does the resultant force meet the line DC produced?

(NI 1983)

Q2.

In the x–y plane, four forces

$$\mathbf{i},\ 2\mathbf{j},\ 2\mathbf{i} \text{ and } \mathbf{i} + \mathbf{j}$$

act at the points whose coordinates are

$$(0, 0),\ (1, 0),\ (1, 1) \text{ and } (0, 1),$$

respectively. This system is equivalent to a single force $\mathbf{F}$. Find the magnitude of $\mathbf{F}$ and the equation of its line of action. (JMB 1986)

Q3. Forces of magnitudes 1, 2, 3 and 5 newtons act along the sides AB, BC, CD and DA, respectively, of a rectangle $ABCD$ in which $AB = 4$ metres and $BC = 3$ metres. Two other forces of magnitudes P and Q newtons act along the diagonals AC and BD respectively. All six forces act in the senses indicated by the order of the letters. Given that the system reduces to a couple, determine the values of P and Q. Determine also the moment of the couple. (JMB 1986)

Q4. Forces $\mathbf{F}_1$ and $\mathbf{F}_2$ act at points whose position vectors, relative to a fixed origin, are $\mathbf{r}_1$ and $\mathbf{r}_2$ respectively, where

$$\mathbf{F}_1 = (4\mathbf{i} + 3\mathbf{j})\text{ N}, \qquad \mathbf{r}_1 = (2\mathbf{i} + \mathbf{j})\text{ m},$$
$$\mathbf{F}_2 = (6\mathbf{i} - 3\mathbf{j})\text{ N}, \qquad \mathbf{r}_2 = (-3\mathbf{j})\text{ m}.$$

Determine the resultant of these forces. Show that the position vector of the point of intersection of the lines of action of the forces $\mathbf{F}_1$ and $\mathbf{F}_2$ is $(-2\mathbf{i} - 2\mathbf{j})$ m and hence write down a vector equation of the line of action of the resultant of these forces.

A third force $\mathbf{F}_3 = (-6\mathbf{i} - 5\mathbf{j})$ N acting at the point with position vector $\mathbf{r}_3 = (-2\mathbf{i} - 2\mathbf{j})$ m is now added to the system. When the system of forces $\mathbf{F}_1$, $\mathbf{F}_2$, $\mathbf{F}_3$, is applied to a lamina it is found that equilibrium can be maintained by applying a force $\mathbf{F}_4$ at the point with position vector $(-\mathbf{i} + 2\mathbf{j})$ m together with a couple G. Find $\mathbf{F}_4$ and G, stating whether G is clockwise or anti-clockwise. (AEB 1986)

SOLUTIONS

S1.

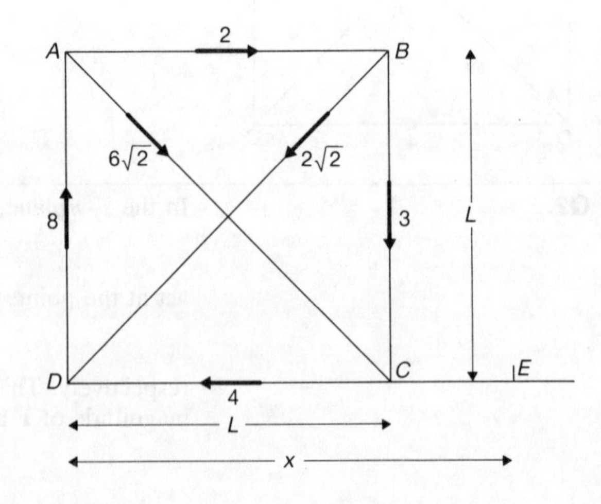

Let the resultant force have components X in the direction DC and Y in the direction DA.
Resolving forces,

$$\rightarrow: \quad X = 2 - 4 + 6\sqrt{2}\left(\frac{1}{\sqrt{2}}\right) - 2\sqrt{2}\left(\frac{1}{\sqrt{2}}\right) = 2.$$

$$\uparrow: \quad Y = 8 - 3 - 6\sqrt{2}\left(\frac{1}{\sqrt{2}}\right) - 2\sqrt{2}\left(\frac{1}{\sqrt{2}}\right) = -3.$$

Therefore the magnitude of the resultant is
$\sqrt{X^2 + Y^2} = \sqrt{2^2 + 3^2} = \sqrt{13}$. ■

The resultant acts in the direction shown in the diagram, in which $\tan\theta = \frac{3}{2}$. [A]
Therefore the angle the resultant makes with DC is arc tan 1·5. ■

Suppose the resultant acts through E on DC produced, where $DE = x$. Then the moment of the resultant about D is equal to the total moment about D.

$$-3x = -2L - 3L - 6\sqrt{2}\left(\frac{L}{\sqrt{2}}\right) = -11L.$$

Hence the distance from D at which the line of action of the resultant force meets DC produced is $DE = x = \dfrac{11L}{3}$. *11m* ■

[A] Draw a separate diagram to show the resultant. If you draw it on the main diagram there is a danger that you might later include it on the wrong side of the moment equation.

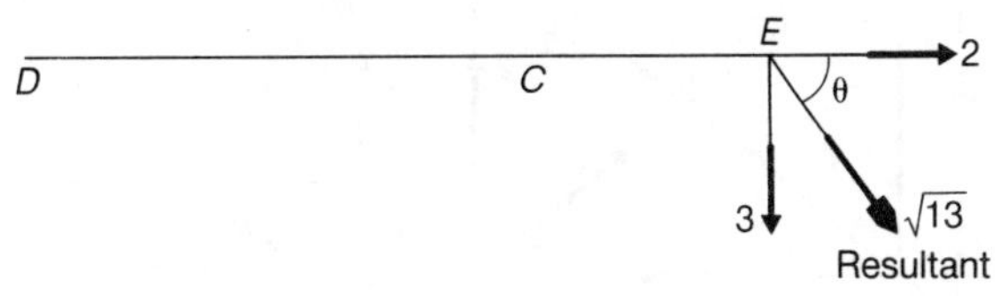

Anticipating part (iii), we have shown the resultant acting at E.

S2.

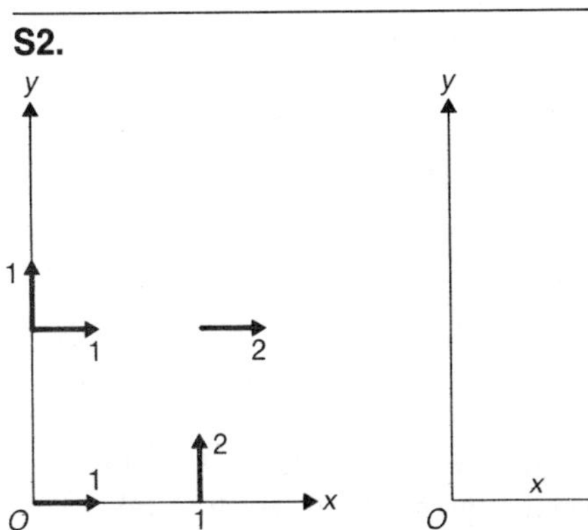

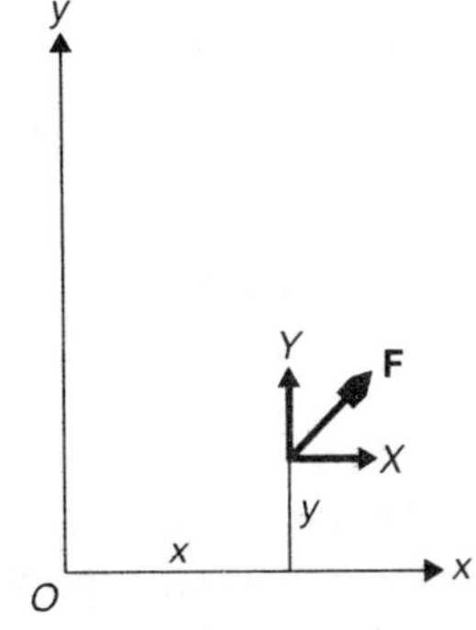

[A]

Let the resultant $\mathbf{F} = X\mathbf{i} + Y\mathbf{j}$, and suppose $\mathbf{F}$ acts at the point $P(x, y)$.

$\rightarrow$: $\quad X = 1 + 2 + 1 = 4.$

$\uparrow$: $\quad Y = 2 + 1 = 3.$

$\therefore$ the magnitude of $\mathbf{F}$ is $\sqrt{X^2 + Y^2} = 5$.
Equate moments about O.

O ⟲: $\quad Yx - Xy = 2 \times 1 - 2 \times 1 - 1 \times 1 = -1.$

$\therefore \quad 3x - 4y = -1.$ $\quad$ (1)[B]

This is the equation of the line of action of the single resultant $\mathbf{F}$. *9m* ■

[A] Draw a diagram showing the components of the forces, and *omitting* **i** and **j** which are not needed in the work and may lead to error if included. Indicate the resultant **F** on a separate diagram to avoid confusion. Then the force systems in the two diagrams are to be equivalent.
[B] Using I5. Alternatively seek the equation in the form $y = mx + c$.
The gradient

$$m = \frac{Y}{X} = \frac{3}{4}.$$

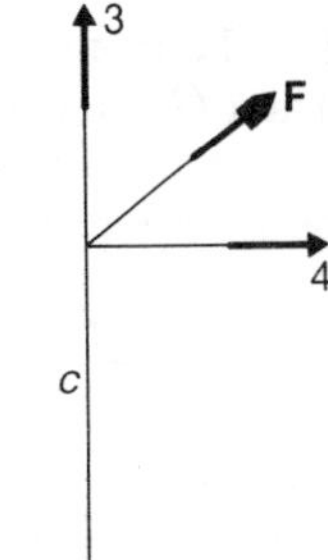

The intercept c is found by equating moments about O.

O ⟲: $\quad -4c = 2 \times 1 - 2 \times 1 - 1 \times 1 = -1.$

$\therefore \quad c = \dfrac{1}{4}.$

$\therefore$ the equation is $y = \dfrac{3}{4}x + \dfrac{1}{4}$, which is the same as (1).

S3.

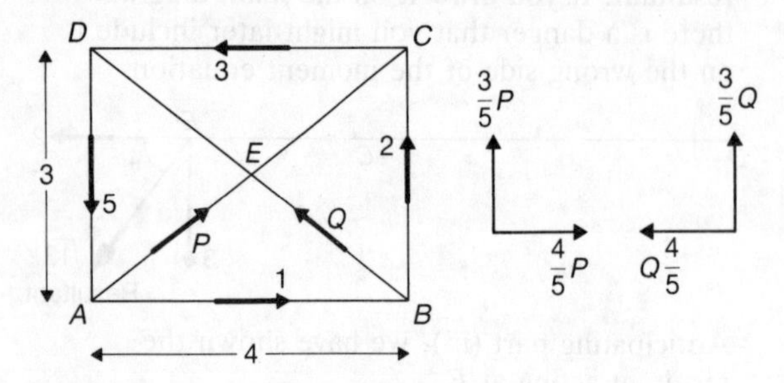

[A] Since the system reduces to a couple, the vector sum of all the forces is zero. [B]

$$\rightarrow: \quad 1 - 3 + \frac{4}{5}P - \frac{4}{5}Q = 0.$$

$$\uparrow: \quad 2 - 5 + \frac{3}{5}P + \frac{3}{5}Q = 0.$$

$$\therefore 4P - 4Q = 10 \quad \text{and} \quad 3P + 3Q = 15.$$

$$\therefore P = \frac{15}{4} \quad \text{and} \quad Q = \frac{5}{4}. \blacksquare$$

Let the moment of the couple be G↺. Then G is the sum of the moments of all the forces about any point. Take moments about the point of intersection, E, of the diagonals.

$$E↺: \quad G = 1 \times \frac{3}{2} + 2 \times 2 + 3 \times \frac{3}{2} + 5 \times 2$$

$$= 20. \quad \text{[C]}$$

$\therefore$ the moment of the couple is anticlockwise in the diagram and its magnitude is 20 N m. [D]

16m ■

[A] You must transfer the information accurately to a diagram. A special hazard in this question is possible confusion between the linear dimensions (4 and 3 metres) and the force magnitudes; keep them well separated.

Also draw diagrams showing the components of P and Q, which you will need. In this problem no trigonometry is needed because you can exploit the (3, 4, 5) triangle.

[B] The case of F8.

[C] In this type of problem, always look for the most convenient point about which to take moments. *Any* point will enable you to reach the result, but here E is especially advantageous because the moment equation involves neither P nor Q. This makes it easier, and also safer, because a mistake made in finding P or Q will not be transmitted to the calculation of G.

[D] Scrupulously state the sense (↺) of the couple, or you may lose marks.

S4.

[A] $\mathbf{F}_1 + \mathbf{F}_2 = 10\mathbf{i}$.

$\therefore$ the resultant is $10\mathbf{i}$ N, acting along a line to be found. ■

The line of action of $\mathbf{F}_1$ passes through the point $\mathbf{r}_1$ and has the direction of $\mathbf{F}_1$, and so its equation may be written

[A] Reading through the question we surmise that the first requirement is just the vector sum $\mathbf{F}_1 + \mathbf{F}_2$. Then there is some vector geometry. When $\mathbf{F}_3$ is added, more vector addition is needed to find $\mathbf{F}_4$. It is only in the last part that we have to take moments, and so it is not until then that a diagram of the force system may be helpful.

$\mathbf{r} = \mathbf{r}_1 + \alpha\mathbf{F}_1 = 2\mathbf{i} + \mathbf{j} + \alpha(4\mathbf{i} + 3\mathbf{j})$. (1) B C

Similarly the line of action of $\mathbf{F}_2$ is represented by the equation

$\mathbf{r} = \mathbf{r}_2 + \beta\mathbf{F}_2 = -3\mathbf{j} + \beta(6\mathbf{i} - 3\mathbf{j})$. (2)

The point of intersection, A, lies on both lines. Substitute $\mathbf{r} = -2\mathbf{i} - 2\mathbf{j}$ in (1), and write down the x and y components of the equation.

$-2 = 2 + 4\alpha$ and $-2 = 1 + 3\alpha$.

Both these equations give $\alpha = -1$. Therefore the point $(-2\mathbf{i} - 2\mathbf{j})$ lies on the line of action of $\mathbf{F}_1$, at the point where $\alpha = -1$.
Substituting $\mathbf{r} = -2\mathbf{i} - 2\mathbf{j}$ in (2) gives

$-2 = 6\beta$ and $-2 = -3 - 3\beta$.

Both equations give $\beta = -\dfrac{1}{3}$.

Hence the position vector of the point of intersection A is $(-2\mathbf{i} - 2\mathbf{j})$ m. D E
The line of action of the resultant passes through A. F
Therefore its equation is

$\mathbf{r} = -2\mathbf{i} - 2\mathbf{j} + t(\mathbf{F}_1 + \mathbf{F}_2)$ (t is a parameter),

$= -2\mathbf{i} - 2\mathbf{j} + 10t\mathbf{i}$. ■

The force $\mathbf{F}_3$ also acts through A. When it is added, the system becomes equivalent to a single force at A equal to

$\mathbf{F}_1 + \mathbf{F}_2 + \mathbf{F}_3 = 10\mathbf{i} - 6\mathbf{i} - 5\mathbf{j} = 4\mathbf{i} - 5\mathbf{j}$.

The force $\mathbf{F}_4$ acts at the point $B(-\mathbf{i} + 2\mathbf{j})$.
We now reduce the system $(\mathbf{F}_1, \mathbf{F}_2, \mathbf{F}_3)$ to a force at B and a couple. The force is $4\mathbf{i} - 5\mathbf{j}$. The couple is the total moment about B, which is the moment about B of $4\mathbf{i} - 5\mathbf{j}$ acting at A. G
The couple is $5 \times 1 + 4 \times 4 = 21$.
The system $(\mathbf{F}_4, G)$ which maintains equilibrium is the exact reverse of the force and couple just found. H

$\therefore \mathbf{F}_4 = -4\mathbf{i} + 5\mathbf{j}$ N

and $G = -21 \circlearrowleft = 21$ N m clockwise. *25m* ■

B Using the parametric vector equation of the line. The position vector $\mathbf{r}$ of any point P on the line is $\mathbf{r} = \overrightarrow{OP} = \overrightarrow{OP_1} + \alpha\mathbf{F}_1$, because $\alpha\mathbf{F}_1$ is a vector in the direction of $\mathbf{F}_1$. As the parameter α varies, P moves along the line.
C Omit the units so that they do not get in the way of the working.
D Alternatively we could have supposed the position vector of the point of intersection to be $p\mathbf{i} + q\mathbf{j}$ and then solved four equations for p, q, α and β. This would take longer and is unnecessary, because we have not been asked to *find* the point. We are only asked to *show* that $-2\mathbf{i} - 2\mathbf{j}$ is on both lines, which is easier.
E Include the unit (m) when presenting the result.
F From F2.
G Using F6. We now need a diagram to help work out the moment.

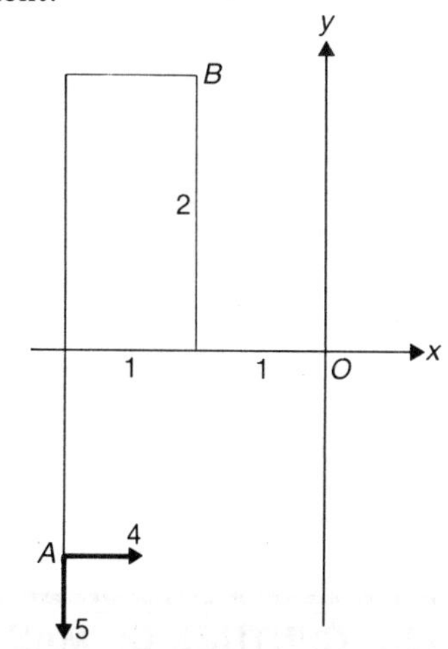

H Questions of this type need to be read very carefully, because it is easy to forget whether $\mathbf{F}_4$ and G are to be found so as to maintain equilibrium, or are to be equivalent to the given system.

Chapter 13 Equilibrium of a rigid body

13.1 GETTING STARTED

In this chapter the general theory of Chapter 12 is applied to problems involving the equilibrium of a particular rigid body which is subject to a system of coplanar forces. If the body is in equilibrium, the system of all the forces which act on it must be a *null* system, that is, equivalent to no force. Conversely, if the body is at rest and a null system of forces is applied to it, it will remain stationary, in equilibrium. In Chapters 13 and 14 the bodies are assumed to be at rest.

The **weight of a body** is the resultant of the system of gravitational forces which act on it. Assuming uniform gravity, as we may in all practical cases involving objects close to a point of the surface of the Earth, the weight of any particle of mass m in the body is $m\mathbf{g}$, where $\mathbf{g}$ is the acceleration due to gravity. The weight of the whole body is thus the resultant of a set of forces which are all in the same direction. This resultant is a single force. The way to determine the line of action of this force is through knowledge of the **centre of mass** of the body. We begin this chapter by studying centres of mass, and then go on to consider equilibrium problems of various types.

13.2 CENTRES OF MASS

ESSENTIAL FACTS

F1. The centre of mass of the system consisting of a particle P_1 with position vector $\mathbf{r}_1$ and mass m_1, and a particle P_2 with position

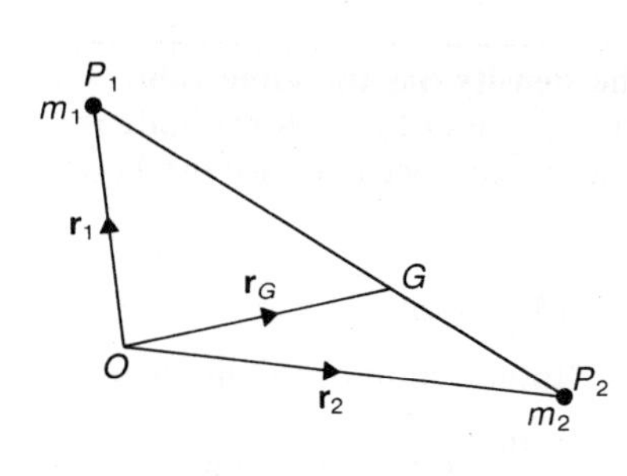

vector $\mathbf{r}_2$ and mass m_2, is the point G with position vector $\mathbf{r}_G$, given by

$$(m_1 + m_2)\mathbf{r}_G = m_1\mathbf{r}_1 + m_2\mathbf{r}_2.$$

The centre of mass G divides the line P_1P_2 in the ratio $m_2:m_1$, that is, in **inverse ratio** to the masses.

F2.

Write all the position vectors in F1 in the form $\mathbf{r} = x\mathbf{i} + y\mathbf{j} + z\mathbf{k}$, and write $m_1 + m_2 = M$. Then the coordinates of G are given by

$$Mx_G = m_1x_1 + m_2x_2, \qquad My_G = m_1y_1 + m_2y_2,$$
$$Mz_G = m_1z_1 + m_2z_2.$$

F3.

Let $P_1, P_2, \ldots, P_i, \ldots, P_n$ be a system of n particles. Let the position vector of the particle P_i be $\mathbf{r}_i = x_i\mathbf{i} + y_i\mathbf{j} + z_i\mathbf{k}$ and let its mass be m_i. Then the position vector $\mathbf{r}_G$ of the mass centre G of the system is given by

$$\mathbf{r}_G\sum_{i=1}^{n} m_i = \sum_{i=1}^{n} m_i\mathbf{r}_i.$$

When expressed in component form, this is often abbreviated:

$$M = \sum m; \quad Mx_G = \sum mx, \quad My_G = \sum my, \quad Mz_G = \sum mz.$$

F4.

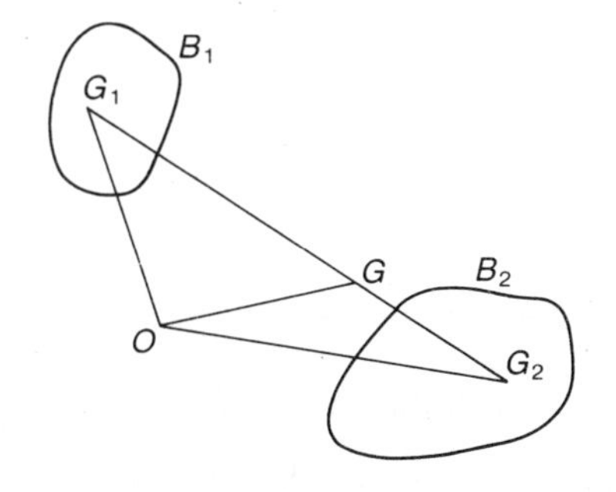

Let B_1 and B_2 be bodies (systems of particles) with masses M_1 and M_2 and centres of mass G_1 and G_2 respectively. Let G be the centre of mass of the system of the two bodies B_1 and B_2. Then

$$(M_1 + M_2)\overrightarrow{OG} = M_1\overrightarrow{OG}_1 + M_2\overrightarrow{OG}_2.$$

Thus G divides the line G_1G_2 in the ratio $M_2:M_1$.

F5.

Suppose a system S possesses a **plane of symmetry**. This means that when the system is reflected in the plane the same distribution of mass is obtained. Then it follows from F3, by choosing the plane $x = 0$ to be the plane of symmetry, that the centre of mass of S lies on the plane of symmetry.

F6. Uniform body

A uniform body is one in which the **density has the same value** everywhere in the body. The centre of mass of a uniform body is the same point as the **centroid** of the region occupied by the body.

F7. Centres of mass of some uniform bodies

Thin straight rod	Mid-point
Triangular lamina	Intersection of the medians
Circular arc, radius r and angle 2θ	$\dfrac{r\sin\theta}{\theta}$ from the centre of the circle
Sector of a circle radius r, angle 2θ	$\dfrac{2r\sin\theta}{3\theta}$ from the centre
Cone of height h	$\dfrac{3h}{4}$ from the vertex
Hemisphere of radius r	$\dfrac{3r}{8}$ from the centre
Hemispherical shell of radius r	$\dfrac{r}{2}$ from the centre

ILLUSTRATIONS

I1.

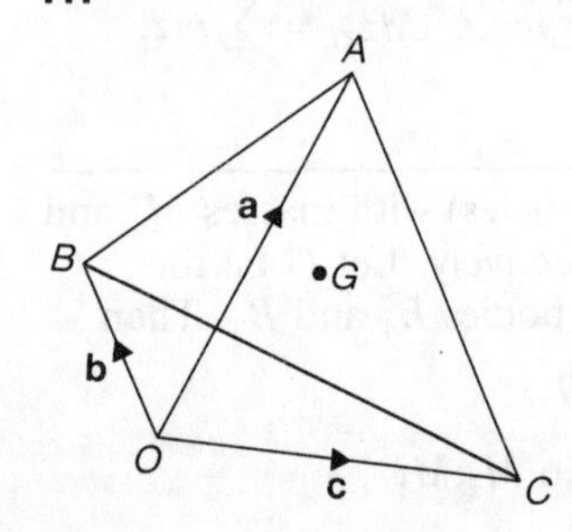

Show that the centre of mass of three identical particles placed at the vertices of a triangle is at the intersection of the medians of the triangle. □

Let the vertices of the triangle be $A(\mathbf{a})$, $B(\mathbf{b})$ and $C(\mathbf{c})$, and let the mass of each particle be m. Then, by F3, the position vector of the mass centre, G, is given by

$$3m\mathbf{r}_G = m\mathbf{a} + m\mathbf{b} + m\mathbf{c}.$$

$$\therefore\ \mathbf{r}_G = \frac{\mathbf{a}+\mathbf{b}+\mathbf{c}}{3}.$$

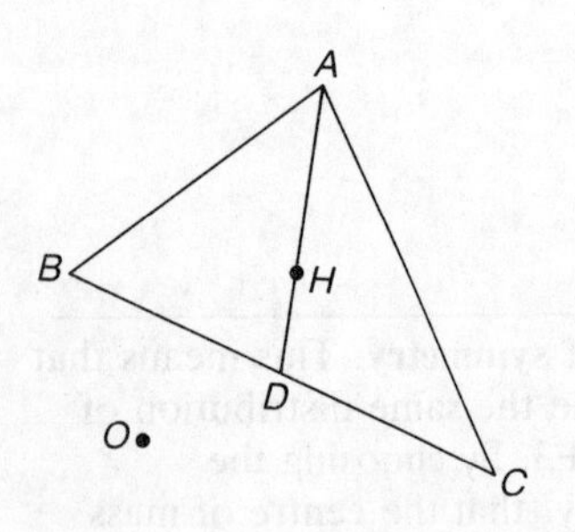

Let D be the mid-point of BC.

Then $\overrightarrow{OD} = \dfrac{\mathbf{b}+\mathbf{c}}{2}$. $\quad\therefore\ \overrightarrow{AD} = \overrightarrow{AO} + \overrightarrow{OD} = -\mathbf{a} + \dfrac{\mathbf{b}+\mathbf{c}}{2}$.

Let H be the point on the median AD such that $AH = 2HD$.

$$\text{Then } \overrightarrow{OH} = \overrightarrow{OA} + \overrightarrow{AH} = \mathbf{a} + \tfrac{2}{3}\overrightarrow{AD} = \mathbf{a} - \frac{2}{3}\mathbf{a} + \frac{2}{3}\,\frac{\mathbf{b}+\mathbf{c}}{2}$$

$$= \frac{\mathbf{a}+\mathbf{b}+\mathbf{c}}{3}.$$

The symmetry of this last expression shows that the other two medians also pass through H. So, having proved that $\mathbf{r}_G = \overrightarrow{OH}$, we have shown that G is at the intersection of the medians. ■

I2. A uniform lamina is bounded by the curve $y = 1 - x^2$ and the line $y = 0$. Find its centre of mass. □

The system is symmetrical about the line Oy. Therefore the centre of mass G lies on Oy.

We have here used F5 (strictly, we could say that there is symmetry about the plane $x = 0$).

Suppose G is the point $(0, y_G)$. To find y_G, we use the formula $My_G = \Sigma my$ from F3. Since the lamina is uniform, the mass per unit area is a constant factor of both sides of the equation, so we omit it from the calculation. So, on the LHS, M may be replaced by the area A of the lamina, and the RHS becomes the **moment of area** with respect to the coordinate y. Thus G is the centroid of the lamina (as defined in *PM*11.8F4), and

$$Ay_G = \int_0^1 \frac{y}{2} y \, dx.$$

This formula is obtained by dividing the region into vertical strips, of which a typical one is shown in the diagram. The area of the strip of width δx and length y is $y\delta x$. The centre of mass of the strip is at its mid-point.

$$A = \int_{-1}^1 y \, dx = \int_{-1}^1 (1 - x^2) \, dx = \left[x - \frac{x^3}{3}\right]_{-1}^1 = \frac{4}{3}.$$

$$Ay_G = \int_{-1}^1 \frac{1 - 2x^2 + x^4}{2} dx = \frac{1}{2}\left[x - \frac{2x^3}{3} + \frac{x^5}{5}\right]_{-1}^1 = \frac{8}{15}.$$

$\therefore \frac{4}{3}x_G = \frac{8}{15}$. $\quad \therefore G$ is the point $(0, \frac{2}{5})$. ■

I3. A uniform lamina occupies the region in which

$$x \geqslant 0, \quad y \geqslant 0 \quad \text{and} \quad y \leqslant 1 - x^2.$$

Find the coordinates of the centre of mass of the lamina. □

Suppose the mass centre is the point $G(x_G, y_G)$. From I2, the area B of the lamina is $\frac{A}{2} = \frac{2}{3}$.

Similarly we see that $y_G = \frac{2}{5}$, as before.

The formula from F3, $\quad Mx_G = \sum mx, \quad$ becomes

$$Bx_G = \int_0^1 xy \, dx = \int_0^1 (x - x^3) \, dx = \left[\frac{x^2}{2} - \frac{x^4}{4}\right]_0^1 = \frac{1}{4}.$$

$\therefore \frac{2}{3}x_G = \frac{1}{4}$. $\quad \therefore G$ is the point $(\frac{3}{8}, \frac{2}{5})$. ■

14. A cone of height k is removed from a uniform solid cone of height h and base radius a by cutting in a plane at a distance k from the vertex. Find the centre of mass of the solid frustum which remains. □

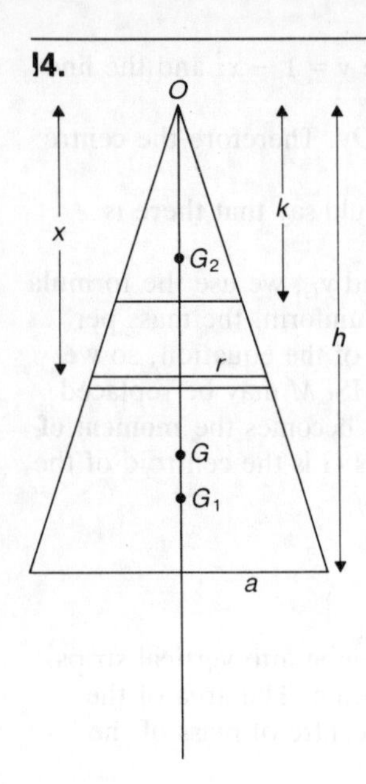

As the frustum is a uniform solid of revolution, every plane containing the axis is a plane of symmetry, and so the mass centre G lies on the axis. Let O be the vertex of the cone and Ox the axis of symmetry.

Method 1: From first principles. Consider a plane section of the cone at a distance x from the vertex. (It is easier to take the vertex of the cone as the origin, rather than the centre of the plane base.)

The plane section is a disc of radius r, where $r = \dfrac{ax}{h}$.

$$\therefore \text{ the volume of the frustum is } V = \int_k^h \pi\left(\frac{ax}{h}\right)^2 \mathrm{d}x = \frac{\pi a^2}{3h^2}(h^3 - k^3).$$

The formula from F3, $\quad Mx_G = \sum mx, \quad$ becomes

$$Vx_G = \int_k^h x\pi r^2\,\mathrm{d}x = \frac{\pi a^2}{h^2}\int_k^h x^3\,\mathrm{d}x = \frac{\pi a^2}{4h^2}(h^4 - k^4).$$

Hence $OG = x_G = \dfrac{3}{4}\left(\dfrac{h^4 - k^4}{h^3 - k^3}\right)$. ■

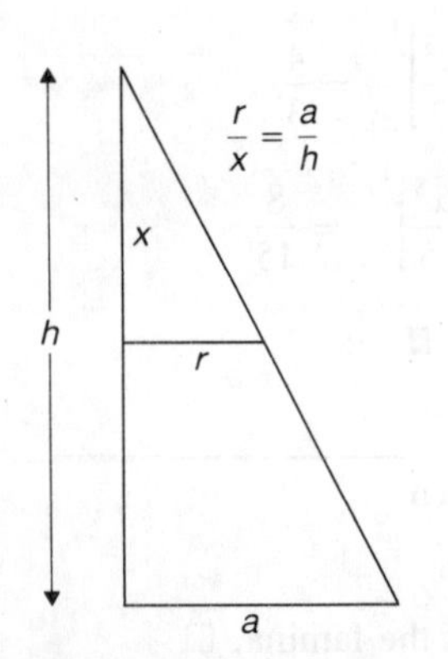

Method 2: Using F4.

	Full Cone =	Removed Cone	+	Frustum
Volume	$V_1 = \frac{1}{3}\pi ha^2$	$V_2 = \frac{1}{3}\pi k\left(\frac{ak}{h}\right)^2$		$V = V_1 - V_2$
Mass centre	$OG_1 = \frac{3}{4}h$	$OG_2 = \frac{3}{4}k$		OG

Now, applying F4, $\quad V_1(OG_1) = V_2(OG_2) + V(OG).$

Removing the common factor $\frac{1}{3}\pi a^2$, we find

$$h\left(\frac{3}{4}h\right) - \frac{k^3}{h^2}\left(\frac{3}{4}k\right) = \left(h - \frac{k^3}{h^2}\right)OG,$$

from which the result follows.

13.3 EQUILIBRIUM CONDITIONS

ESSENTIAL FACTS

In the following Facts a rigid body is subject to a system of coplanar forces. It is assumed that the body is at rest, so that if the system of forces is equivalent to zero force and zero couple the body will be in equilibrium.

F1. Weight

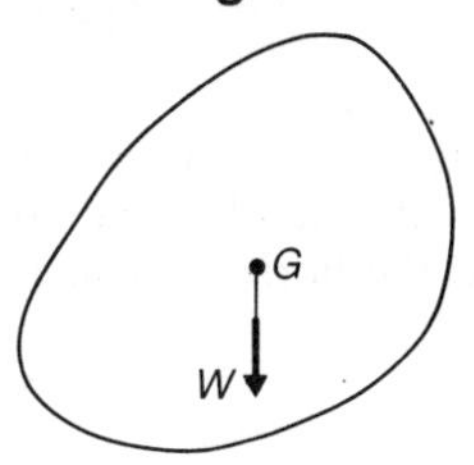

The weight of a rigid body is equivalent to a single force W acting at the centre of mass G of the body.

F2. Resolving and taking moments

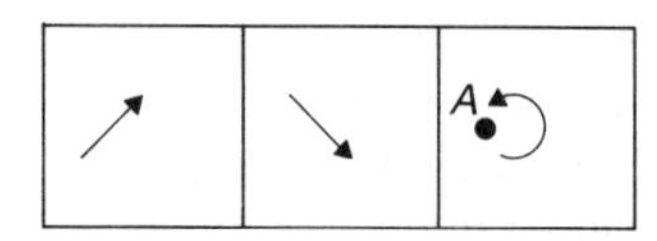

The **general condition for equilibrium** is that

Sum of components of all the forces in **two** directions is zero
and
Sum of moments of all the forces about a point is zero.

F3. Two forces

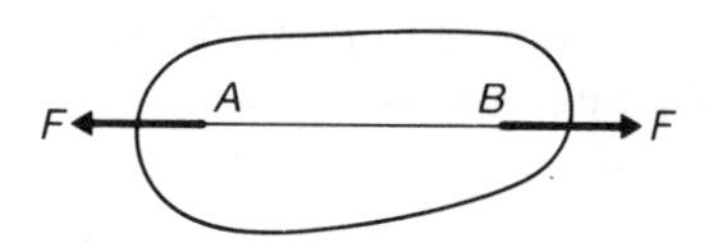

The condition for equilibrium under the action of two forces is that the forces must be equal in magnitude and opposite in direction, and act along the same line.

F4. Three forces

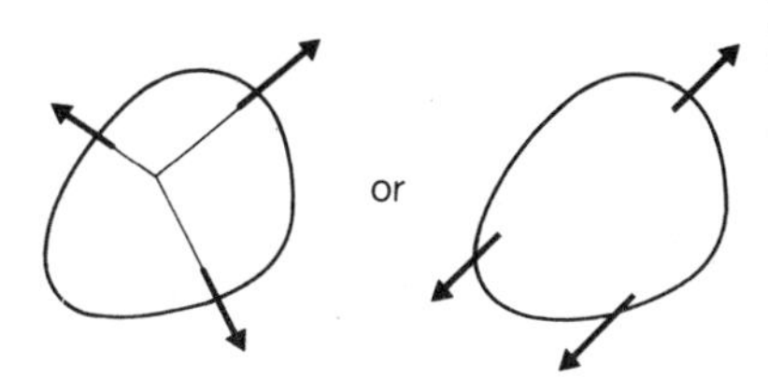

If a body is in equilibrium under the action of three non-zero forces, the lines of action of the forces must be either concurrent or parallel.

F5.

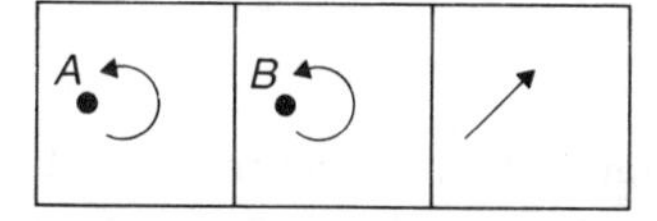

The combination
Zero total moment about each of two points A and B
and
zero total components in a direction *not perpendicular* to AB

ensures the equilibrium of a rigid body.

F6.

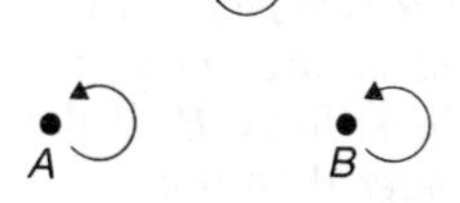

The combination
Zero total moment about each of three
non-collinear points A, B and C

ensures equilibrium.

I1. A uniform rectangular solid S of weight W is formed by joining two cubes, each of edge $2a$. It rests in equilibrium with one of its square faces on a rough plane inclined at an angle θ to the horizontal and with four of its edges horizontal.

(a) Find the greatest possible value, α, of θ.

(b) If $\theta > \alpha$, find
  (i) the least couple,
  (ii) the least horizontal force, applied to the highest point of the solid,
required to maintain equilibrium. □

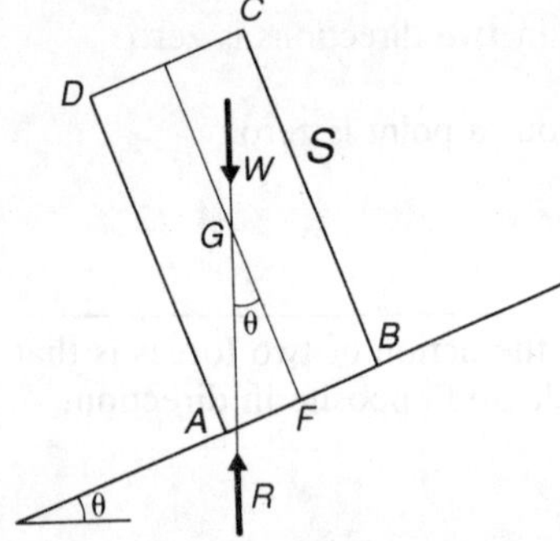

[A] The direction and line of action of R follow from F3. Also $R = W$, but we do not need this here. Equilibrium would be impossible on a smooth inclined plane, because then R would have to act normal to the plane. Note that there is no need to analyse the friction in this type of problem; the plane is assumed to be sufficiently rough to prevent slipping, and no more information is required.

By symmetry, the centre of mass G of the solid is at distance $2a$ from each square face and a from each of the other faces.
The forces acting on S are:

Its weight $\quad W \downarrow$ at G

The reaction of the plane $\quad R \uparrow$ through G [A]

(a) Now R can act only at a point of contact, and so its line of action must meet the plane at a point E which is between A and B in the diagram. Let F be the mid-point of AB. Then $\theta \leqslant \angle AGF = \arctan \frac{1}{2}$. Therefore the greatest possible value of θ is $\alpha = \arctan \frac{1}{2}$. ■

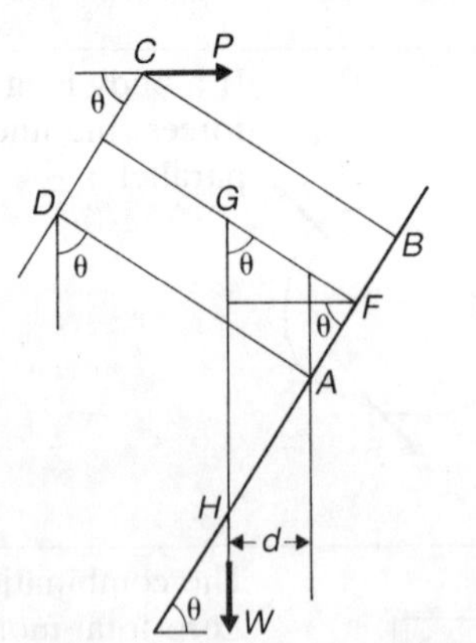

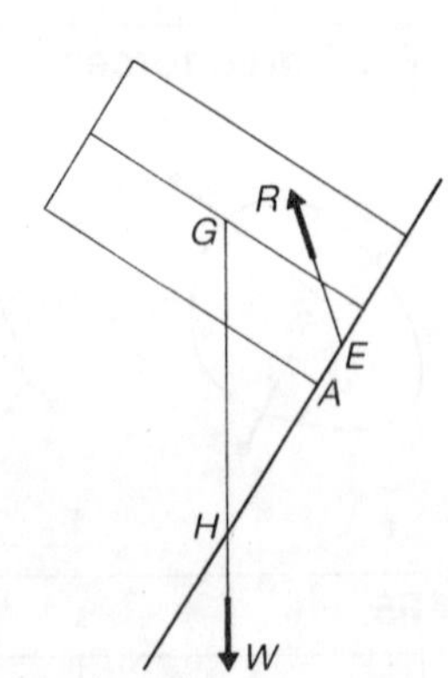

(b) When $\theta > \alpha$ the weight acts along a line that meets the plane at a point H not in AB.
The moment of the weight about A is

$$Wd = W(GF \sin\theta - FA\cos\theta)$$
$$= Wa(2\sin\theta - \cos\theta). \quad (1)$$

(i) The least couple required to maintain equilibrium has a moment equal and opposite to (1), and is therefore

$$G = Wa(2\sin\theta - \cos\theta).$$ [B] ■

[B] We have here assumed that R acts through A. Why? The reason is as follows. Suppose R acts at E in AB. Then the moment of R about A is ↺, as is the moment of W. Therefore the least magnitude of the total moment of W and R is attained when the moment of R is zero, that is, when R acts through A. The same argument applies in (ii) for finding the least force P. If the couple G or the force P is *greater* than the

(ii) The moment about A of the least horizontal force P applied at C to maintain equilibrium must also be equal and opposite to (1).

$$\therefore P(CD\sin\theta + DA\cos\theta) = Wa(2\sin\theta - \cos\theta).$$

$$\therefore P = W\frac{2\sin\theta - \cos\theta}{2\sin\theta + 4\cos\theta}.$$ C ■

required minimum, then the point of application of R will be at a point E such that the moment of R just balances the moments of the other forces.
C In this case it was easy to draw a diagram of the body in the tilted position. When the geometry is less simple, time can be saved in this type of problem by drawing an oblique line representing the vertical, as shown below.

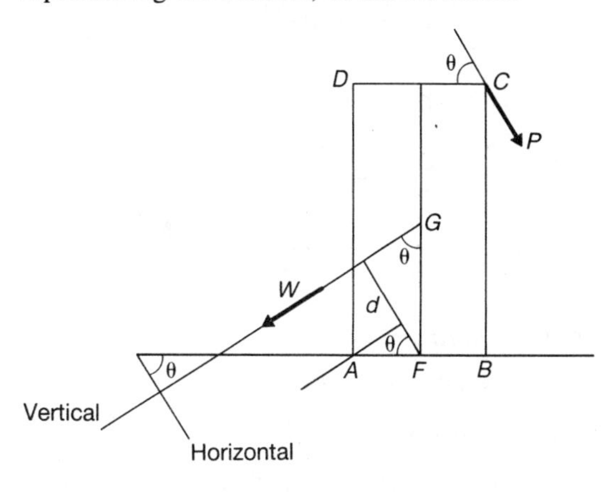

12.

Show that, if a body is in equilibrium under the action of three coplanar non-zero forces, the lines of action of the forces are either concurrent or parallel. (This is a partial statement of F4. It assumes the forces to be coplanar.) □

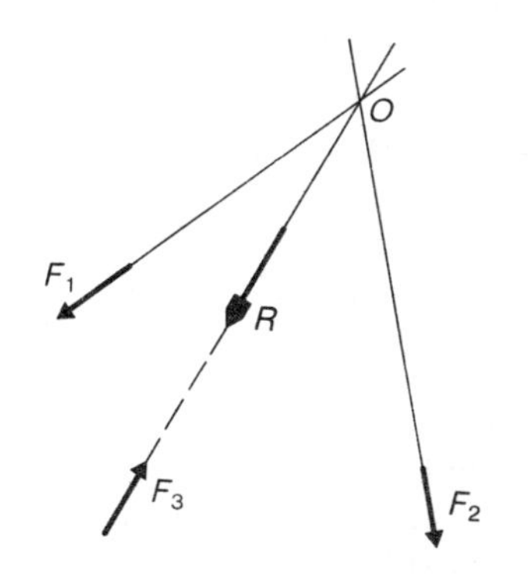

Let the forces be $\mathbf{F}_1$, $\mathbf{F}_2$ and $\mathbf{F}_3$. Suppose the lines of action are not all parallel. Then two of the lines meet in a point O. The forces $\mathbf{F}_1$ and $\mathbf{F}_2$ (say) which act along these lines are equivalent to a single force $\mathbf{R} = \mathbf{F}_1 + \mathbf{F}_2$ acting at O. Since the forces $\mathbf{F}_1$, $\mathbf{F}_2$, $\mathbf{F}_3$ are in equilibrium, then so is the force system consisting of $\mathbf{R}$ and $\mathbf{F}_3$. Therefore $\mathbf{F}_3$ has the same line of action as $\mathbf{R}$. So $\mathbf{F}_3$ also acts through O, and hence the lines of action are concurrent. ■

Therefore **either** the lines are parallel **or** they are concurrent.

Note that we do not need to enquire what happens when the lines are all parallel. We have proved

'If (not all parallel) then (concurrent).'

This is exactly the same proposition as

'Lines must be parallel or concurrent.'

It would be possible to start from the assumption 'not concurrent' and then prove that they are parallel, but it is less easy because there are more cases to consider if that approach is adopted.

I3. A uniform rod AB of length $4a$ and weight W is at rest in a horizontal position. It is supported on a smooth peg C, where $AC = a$, and is in contact at A with a smooth plane ceiling sloping at an angle θ to the horizontal. Equilibrium is maintained by a force F applied at B in a direction parallel to the ceiling. Find F and the reactions at the peg and the ceiling. [A]

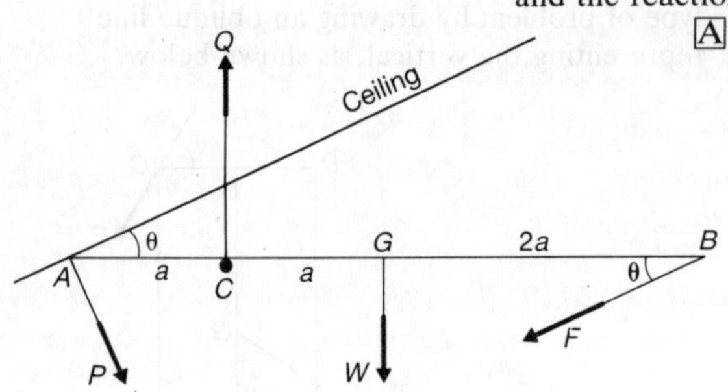

[A] Draw the horizontal rod first. Then mark $W\downarrow$ at the mid point G. Mark $Q\uparrow$, because the smooth peg is 'supporting' the rod. Mark $P\searrow$ because the smooth ceiling exerts an outward force perpendicular to the ceiling. Then mark in F parallel to the plane. F acts in the direction shown because P is the only other force with a horizontal component, and it is pushing to the right.

Resolving horizontally,

$\rightarrow$: $\quad P\sin\theta - F\cos\theta = 0. \quad (1)$

Resolving vertically,

$\uparrow$: $\quad Q = P\cos\theta + W + F\sin\theta. \quad (2)$

Taking moments about A,

$A\circlearrowleft$: $\quad Qa - W(2a) - F\sin\theta\,(4a) = 0. \quad (3)$ [B]

Eliminate Q from (2) and (3).

$P\cos\theta - 3F\sin\theta = W.$

From (1), $\quad F = P\dfrac{\sin\theta}{\cos\theta}.$

$\therefore P\cos\theta - \dfrac{3P\sin^2\theta}{\cos\theta} = W.$

$\therefore P = \dfrac{W\cos\theta}{1-4\sin^2\theta}$, whence $\quad F = \dfrac{W\sin\theta}{1-4\sin^2\theta}.$

Substituting for F in (3), we obtain

$Q = 2W\dfrac{1-2\sin^2\theta}{1-4\sin^2\theta}.$

P, Q and F act in the directions shown. The result is valid for $1 - 4\sin^2\theta > 0$, that is, for $\theta < 30°$. Equilibrium is not possible for $\theta \geqslant 30°$ because the ceiling cannot *pull*, and so P cannot be negative. ■

[B] There are many other equations which can be written, corresponding to resolution in different directions and moments about different points. Some of them may be more convenient than the ones we have chosen. Thus instead of (2), which contains all three unknowns, we could take

$C\circlearrowleft$: $\quad Pa\cos\theta - Wa - 3Fa\sin\theta = 0, \quad (4)$

or $B\circlearrowleft$: $\quad 4Pa\cos\theta - 3Qa + 2Wa = 0. \quad (5)$

But be careful not to have too many equations. Only three are needed, and they must be *independent* equations. Thus (2), (3) and (4) are not independent, but (1), (3) and (4) are (see F5). Similarly (3), (4) and (5) are not independent, because A, C and B are collinear (see F6). Likewise, having resolved in two directions to obtain equations (1) and (2), it is no use resolving parallel to the ceiling,

? $\nearrow$: $\quad Q\sin\theta - W\sin\theta - F = 0.$ **?**

because this equation can already be deduced from (1) and (2).

14.

A uniform rod of weight W is freely pivoted at one end A to a fixed point, and rests in equilibrium with its other end B touching a fixed smooth vertical wall. The horizontal distance from A to the wall is $2a$, and B is at a height $3a$ above the level of A. Find the reactions exerted by the pivot and the wall. □

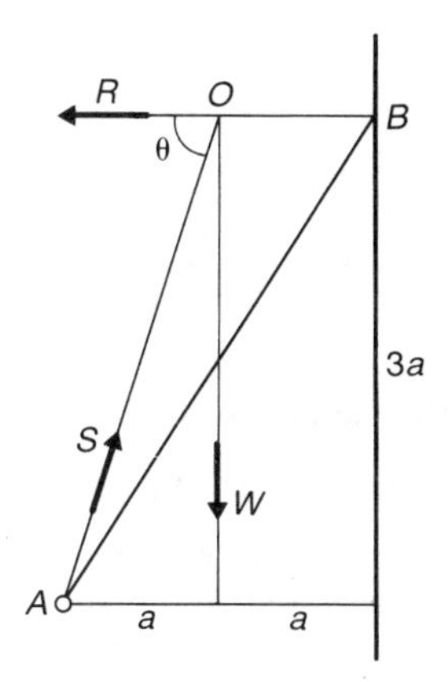

The wall is **smooth**, and so it exerts a force R perpendicular to its surface. The pivot is 'free', which means that its action on the rod consists of a single force S acting through A. (Action at pivots is discussed in detail in 14.2.)

The only other force is the weight $W\downarrow$ acting at the mid-point G of AB. This is a *three-force* problem and lines of the action of W and R are certainly not parallel. By F4, the lines of action of the forces meet in a point O. OG must be vertical, and OB perpendicular to the wall, so the rod and the three forces are all in a vertical plane perpendicular to the wall.

Let S act at an angle θ to the horizontal. Since S acts through O, $\tan\theta = \dfrac{3a}{a} = 3.$

Resolving vertically, $\quad S\sin\theta = W. \quad \therefore S = W\operatorname{cosec}\theta = \frac{1}{3}W\sqrt{10}.$

Resolving horizontally, $\quad S\cos\theta = R. \quad \therefore R = W\cot\theta = \frac{1}{3}W.$ ■

13.4 EXAMINATION QUESTIONS AND SOLUTIONS

Q1.

Show by integration that the centre of mass of a uniform triangular lamina PQR is at a distance $\frac{1}{3}h$ from QR, where h is the length of the altitude through P.

A uniform lamina $ABCD$ is in the form of a trapezium in which $AB = AD = a$, $CD = 2a$ and $\angle BAD = \angle ADC = 90°$. Find the distance of the centre of mass of the lamina from AD and from AB.

The lamina stands with the edge AB on a plane inclined at an angle α to the horizontal with A higher than B. The lamina is in a vertical plane through a line of greatest slope of the plane. If the lamina is on the point of overturning about B, find the value of $\tan\alpha$. (LON 1986)

Q2. The diagram shows a plane section through the axis of a uniform solid cone of height h and base radius $2h$ from which a coaxial cone of height a and base radius h has been removed.

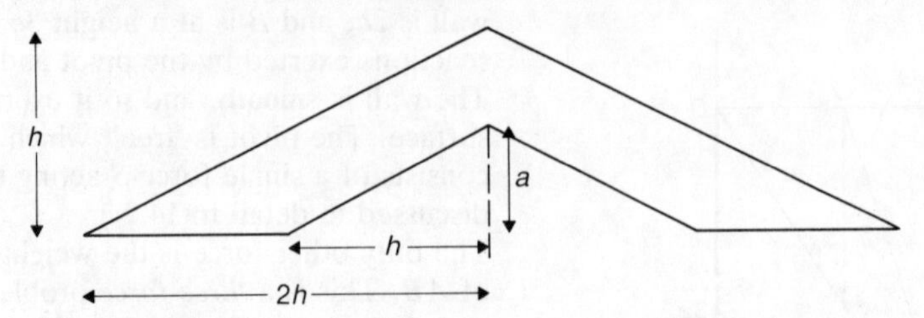

Given that the centre of gravity of the resulting solid S is at a distance $\frac{4h}{15}$ from its plane base, find the possible values of a.

The solid S, which is of weight W, is now placed with its flat base on a plane inclined at an angle θ to the horizontal, the plane being sufficiently rough to prevent slipping. Find the minimum value of $\tan\theta$ such that S cannot stay unsupported on the plane and, for such values of θ, obtain the magnitude of the least force, in terms of W and θ, applied at the vertex V, that will maintain equilibrium. (AEB 1984)

Q3.

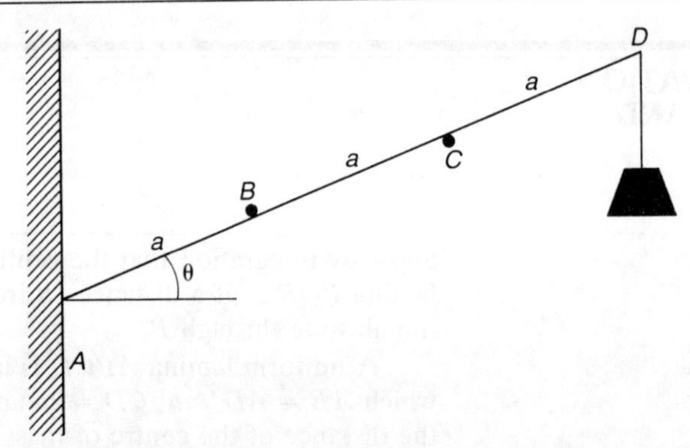

The diagram shows a thin light rod AD resting at an angle θ to the horizontal with its lower end A in contact with a smooth vertical wall which is perpendicular to the vertical plane containing the rod. The rod passes under a peg B and over a peg C, both pegs being fixed, smooth, horizontal and parallel to the wall. A weight W is suspended from D and the rod rests in equilibrium. Given that $AB = BC = CD = a$, find the reactions at B and C and show that

$$3\cos^2\theta \geqslant 2.$$

(JMB 1983)

Q4.

A straight uniform rod $CDEF$, of length $6a$ and mass M, is suspended from a fixed point A by two light inextensible strings AD and AE, each of length $2a$, where D and E are the points of trisection of the rod. A particle of mass m is attached to the rod at a point distant x from the end C and the system hangs in equilibrium. Given that the tension in the string AD is twice the tension in the string AE, show that the rod is inclined to the horizontal at an angle θ, where $\tan\theta = 1/(3\sqrt{3})$.
Hence find, in terms of M, m, a, and g,

(a) the tensions in the strings,
(b) the distance x. (LON 1984)

SOLUTIONS

S1.

Let M be a point on the altitude PN and let $QR = a$. The length of the intercept SMT parallel to QR is $\dfrac{ax}{h}$. Let the mass centre G be at a distance $h - x_G$ from QR. Let the mass per unit area be σ. Then, by equating moments of mass with respect to x,

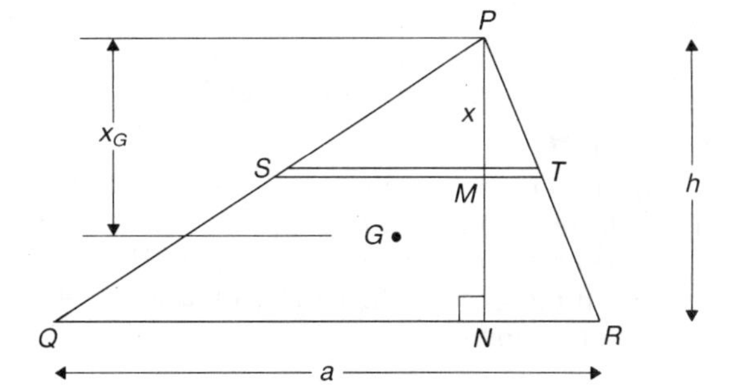

$$\left(\frac{\sigma a h}{2}\right)x_G = \int_0^h x\sigma \frac{ax}{h}\,\mathrm{d}x = \frac{\sigma a}{h}\left[\frac{x^3}{3}\right]_0^h = \frac{\sigma a h^2}{3}.$$

Hence $x_G = \frac{2}{3}h$. It follows that the distance of G from QR is $h - x_G = \frac{1}{3}h$. ■

Choose A as the origin, AB as the x-axis and AD as the y-axis. [A]
Mark E the mid-point of DC. Let the mass of the triangle BEC be m. Then the mass of the square $ABED$ is $2m$.
The mass centres of the square and the triangle are

$$G_1\left(\frac{a}{2}, \frac{a}{2}\right) \quad \text{and} \quad G_2\left(\frac{4a}{3}, \frac{2a}{3}\right).$$

Let the mass centre of the lamina be $G\ (x, y)$.

Then $\quad (2m + m)x = 2m\left(\frac{a}{2}\right) + m\left(\frac{4a}{3}\right)$

and $\quad (2m + m)y = 2m\left(\frac{a}{2}\right) + m\left(\frac{2a}{3}\right)$. [B]

$\therefore x = \frac{7a}{9}$ and $y = \frac{5a}{9}$. [C]

$\therefore G$ is at a distance $\frac{7a}{9}$ from AD and $\frac{5a}{9}$ from AB. ■

If the vertical line through G passes between A and B, the lamina can be supported on AB. [D] It will be on the point of overturning if GB is vertical. If this occurs when AB is on a plane of inclination α, then

the angle $ABG = \frac{\pi}{2} - \alpha$.

$\therefore \tan\alpha = \frac{2a/9}{5a/9} = \frac{2}{5}$. 25m ■

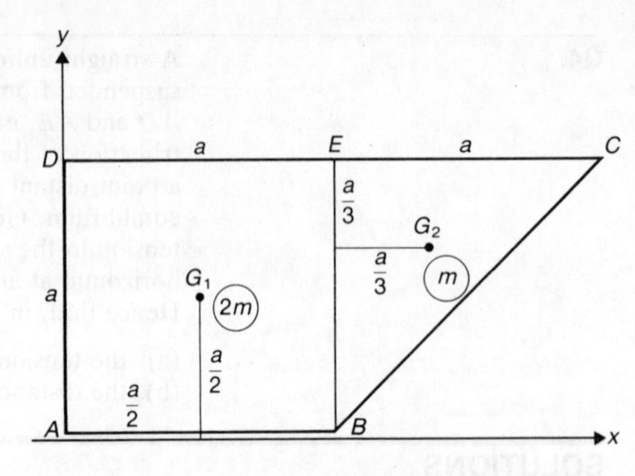

[A] Start by choosing a co-ordinate system: the wording of the question suggests this one.
[B] Using 13.2F4, and expressing the vectors in component form.
[C] *Check*: If you have time, confirm that G divides G_1G_2 in the ratio 1:2.

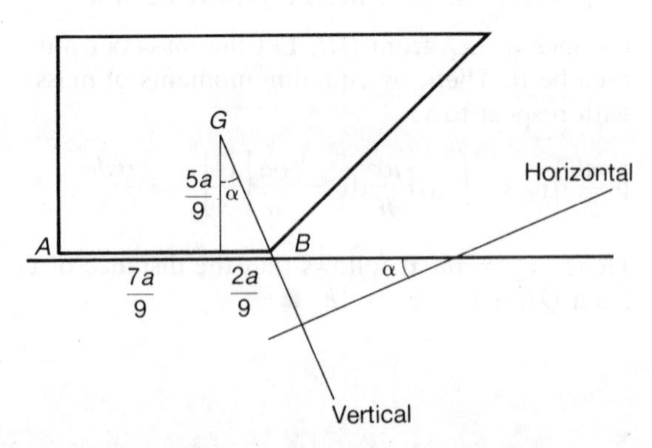

[D] As in 13.3I1C, to answer this part of the question you can avoid spending time (and risking error) on drawing a diagram showing the lamina in a tilted position.

S2.
Let the density of the material be ρ.

Full cone $=$ removed cone $+$ remainder.

Mass $M_1 = \frac{1}{3}\rho\pi(2h)^2h$, $\qquad M_2 = \frac{1}{3}\rho\pi h^2 a$,

$M = M_1 - M_2$

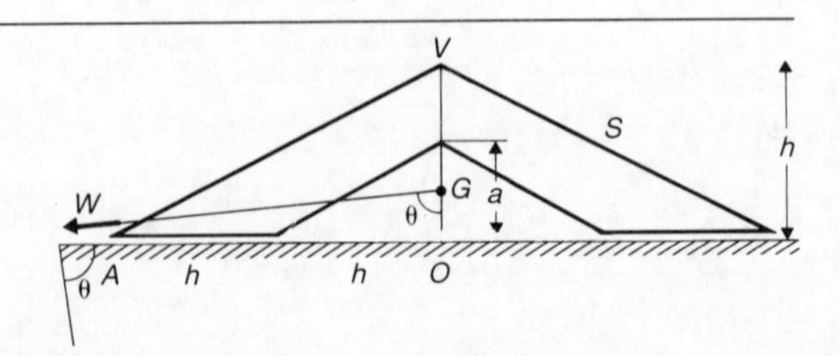

Mass centre

$$OG_1 = \frac{1}{4}h, \quad OG_2 = \frac{1}{4}a, \quad OG = \frac{4h}{15}$$

Equate moments of mass.

$$M_1\,(OG_1) = M_2\,(OG_2) + M\,(OG).$$

$$\therefore 4h\left(\frac{1}{4}h\right) = a\left(\frac{1}{4}a\right) \qquad + \left(4h - a\right)\frac{4h}{15}.$$

$$\therefore 15a^2 - 16ha + 4h^2 = 0.$$

$$\therefore (5a - 2h)(3a - 2h) = 0.$$

$$\therefore a = \frac{2h}{5} \quad \text{or} \quad \frac{2h}{3}. \ \blacksquare$$

When S is placed on the inclined plane, it can stay unsupported if the vertical line through G cuts the plane within the circular plane base of the outer cone. Let A be the lowest point of the circumference of the circle. [A]
Then, if S cannot stay unsupported, $\theta > \angle AGO$.

$$\therefore \tan\theta > \tan\angle AGO = 2h\left(\frac{15}{4h}\right) = \frac{15}{2}.$$

Hence the minimum value of $\tan\theta$ is $\frac{15}{2}$. [B]

For equilibrium, the total moment of force about A must be zero. In our diagram the weight W has an anticlockwise moment about A. [C]
Referring to the diagram, the magnitude of this moment is

$$W\,(AM) = W(GO\sin\theta - OA\cos\theta)$$

$$= W\left(\frac{4h}{15}\sin\theta - 2h\cos\theta\right).$$

The moment about A of the least force applied at V to maintain equilibrium is equal to this in magnitude, and opposite in direction. [D]
Now, in order to obtain this moment with the *least possible* force F acting at V, the direction of F should be perpendicular to VA.
The moment of F about A is then

$$F(VA) = F\sqrt{5}h \ \circlearrowleft.$$

Hence the least force is

$$F = \frac{W}{\sqrt{5}}\left(\frac{4\sin\theta}{15} - 2\cos\theta\right). \qquad \textit{25m} \ \blacksquare$$

[A] As in 13.3I1C, the above diagram can be used for this part of the question.
[B] As can be seen from 13.5I1, the coefficient of friction μ required to prevent slipping is greater than 7·5. This shows that the assertion made by some authors that '$\mu < 1$ always' is certainly not true in all examination questions!

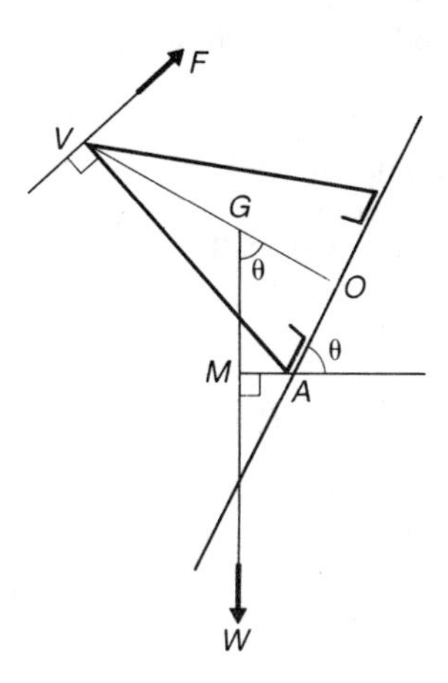

[C] This is an example of a diagram which is difficult to draw if you try to make the scale correspond to the data. The angle of inclination ($\theta > \arctan 7{\cdot}5$) is very steep, and so the lines become confused. In our sketch we have made the cone much narrower so as to obtain a clear diagram. Alternatively, work with an untilted diagram which is easier to draw.

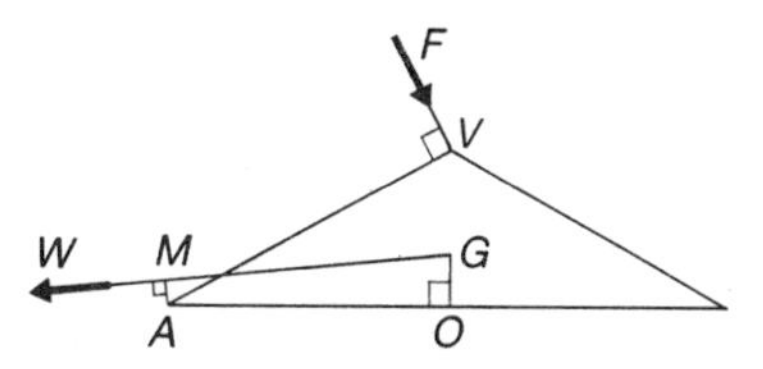

[D] The reaction of the plane may be assumed to act at A when the *least* force is applied, as explained in 13.3I1.

S3.

The forces acting on the rod are:

The reaction of the wall at A,	$P\rightarrow$
The reaction at B, perpendicular to the rod	$Q\searrow$
The reaction at C, perpendicular to the rod	$R\nwarrow$
The weight at D	$W\downarrow$ [A]

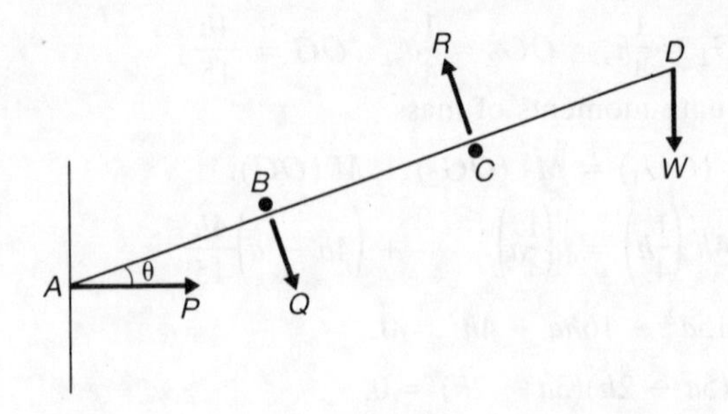

The value of P is not demanded, and so we look for equations which do not contain P. These are found if we resolve vertically, and take moments about A.

$\uparrow$: $\quad -Q\cos\theta + R\cos\theta - W = 0.$

$A\circlearrowleft$: $\quad -Qa + 2Ra - 3Wa\cos\theta = 0.$

We now have two equations for the two unknowns Q and R. Simplifying:

$$-Q + R = W\sec\theta. \quad (1)$$

$$-Q + 2R = 3W\cos\theta.$$

$$\therefore \quad Q = 3W\cos\theta - 2W\sec\theta$$

and $\quad R = 3W\cos\theta - W\sec\theta.$

Now the pegs cannot *pull*, so $Q \geqslant 0$ and $R \geqslant 0$. From (1), $R > Q$. Therefore both conditions are satisfied if $Q \geqslant 0$.

$\therefore 3\cos\theta \geqslant 2\sec\theta. \qquad \therefore 3\cos^2\theta \geqslant 2.$ *13m* ■

[A] Draw your own diagram – don't rely on the one on the question paper. Let the examiner see that you have correctly assigned the forces. This depends on noticing some important words in the question.
Smooth defines the direction of the forces at A, B and C. **Under** and **over**: be careful to mark Q downwards and R upwards.
Light means that there is no weight acting at the mid-point of the rod.
In particular, if you forget the implication of 'smooth' at A, which is a mistake easily made, you will draw a diagram with a vertical force component at A. Then there will be 4 unknowns and only 3 equations, and no progress can be made.

S4.

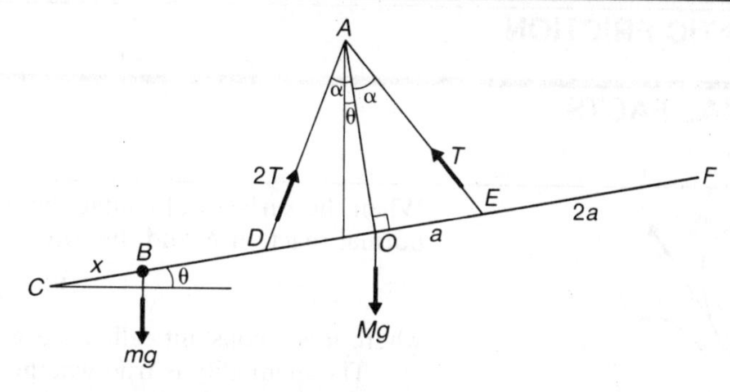

The forces acting on the rod are:

The tension T acting along EA.
The tension $2T$ acting along DA.
The weight Mg at the centre O.
The weight mg at B, where $CB = x$.

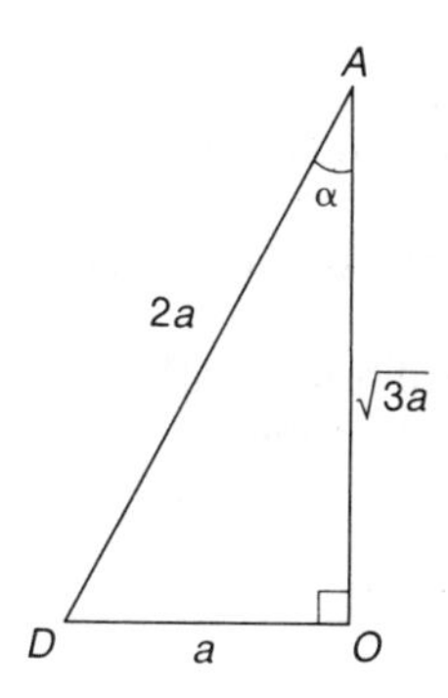

Resolving horizontally,

$$2T\sin(\alpha - \theta) = T\sin(\alpha + \theta).$$

$$\therefore\ 2\sin\alpha\cos\theta - 2\cos\alpha\sin\theta = \sin\alpha\cos\theta + \cos\alpha\sin\theta.$$

$$\therefore\ 3\tan\theta = \tan\alpha. \qquad \therefore\ \tan\theta = \frac{\tan\alpha}{3} = \frac{1}{3\sqrt{3}}.\ ■$$

Resolving vertically,

$$Mg + mg = 2T\cos(\alpha - \theta) + T\cos(\alpha + \theta)$$

$$= T(3\cos\alpha\cos\theta + \sin\alpha\sin\theta)$$

$$= T\left(3\,\frac{\sqrt{3}}{2}\,\frac{3\sqrt{3}}{\sqrt{28}} + \frac{1}{2}\,\frac{1}{\sqrt{28}}\right) = T\frac{28}{2\sqrt{28}} = T\sqrt{7}.$$

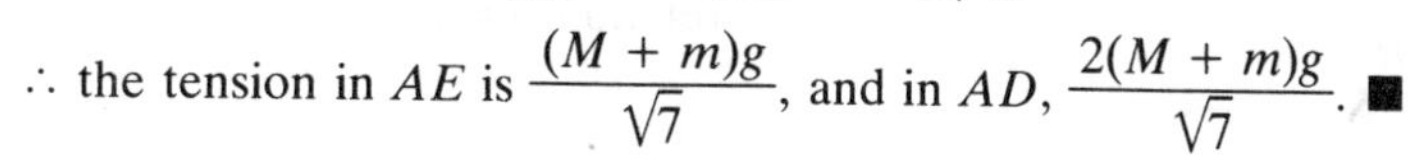

$$\therefore\ \text{the tension in } AE \text{ is } \frac{(M + m)g}{\sqrt{7}}, \text{ and in } AD, \frac{2(M + m)g}{\sqrt{7}}.\ ■$$

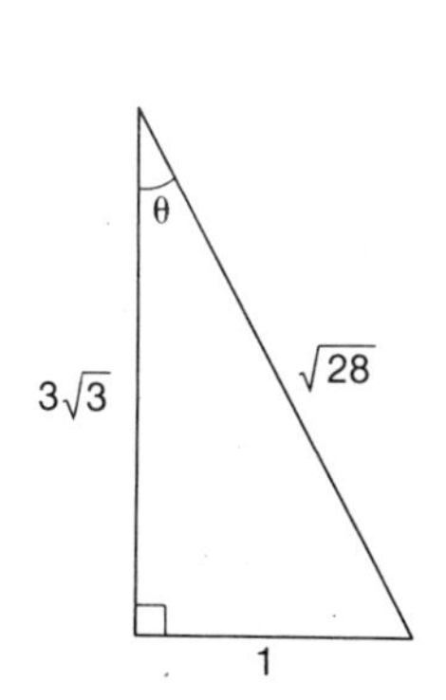

Taking moments about A,

$$mg(BO\cos\theta - OA\sin\theta) = Mg\,OA\sin\theta.$$

$$\therefore\ BO - OA\tan\theta = \frac{M}{m}OA\tan\theta.$$

$$\therefore\ 3a - x = OA\tan\theta\left(1 + \frac{M}{m}\right) = \frac{a\sqrt{3}}{3\sqrt{3}}\left(1 + \frac{M}{m}\right).$$

$$\therefore\ x = 3a - \frac{a}{3}\left(1 + \frac{M}{m}\right) = \frac{a}{3}\left(8 - \frac{M}{m}\right).$$

25m ■

13.5 STATIC FRICTION

ESSENTIAL FACTS

F1.

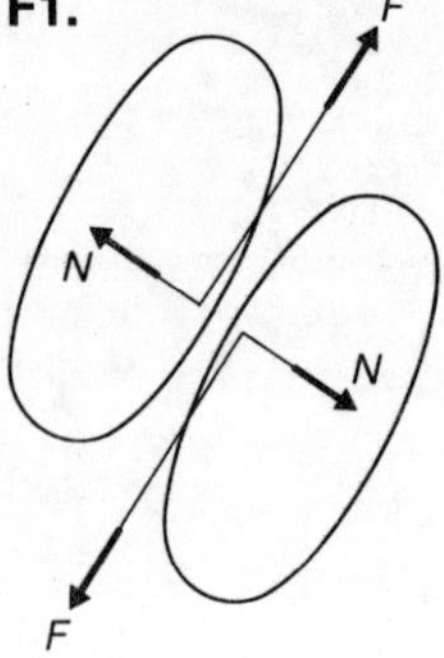

When the surfaces of contact between two bodies are **rough**, the normal reaction **N** and the force of friction **F** satisfy the relation

$$|\mathbf{F}| \leqslant \mu\,|\mathbf{N}|,$$

where μ is a constant called the **coefficient of static friction**.

The inequality is true whether or not there is slipping at the point of contact. In the case of slipping, $|\mathbf{F}| = \mu'|\mathbf{N}|$, where μ' is the coefficient of **dynamic** friction and $\mu' \leqslant \mu$ (see 9.4F3).

F2. Limiting friction

When $|\mathbf{F}| = \mu|\mathbf{N}|$ and there is no slipping, the friction is said to be limiting.

The phrases 'about to slip' and 'on the point of slipping' mean 'in limiting friction'.

A body, or system of bodies, which is in equilibrium with limiting friction at a contact, is called a **system in limiting equilibrium**.

F3. Angle of friction

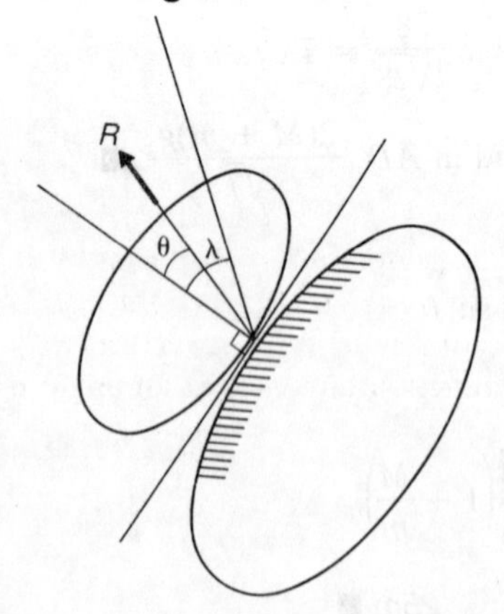

The angle of friction λ is defined by $\lambda = \arctan \mu$.
The acute angle θ between the direction of the **total reaction R** and the normal to the surface at the point of contact satisfies the relation $\theta \leqslant \lambda$.

F4. Coplanar force problems

When the forces acting are coplanar, and represented by scalars associated with directions shown in a diagram, the normal reaction exerted **by** a surface is represented by a **positive** scalar N associated with the direction of the **outward** normal to the surface. The force of friction is represented by a scalar F, which may be positive or negative, associated with a direction along the tangent to the surface. The friction relation (F1) takes the form

$$|F| \leqslant \mu N.$$

Equivalently, $-\mu N \leqslant F \leqslant \mu N$.

ILLUSTRATIONS

In these Illustrations the bodies are in equilibrium under systems of coplanar forces.

I1.

A particle rests in equilibrium on a plane inclined at an angle α to the horizontal. Find the least value required of the coefficient of friction μ between the particle and the plane. □

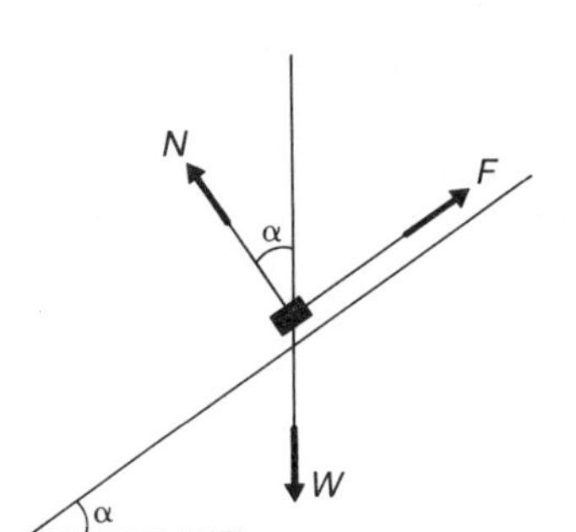

The forces acting on the particle are

Its weight	$\downarrow W$
The normal reaction exerted by the plane along the outward normal	$\nwarrow N$
The force of friction along the line of greatest slope	$\searrow F$

$\rightarrow$: $\quad N\sin\alpha = F\cos\alpha.$

$\therefore \quad F = N\tan\alpha.$ $\quad$ Now $N\tan\alpha$ is positive.

$\therefore$ F is **positive**, and so the friction acts **up** the slope.

Then, by F4, $\quad F \leqslant \mu N.$

$$\therefore \mu \geqslant \frac{F}{N} = \tan\alpha.$$

$\therefore$ the least value of μ is $\tan\alpha$. ■

Alternatively we can use the angle of friction.

The particle is in equilibrium under the action of its weight and the total reaction. Therefore these two forces are equal in magnitude and opposite in direction, and so the total reaction is a vertically upward force R, where $R = W$. The reaction thus acts at an angle α to the normal.

Therefore, by F3, $\quad \alpha \leqslant \lambda.$

$\therefore \tan\alpha \leqslant \tan\lambda = \mu,$ as before.

I2.

A particle of weight W rests on a rough horizontal plane, the coefficent of friction being μ. A force of magnitude P is applied at an angle θ to the plane, and pulling away from the plane.
(a) Find the greatest possible value of P consistent with equilibrium.
(b) Find the least force, pulling away from the plane, at any angle, that will cause the particle to be in limiting equilibrium. □

Analytical method

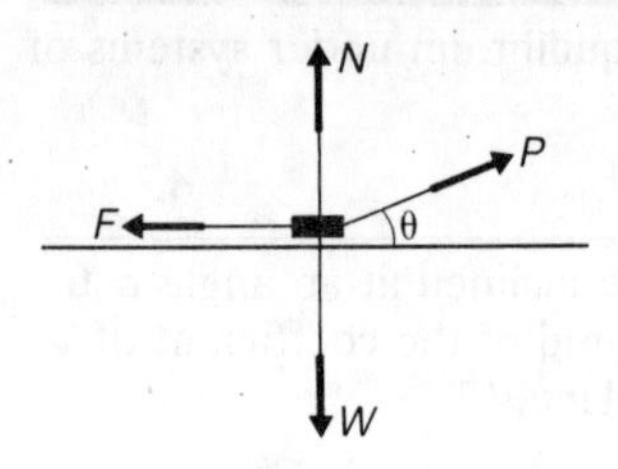

(a) Let the normal reaction be N and the friction F, acting as shown in the diagram

$\rightarrow$: $\quad F = P\cos\theta$

$\uparrow$: $\quad N = W - P\sin\theta.$

Now $P\cos\theta$ is positive and so F is positive.

$\therefore F \leqslant \mu N$ (from F4).

$\therefore P\cos\theta \leqslant \mu W - \mu P\sin\theta.$

$\therefore P(\cos\theta + \mu\sin\theta) \leqslant \mu W.$

$$\therefore P \leqslant \frac{\mu W}{\cos\theta + \mu\sin\theta}. \quad (1)$$ ■

Check: $P\sin\theta \leqslant \dfrac{W}{1 + \mu^{-1}\cot\theta} < W,$

as we expect, for necessarily $N \geqslant 0$.

Note that this question could have been stated, equivalently:
Find the condition that P will cause the particle to move.
(Answer the same, but with $>$ instead of $\leqslant$.)

(b) From (1), we see that the *greatest* value of $\cos\theta + \mu\sin\theta$ is required. By the formula on page 244, this is $\sqrt{1+\mu^2}$.

Therefore the least force required is $\dfrac{\mu W}{\sqrt{1+\mu^2}}$. (3)

■

Geometrical method

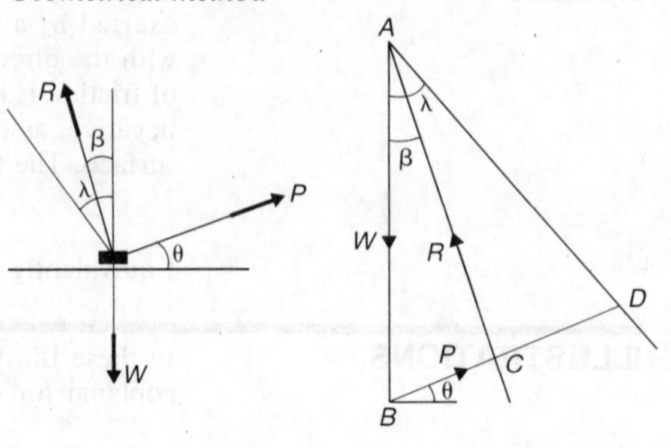

(a) Let the angle of friction be λ. The total reaction R acts at an angle $\beta \leqslant \lambda$ to the normal. Draw a triangle of forces ABC showing the vector sum of the three forces W, P and R to be zero. Then the friction condition (F3) is that the angle $BAC \leqslant \lambda$. Let D be the intersection of BC produced with a line through A at an angle λ to AB. Then if $\overrightarrow{BC}$ represents the force ($P\nearrow$), we have $BC \leqslant BD = \dfrac{AB}{\sin\angle ADB}\sin\lambda$, where

$$\angle ADB = \pi - \lambda - \left(\frac{\pi}{2} - \theta\right) = \frac{\pi}{2} - \lambda + \theta.$$

$$\therefore P \leqslant \frac{W\sin\lambda}{\cos(\lambda - \theta)}, \quad (2)$$

which is the same as (1).

Note: Be careful to put R on the correct side of the vertical.
Equilibrium is impossible under a force system such as

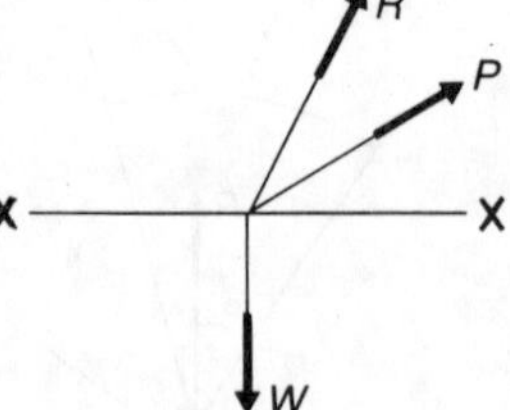

which may lead to an incorrect triangle of forces like

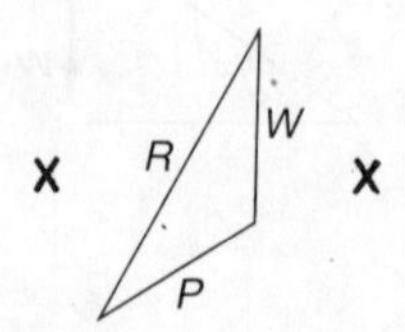

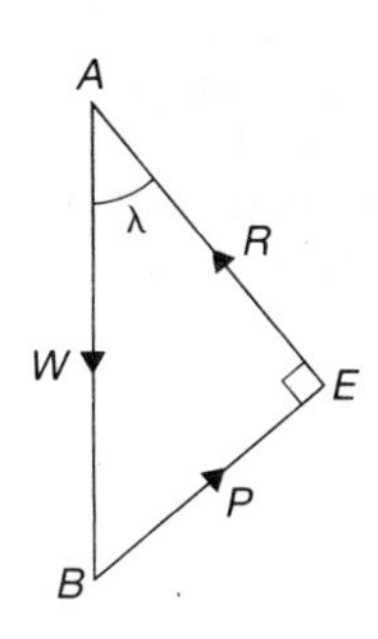

(b) From (2), since $\cos(\lambda - \theta) \leqslant 1$, the least force is $W \sin \lambda$, which is the same as (3).
Note that the least force is represented by BE, where E is the foot of the perpendicular from B to AD, and $BE = AB \sin \lambda$. So the last result can be found directly without using the preceding work.

---

13. A particle of weight W rests on a rough horizontal plane, the coefficient of friction being μ. A force of magnitude P is applied at an angle θ to the plane and pushing into the plane. Find the greatest possible value of P consistent with equilibrium. □
Let the normal reaction N and the friction F act as shown in the diagram.

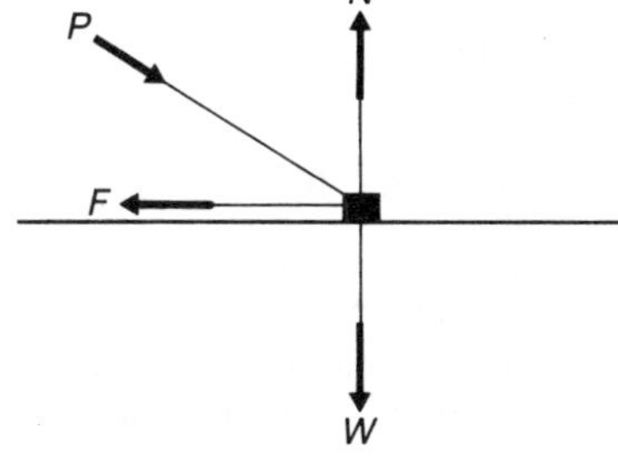

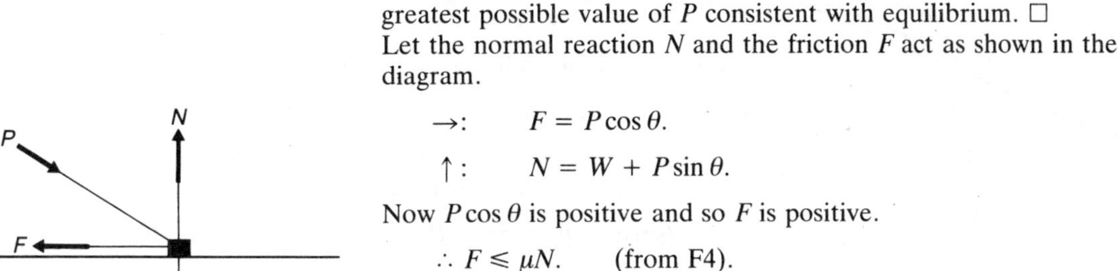

$\rightarrow$: $\quad F = P\cos\theta.$

$\uparrow$: $\quad N = W + P\sin\theta.$

Now $P\cos\theta$ is positive and so F is positive.

$\therefore\ F \leqslant \mu N.$ (from F4).

$\therefore\ P\cos\theta \leqslant \mu W + \mu P \sin\theta.$

$$\therefore\ P(\cos\theta - \mu\sin\theta) \leqslant \mu W. \qquad (1)$$

Now, **pause** to consider the sign of the LHS before attempting to obtain an explicit inequality for P.
(i) In the case when $\cot\theta \leqslant \mu$, the LHS $\leqslant 0$, and so the relation (1) is true for all values of P.
(ii) In the case when $\cot\theta > \mu$, $\cos\theta - \mu\sin\theta$ is positive, and so we may conclude that $P \leqslant \dfrac{\mu W}{\cos\theta - \mu\sin\theta}$.

Thus, for $\cot\theta \leqslant \mu$ there is *no* greatest value of P, and for $\cot\theta > \mu$ the greatest value of P is $\dfrac{\mu W}{\cos\theta - \mu\sin\theta}$. ■

The reason for the two cases (i) and (ii) can be seen by using the angle of friction λ.

Case (i) $\cot\theta \leqslant \mu$ is equivalent to $\dfrac{\pi}{2} - \theta \leqslant \lambda$. The resultant R of P and W acts at an angle β to the vertical, where $\beta < \dfrac{\pi}{2} - \theta$, for all values of P. Hence $\beta \leqslant \lambda$ and so, by F3, equilibrium can be maintained by a *total reaction* of R acting at an angle $\beta(\leqslant \lambda)$ to the normal, for all values of P.

But (case ii), if $\dfrac{\pi}{2} - \theta > \lambda$, then $\beta < \lambda$ only for *some* values of P.

I4. A particle of weight W is maintained in equilibrium on a rough plane inclined at an angle α to the horizontal by a force of magnitude P acting up the line of greatest slope. The coefficient of friction is μ, and $\mu < \tan\alpha$. Find the range of values that P can take. □

Let the normal reaction be N and the force of friction F.

Resolving up the slope: $F = W\sin\alpha - P$

Resolving perpendicular to the slope: $N = W\cos\alpha$

Now we do not know whether $W\sin\alpha - P$ is positive or negative. We therefore write the friction relation (F4) in the form

$$-\mu N \leqslant F \leqslant \mu N.$$

$$\therefore -\mu W\cos\alpha \leqslant W\sin\alpha - P \leqslant \mu W\cos\alpha.$$

From the LH inequality, $P \leqslant W(\sin\alpha + \mu\cos\alpha)$.

From the RH inequality, $P \geqslant W(\sin\alpha - \mu\cos\alpha)$.

Now $\sin\alpha - \mu\cos\alpha > 0$ because $\mu < \tan\alpha$.

$\therefore$ the range of values that P can take is given by

$$W(\sin\alpha - \mu\cos\alpha) \leqslant P \leqslant W(\sin\alpha + \mu\cos\alpha). \blacksquare$$

Note: In the case when $\mu \geqslant \tan\alpha$, $\sin\alpha - \mu\cos\alpha \leqslant 0$, so that P could be zero, that is, the particle could rest on the plane without the help of the force P. Indeed, it would stay in equilibrium even if subject to a force *down* the plane, of magnitude no greater than $W|\sin\alpha - \mu\cos\alpha|$.

I5. Sliding and toppling

A uniform cylinder of radius r and height h rests with its plane base on a rough horizontal plane, the coefficient of friction being μ. The inclination of the plane to the horizontal is slowly increased. Find the range of values of μ, in terms of r and h, for which equilibrium will be broken by the cylinder toppling over. □

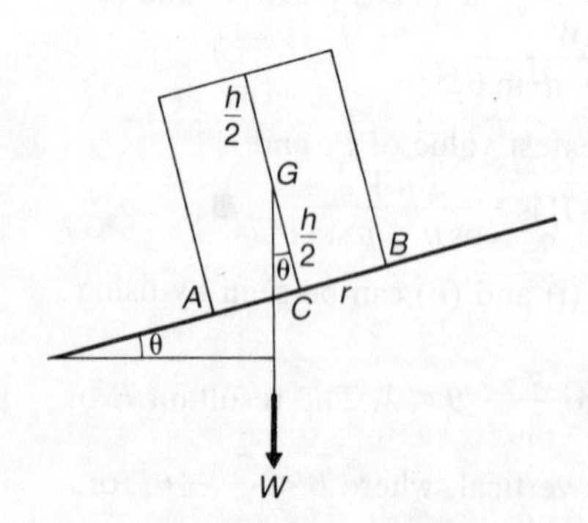

No toppling: The cylinder can be in equilibrium only if the vertical line through the centre of mass G passes through the plane base, that is, between A and B in the diagram (as in 13.3I1). Let C be the centre of the plane base. Then this necessary condition for equilibrium is given by

$$\tan\theta \leqslant \frac{CA}{GC} = \frac{r}{h/2} = \frac{2r}{h}. \qquad (1)$$

No sliding: We know from the argument in I1 that

$$\tan\theta \leqslant \mu \qquad (2)$$

is also a necessary condition for equilibrium.

When $\theta = 0$, conditions (1) and (2) are both satisfied. As θ increases equilibrium will continue until one or other of the

inequalities ceases to be true.
If equilibrium is broken by toppling, (1) is the first relation to fail.

$$\therefore \text{ the } \textit{toppling first} \text{ condition is } \mu > \frac{2r}{h},$$

and this inequality defines the required range of values of μ. ■

Similarly, the sliding first condition is $\mu < \dfrac{2r}{h}$.

I6.

A uniform ladder of weight W stands on rough ground and leans against a smooth wall. A man of weight $4W$ stands three-quarters of the way up the ladder. The coefficient of friction between the ladder and the ground is $\frac{1}{5}$. Find the greatest value of θ, the inclination of the ladder to the vertical, which is consistent with equilibrium. □

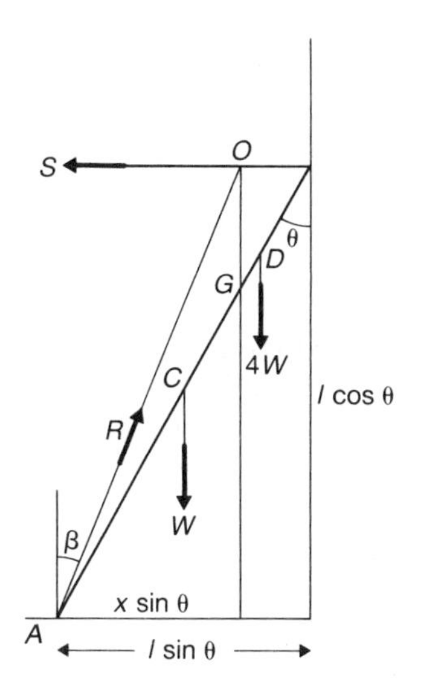

Let the length of the ladder AB be l, and suppose the centre of mass of the (man + ladder) is the point G. The system is in equilibrium under the action of three forces.

Its total weight	$5W\downarrow$ through G
The normal reaction of the smooth wall	$S\leftarrow$ at B
The total reaction of the ground	$R\nearrow$ at A

The lines of action of the three forces are concurrent (13.3F4). They meet at O, the intersection of the vertical line through G with the horizontal line through B. Let $AG = x$.

Then $\tan\beta = \dfrac{x\sin\theta}{l\cos\theta} = \dfrac{x}{l}\tan\theta.$

Let the angle of friction be λ.

By F3, $\beta \leqslant \lambda$. $\quad\therefore \dfrac{x}{l}\tan\theta \leqslant \tan\lambda. \quad \therefore \tan\theta \leqslant \dfrac{1}{5}\dfrac{l}{x}.$

Let C be the mid-point of the ladder and let D be the man.
Now, by 13.2F4, G divides CD in the ratio 4:1.

$$\therefore x = AG = \frac{l}{2} + \frac{4}{5}\frac{l}{4} = \frac{7l}{10}.$$

$$\therefore \tan\theta \leqslant \frac{2}{7}. \qquad \therefore \text{ the maximum inclination is } \arctan\frac{2}{7}. \ ■$$

I7.

A particle P of weight W is maintained in equilibrium on a rough plane inclined at an angle α to the horizontal by a force of magnitude Q. The force acts in the vertical plane containing the line of greatest slope through P, and at an angle θ above the upward direction of this line. The angle of friction is λ ($<\alpha$). Find Q

(i) when P is about to slip down the plane,
(ii) when P is about to slip up the plane.□

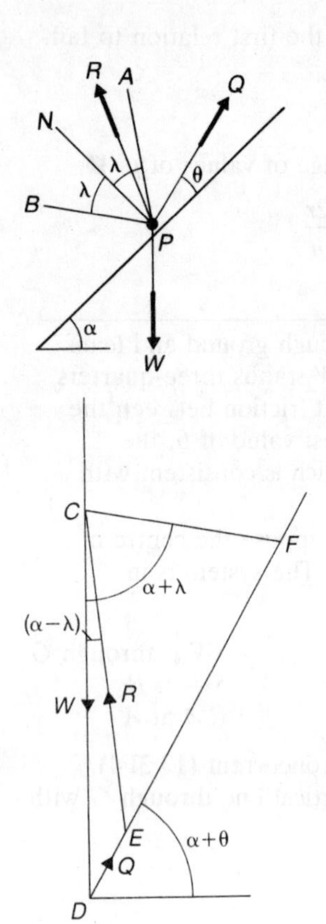

The particle is in equilibrium under the action of three forces.

Its weight $W\downarrow$

The force $Q\nearrow$

The total reaction $R\nwarrow$

Let PN be the normal at P to the plane, and draw PA, PB each making an angle λ with PN. Then, by F3, R acts within the angle APB. Since $\lambda < \alpha$, R cannot act vertically, and so P cannot be in equilibrium in the absence of Q, (see I1) but requires Q to maintain equilibrium.

(i) 'P is about to slip down the plane', means that the friction is limiting and acts *up* the plane. R therefore acts along PA. Construct a triangle CDE of forces. CD representing $W\downarrow$, DE representing $Q\nearrow$ and EC representing $R\nwarrow$ acting along PA.

The angle CED is $\pi - (\alpha - \lambda) - \left(\dfrac{\pi}{2} - \alpha - \theta\right) = \dfrac{\pi}{2} + \lambda + \theta$.

Using the sine rule,

$$\frac{Q}{\sin(\alpha - \lambda)} = \frac{W}{\sin(\pi/2 + \theta + \lambda)} = \frac{W}{\cos(\theta + \lambda)}.$$

$$\therefore\ Q = \frac{W\sin(\alpha - \lambda)}{\cos(\theta + \lambda)}. \blacksquare$$

(ii) 'P is about to slip *up* the plane' means that the friction is limiting and acts *down* the plane. In this case R acts along PB. Construct the triangle CDF of forces. The only change from (i) is that R is now represented by FC, where $\angle DCF = \alpha + \lambda$.

$$\therefore \text{ angle } CFD \text{ is } \pi - (\alpha + \lambda) - \left(\frac{\pi}{2} - \alpha - \theta\right) = \frac{\pi}{2} - \lambda + \theta.$$

$$\therefore\ Q = \frac{W\sin(\alpha + \lambda)}{\cos(\theta - \lambda)}. \blacksquare$$

13.6 EXAMINATION QUESTIONS AND SOLUTIONS

Q1. A heavy uniform wire is bent into the form of an equilateral triangle ABC. The midpoint of BC is M, and the mass centre of the wire is at the point G. Show that $GM = \frac{1}{3}AM$.

The wire is in a vertical plane with A vertically below C and is supported by a rough peg at M, as shown in the diagram. Equilibrium is maintained by a force P which is applied at A and acts in the direction parallel to BC. Given that the wire is on the point of slipping at M, find the coefficient of friction between the wire and the peg. (JMB 1986)

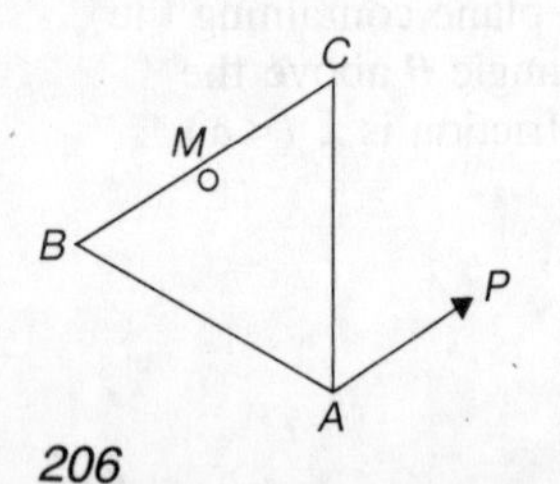

Q2.

A uniform rod AB, of length $2a$ and mass m, rests in equilibrium with its lower end A on a rough horizontal floor. Equilibrium is maintained by a horizontal elastic string, of natural length a and modulus λ. One end of the string is attached to B and the other end to a point vertically above A. Given that θ, where $\theta < \pi/3$, is the inclination of the rod to the horizontal, show that the magnitude of the tension in the string is $\frac{1}{2}mg\cot\theta$.
Prove also that

$$2\lambda = (mg\cot\theta)/(2\cos\theta - 1).$$

Given that the system is in limiting equilibrium and that the coefficient of friction between the floor and the rod is 2/3, find $\tan\theta$. Hence show that $\lambda = 10mg/9$. (LON 1985)

Q3.

A smooth horizontal rail is fixed at a height $3h$ above rough horizontal ground. A uniform rod AB, of mass M and length $6h$, is placed in a vertical plane perpendicular to the rail with the end A resting on the ground. The distance $AC = 5h$, where C is the point of contact between the rail and the rod. Show that the force exerted by the rail on the rod is of magnitude $12Mg/25$.

Given that equilibrium is limiting, find the coefficient of friction between the rod and the ground and show that the force exerted by the ground on the rod is of magnitude $17Mg/25$.

Find, in terms of M and g, the greatest magnitude of the horizontal force which could be applied to the rod at A without disturbing equilibrium. (LON 1986)

Q4.

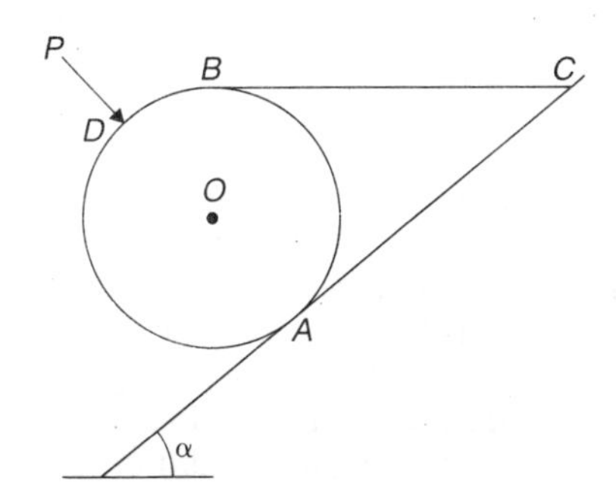

A uniform sphere of weight W is in contact at A with a rough plane which is inclined at an angle α to the horizontal. The coefficient of friction between the sphere and the plane is μ, where $\mu < \tan\frac{\alpha}{2}$. A light inextensible taut string joins the highest point B of the sphere to a point C of the plane; BC is horizontal and C lies on the line of greatest slope of the plane passing through A. Equilibrium of the system is maintained by a force of magnitude P which is directed towards A and applied to the sphere at the point D on the diameter through A, as shown in the diagram.

Show that the tension in the string is $W\tan\frac{\alpha}{2}$.

Show also that

$$\mu P \geqslant W\left(\tan\frac{\alpha}{2} - \mu\right).$$

Explain the significance of the given inequality

$$\mu < \tan\frac{\alpha}{2}.$$

(JMB 1986)

S1.
The line AM is a line of symmetry, that is, when the triangle is reflected in the line the same distribution of mass is obtained. Hence, if we take AM to be the line $x = 0$, it follows from the moment equation $(\Sigma m)x_G = \Sigma mx$ that $x_G = 0$, where x_G is the x-coordinate of G. So G lies on AM.

Similarly, G also lies on the other medians of the triangle. Thus G is the point of intersection of the medians. That is, G is one-third of the distance AM from M.
Therefore $GM = \frac{1}{3}AM$. [A] ■

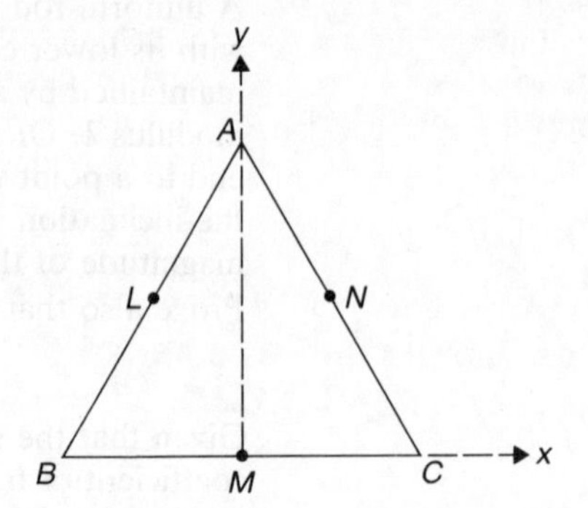

[A] Alternatively, take axes with M as origin, and MC and MA as the x-axis and y-axis. Replace each side of the triangle by a particle of mass m situated at its mid-point.
Then, if $AB = BC = CA = 2a$, we have a mass m at each of the points

$$L\left(-\frac{a}{2}, \frac{a\sqrt{3}}{2}\right), \qquad M\ (0, 0) \qquad \text{and} \quad N\left(\frac{a}{2}, \frac{a\sqrt{3}}{2}\right).$$

Hence the coordinates of G are given by x_G, y_G,

where $$3mx_G = m\left(-\frac{a}{2}\right) + m(0) + m\left(\frac{a}{2}\right) = 0$$

and $$3my_G = m\left(\frac{a\sqrt{3}}{2}\right) + m(0) + m\left(\frac{a\sqrt{3}}{2}\right) = ma\sqrt{3}$$

$$\therefore x_G = 0 \quad \text{and} \quad y_G = \frac{a}{\sqrt{3}}.$$

Thus G lies on AM and $GM = \dfrac{a}{\sqrt{3}}$.
But $AM = a\sqrt{3}$. $\therefore GM = \frac{1}{3}AM$.

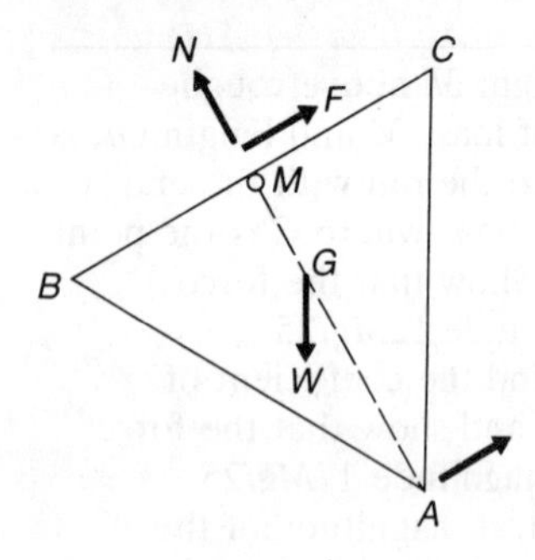

Let the forces acting on the wire be as indicated in the diagram. [B]
Resolve along AM.

$$N = W\cos 30° = \frac{\sqrt{3}}{2}W.$$

$A\circlearrowleft$: $F\,(AM) = W\sin 30°(AG)$.

$$\therefore F = W \times \tfrac{1}{2} \times \tfrac{2}{3}, \text{ as } AG = \tfrac{2}{3}AM.$$

$$\therefore F = \frac{W}{3}$$

As $F > 0$ and the triangle is on the point of slipping, $F = \mu N$.

$$\therefore \mu = \frac{2}{3\sqrt{3}}.$$ *18m* ■

[B] As the triangle is in limiting equilibrium it is tempting to denote the force of friction by μN. However it is not too obvious how equilibrium would be broken, so it is better to call the friction force F. Then, if we have put it in the wrong direction we will obtain a value $F < 0$. In this case we would use $|F| = \mu N$ as the condition for limiting equilibrium.

S2.

Let the forces acting on the rod be as indicated in the diagram. [A]

Let ϕ be the inclination of the reaction at A to the upward vertical.

$A\circlearrowleft$: $\quad T \times 2a\sin\theta = mga\cos\theta.$

$\therefore T = \frac{1}{2}mg\cot\theta.$ ■

Now $T = \dfrac{\lambda}{a}(CB - a).$ [B]

$$\therefore \frac{1}{2}mg\cot\theta = \frac{\lambda}{a}(2a\cos\theta - a).$$

$$\therefore 2\lambda = \frac{mg\cot\theta}{2\cos\theta - 1}. \qquad (1)■$$

$$\tan\phi = \frac{CD}{CA} = \frac{a\cos\theta}{2a\sin\theta} = \frac{1}{2}\cot\theta.$$

In limiting equilibrium $\tan\phi = \frac{2}{3}$. [C]

$$\therefore \tan\theta = \frac{1}{2}\frac{3}{2} = \frac{3}{4}.$$

Substituting in (1):

$$2\lambda = \frac{mg(4/3)}{2(4/5) - 1} = \frac{20mg}{9}.$$

$$\therefore \lambda = \frac{10mg}{9}.$$ 25m ■

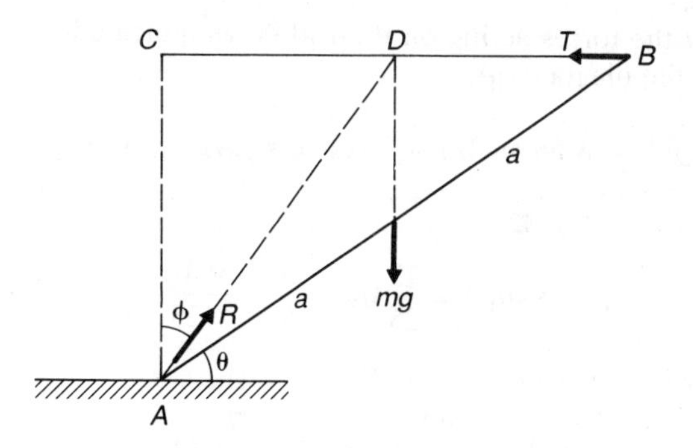

[A] There are three forces acting on the rod; the tension T in the string, the weight mg and the reaction R at A. As the lines of action of T and mg meet at D, the mid-point of CB, the line of action of R must also pass through D.

[B] Using Hooke's Law.

[C] In limiting equilibrium ϕ is equal to λ, the angle of friction. Then, as $\tan\lambda$ is equal to the coefficient of friction, it follows that $\tan\phi = \frac{2}{3}$.

S3.

Let the forces acting on the rod be as indicated on the diagram. [A]

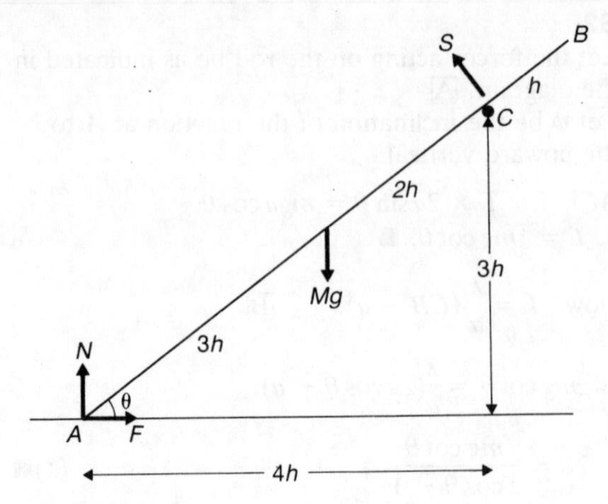

$A\circlearrowleft$: $\quad S.5h = Mg \times 3h\cos\theta = Mg \times 3h \times \frac{4}{5}$.

$\therefore S = \frac{12}{25}Mg$. ■

$\rightarrow$: $\quad F = S\sin\theta = \frac{12}{25}Mg \times \frac{3}{5} = \frac{36Mg}{125}$.

$\uparrow$: $\quad N = Mg - S\cos\theta$

$$= Mg - \frac{12}{25}Mg \times \frac{4}{5} = \frac{77}{125}Mg.$$

As $F > 0$ and the equilibrium is limiting, we have $F = \mu R$.

$\therefore \mu = \frac{36}{77}$. ■

The force exerted by the ground on the rod has magnitude

$$\sqrt{N^2 + F^2} = \frac{Mg}{125}\sqrt{77^2 + 36^2}$$

$$= \frac{Mg}{125} \times 85 = \frac{17Mg}{25}. \ ■$$

When we apply a horizontal force P at A the force S, at C, is unchanged. [B]

Hence, for equilibrium, the force at A must consist of a normal reaction $\frac{77}{125}Mg$ and a horizontal force $\frac{36}{125}Mg$.

Let P act as shown in the diagram.

Then the force of friction will change to a value F_1 such that $P + F_1 = F = \frac{36}{125}Mg$.

Now $|F_1| \leqslant \mu \times \frac{77}{125}Mg = \frac{36}{125}Mg$.

$$\therefore -\frac{36}{125}Mg \leqslant F_1 \leqslant \frac{36}{125}Mg.$$

$$\therefore P - \frac{36}{125}Mg \leqslant P + F_1 \leqslant P + \frac{36}{125}Mg.$$

[A] As the rail is smooth the reaction S at C will be perpendicular to the rod. As the equilibrium is limiting you may be tempted to call the friction μN. However, you may have put it in the wrong direction, so it is safer to call it F. Then, when needed, you can use $|F| = \mu N$ as the condition for limiting equilibrium.

[B] Taking moments about A does not involve the force P and so the same value of S is obtained.

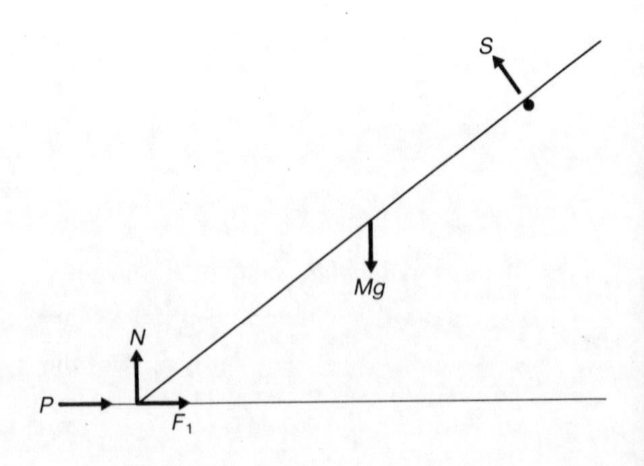

$$\therefore\ P - \frac{36}{125}Mg \leqslant \frac{36}{125}Mg \leqslant P + \frac{36}{125}Mg.$$

$$\therefore\ 0 \leqslant P \leqslant \frac{72}{125}Mg.$$

Thus the maximum magnitude of the horizontal force which could be applied at A is $\frac{72}{125}Mg$. [C] *25m* ■

[C] The value $P = \frac{72}{125}Mg = 2F$ corresponds to the situation when the point A is just about to slip to the right. Hence the friction force has its maximum magnitude and is directed to the left.

S4.

Let the forces acting on the sphere be as indicated in the diagram and let the sphere have radius a and centre O.

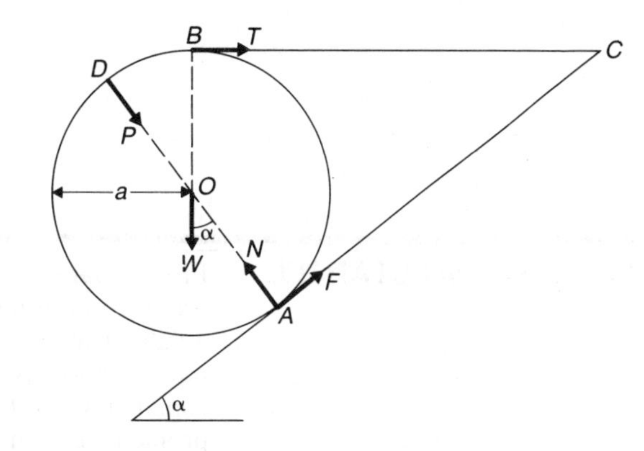

A↺: $\quad T(a + a\cos\alpha) = Wa\sin\alpha.$ [A]

$$\therefore\quad T\left(2\cos^2\frac{\alpha}{2}\right) = W \times 2\sin\frac{\alpha}{2}\cos\frac{\alpha}{2}.$$

$$\therefore\ T = W\tan\frac{\alpha}{2}.\ ■$$

O↺: $\quad Ta = Fa. \qquad \therefore\ F = W\tan\frac{\alpha}{2}.$

C↺: $\quad (N - P)AC = W(BC).$ [B]

$\therefore\ N - P = W, \quad$ as $\quad AC = BC.$

$\therefore\ N = P + W.$

For equilibrium, $F \leqslant \mu N$.

$$\therefore\ W\tan\frac{\alpha}{2} \leqslant \mu P + \mu W.$$

$$\therefore\ \mu P \geqslant W\left(\tan\frac{\alpha}{2} - \mu\right).\ ■$$

For any value of $\mu < \tan\frac{\alpha}{2}$, the RHS is positive, and so there is a *minimum* value of P which is needed to maintain equilibrium. But if $\mu \not< \tan\frac{\alpha}{2}$, then any value of P (including zero) satisfies the inequality. Therefore equilibrium is possible *without* the force P. *16m* ■

[A] We are first asked to find T, the tension in the string. Thus we look for a point which is on the line of action of all the other unknown forces.

[B] If you do not spot that N may be found directly by taking moments about C, the same result may be obtained by resolving perpendicular to AC.

Thus

$$N = P + W\cos\alpha + T\sin\alpha$$

$$= P + W\left(\cos\alpha + 2\sin^2\frac{\alpha}{2}\right)$$

$$= P + W\left(1 - 2\sin^2\frac{\alpha}{2} + 2\sin^2\frac{\alpha}{2}\right)$$

$$= P + W.$$

Chapter 14

Equilibrium of systems of rigid bodies

14.1 GETTING STARTED

In this chapter we consider systems in which two or more bodies are in equilibrium. The bodies may be pivoted together, joined by a light string or simply rest in contact with each other. We restrict ourselves to cases in which the forces acting on the system form a coplanar set of forces and so the method of solution of the problems is to use the general facts stated in Chapters 12 and 13. Each body of the system is assumed to be subject to a set of coplanar forces, with zero vector sum and zero total moment about any point. However an **essential** preliminary to the solution is that we first draw clear diagrams of the system and indicate all the forces acting. In many problems it is advisable to draw two diagrams: one in which only the external forces are indicated and a second 'exploded' diagram in which both external forces and the mutual reactions of the bodies are indicated. When inserting the mutual reactions we must remember Newton's Third Law, which tells us that when two bodies are in contact the reaction experienced by one body is equal in magnitude and opposite in direction to the reaction experienced by the other body.

14.2 FRAMEWORKS

A framework is a set of uniform, light, rigid rods which are smoothly pivoted together.

ESSENTIAL FACTS

F1.

Forces on joints at A and B

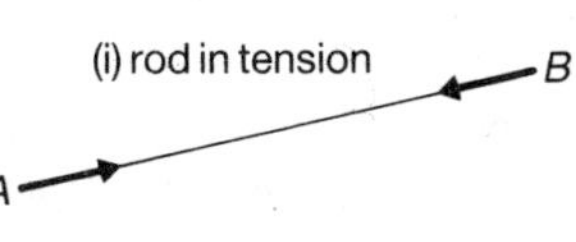

(ii) rod in thrust

A

B

Each rod in the framework is in tension or in thrust. Thus at each end of the rod the force exerted by the rod on the pivot, or joint, is equal in magnitude to the tension, or thrust, in the rod.

F2.

As the pivots are smooth there are no couples exerted by the pivots. Hence the action of a pivot on a body is a force through the centre of the pivot.

F3.

The set of forces acting on the whole of the system, or on any part of the system, forms a set of forces in equilibrium.

F4.

When a number of rods are pivoted at a point on a rigid body the reaction of the pivot on the rigid body is equal in magnitude and opposite in direction to the vector sum of the forces exerted by the rods on the pivot.

ILLUSTRATION

I1.

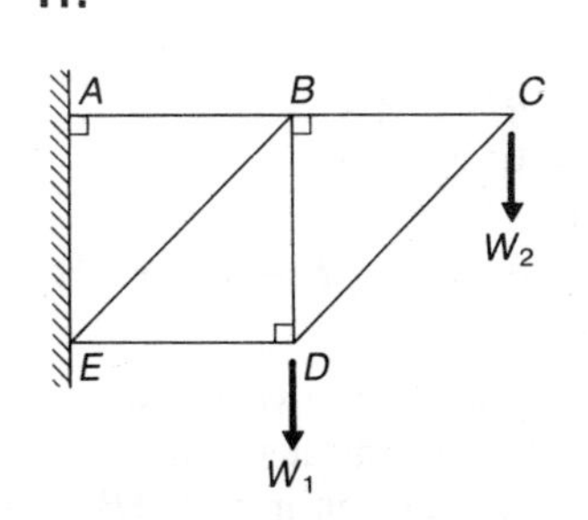

The diagram shows a framework of 6 smoothly jointed light rods in which triangles ABE, DEB and BCD are all right angled and isosceles. Loads of weight W_1 and W_2 are hung from C and D respectively. Find the forces in the rods, stating whether they are in tension or compression.
Find also the horizontal and vertical components of the reactions at A and E. □

Let the tensions in the rods be $T_1, T_2, \ldots, T_6$ as shown. [A]
Let the horizontal and vertical components of the reactions at A and E be P and Q and X and Y respectively.
Consider the forces at C. [B]

$\uparrow$: $T_6 \cos 45° + W_2 = 0. \quad \therefore T_6 = -W_2\sqrt{2}.$

$\rightarrow$: $T_5 + T_6 \cos 45° = 0. \quad \therefore T_5 = W_2.$

Consider the forces at D. [C]

$\uparrow$: $T_4 + T_6 \cos 45° = W_1. \quad \therefore T_4 = W_1 + W_2.$

$\rightarrow$: $T_3 - T_6 \cos 45° = 0. \quad \therefore T_3 = -W_2.$

Consider the forces at B. [D]

$\uparrow$: $T_2 \cos 45° + T_4 = 0. \quad \therefore T_2 = -(W_1 + W_2)\sqrt{2}.$

$\rightarrow$: $T_1 + T_2 \cos 45° = T_5. \quad \therefore T_1 = W_1 + 2W_2.$

Thus AB is in tension of $W_1 + 2W_2$,
BE is in compression of $(W_1 + W_2)\sqrt{2}$,
ED is in compression of W_2,
BD is in tension of $W_1 + W_2$,
BC is in tension of W_2,
CD is in compression of $W_2\sqrt{2}$. ■
Consider the forces at A.

$\uparrow$: $Q = 0$

$\rightarrow$: $P = T_1 = W_1 + 2W_2.$

So the reaction at A is a horizontal force of $W_1 + 2W_2$ in the direction from B to A.
Consider the forces at E.

$\uparrow$: $Y + T_2 \cos 45° = 0. \quad \therefore Y = W_1 + W_2.$

$\rightarrow$: $X = T_2 \cos 45° + T_3. \quad \therefore X = -(W_1 + 2W_2).$

So the reaction at E has a horizontal component of $W_1 + 2W_2$ in the direction from E to D and a vertically upward component of $W_1 + W_2$. [E] ■

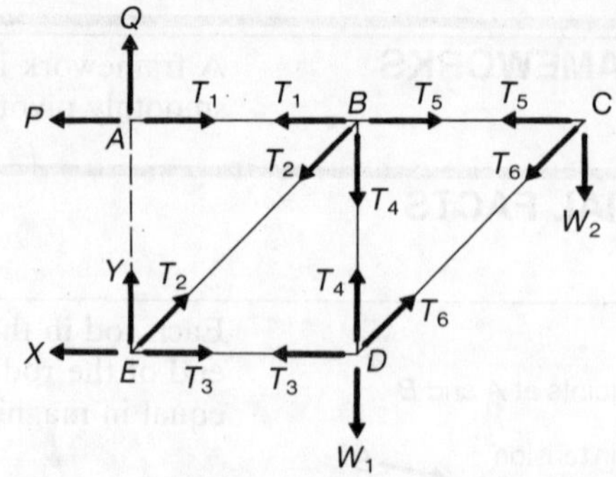

[A] At this stage there is no need to try to decide whether an individual rod is in tension or thrust. We assume them all to be in tension and interpret a negative tension as a thrust.
[B] The forces at C (or any other joint) are in equilibrium. We chose to start at C because it is the only joint with just two unknown tensions. Consequently, resolving in two perpendicular directions will provide sufficient equations to determine the tensions.
[C] As T_6 is known there are now just two unknown tensions at D.
[D] Again, there are now just two unknown tensions at B.
[E] The reactions at A and E may also be found by considering the whole system. If you have time this method may be used to *check* your previous results.

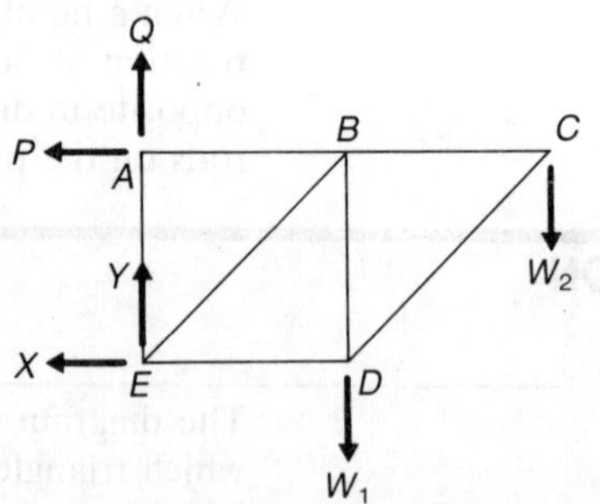

Let $AB = BC = ED = BD = AE = a$.

$A\circlearrowleft$: $Xa + W_1 a + W_2 2a = 0.$
$\therefore X = -(W_1 + 2W_2).$
$E\circlearrowleft$: $Pa = W_1 a + W_2 2a. \quad \therefore P = W_1 + 2W_2.$
$\uparrow$: $Y + Q = W_1 + W_2.$

The total reaction at A (that is, the vector sum of P and Q) must be equal in magnitude and opposite in direction to the tension in rod AB. As this tension is horizontal, the reaction must also be horizontal. Hence $Q = 0$ and $Y = W_1 + W_2$.

14.3 EQUILIBRIUM OF SEVERAL HEAVY BODIES

In this section a heavy body is assumed to be rigid, so that its shape is unaltered by any externally impressed forces. In some circumstances the physical dimensions of the body may be neglected, in which case it is represented by a particle.

ESSENTIAL FACTS

F1. When a set of heavy bodies is in equilibrium the external forces acting on the bodies form a system of forces in equilibrium.

F2. When a set of heavy bodies is in equilibrium the system of forces, consisting of the external forces and internal reactions, acting on a single body (or collection of bodies) in the set form a system of forces in equilibrium.

F3. When two bodies are smoothly pivoted (or smoothly hinged) together the reaction on each body consists of a single force R. The reactions on the bodies are in opposite directions.

ILLUSTRATIONS

I1. Two uniform ladders AB and AC of equal length $2a$ and weights $3W$ and W respectively are smoothly jointed at A and rest in a fixed vertical plane with B and C in contact with a rough horizontal floor. The coefficients of friction at B and C are $\frac{1}{2}$ and $\frac{6}{7}$ respectively and the angle BAC is 2θ. Given that the ladders are in equilibrium, find in terms of W and θ, the horizontal and vertical components of the force exerted on the rod AB at A. The value of θ is gradually increased until equilibrium is broken by one of the ladders slipping. Determine whether slipping first occurs at B or at C. □

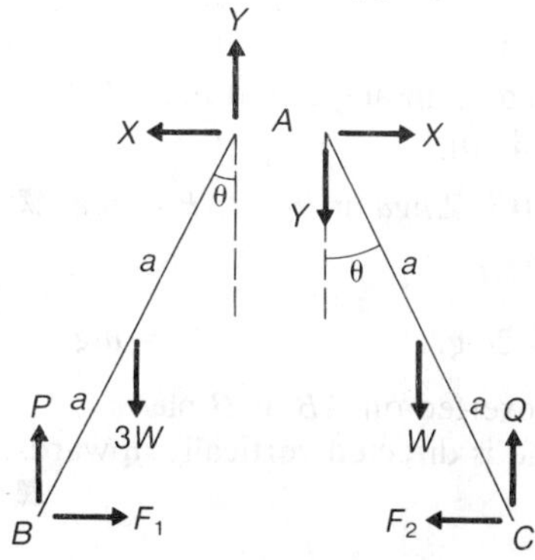

Let the forces acting on the ladders be as indicated in the diagram. [A]
Consider the system as a whole.

$\rightarrow$: $\quad F_1 = F_2$.

$C\circlearrowleft$: $\quad P \times 4a\sin\theta = 3W \times 3a\sin\theta + Wa\sin\theta$.

$\therefore P = \frac{5}{2}W$.

$\uparrow$: $\quad P + Q = 4W. \qquad \therefore Q = \frac{3}{2}W.$ [B]

Consider the ladder AB.

$\rightarrow$: $X = F_1$

$\uparrow$: $P + Y = 3W$. $\therefore Y = \frac{1}{2}W$.

$B\circlearrowleft$: $X \times 2a\cos\theta + Y \times 2a\sin\theta = 3Wa\sin\theta$.

$\therefore X = W\tan\theta$.

Hence the force on AB at A has a vertically upward component of $\frac{1}{2}W$ and a horizontal component, directed away from AC, of $W\tan\theta$. [C] ■

For equilibrium at B, $F_1 \leqslant \frac{1}{2}P$.

$\therefore \tan\theta \leqslant \frac{5}{4}$.

For equilibrium at C, $F_2 \leqslant \frac{6}{7}Q$.

$\therefore \tan\theta \leqslant \frac{9}{7}$.

As $\frac{9}{7} > \frac{5}{4}$ equilibrium will first be limiting at the point B.

Hence equilibrium will be broken by the rod AB slipping at B. [D] ■

[A] As the question asks for the force on AB at A, we draw an 'exploded' diagram giving the external forces on each ladder as well as the reactions at A. Remember that the reaction on AB at A is equal and opposite to the reaction on AC at A.

[B] Alternatively, Q may be found by taking moments about B for the whole system. This may be used as a check on your previous result.

[C] Be careful to specify the directions of the horizontal and the vertical components of the force.

[D] In this question we have modelled the ladders by thin uniform rods. In practice ladders are not uniform, due to the rungs being at intervals along the length of the ladder. However the approximation used is the one conventionally assumed in examinations.

12.

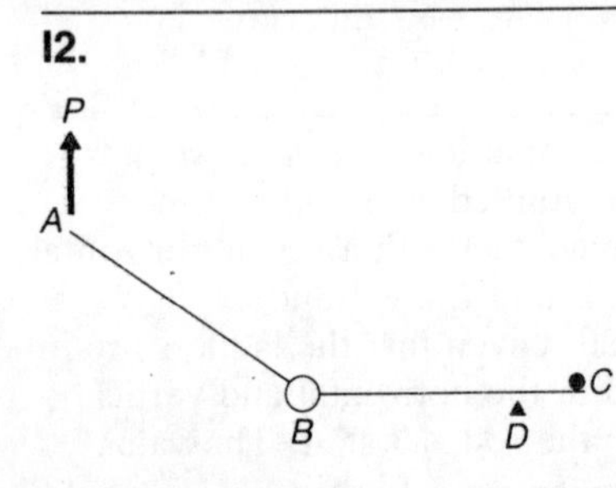

The diagram shows two uniform rods, AB, of length $2a$ and mass $2m$, and BC, of length $2a$ and mass m, smoothly hinged together at B.

The rod BC rests on a small smooth peg fixed at a point D and a particle of mass $4m$ is attached to the rod at C. A vertical force P is applied at A and the value of P is such that equilibrium is maintained with the rod BC horizontal. Find

(i) the value of P,

(ii) the magnitude and direction of the force exerted on AB at B,

(iii) the distance BD. □

Let the forces acting on each rod be as shown in the diagram. [A]

Consider the rod AB.

Let AB be inclined at an angle θ to the downward vertical. [B]

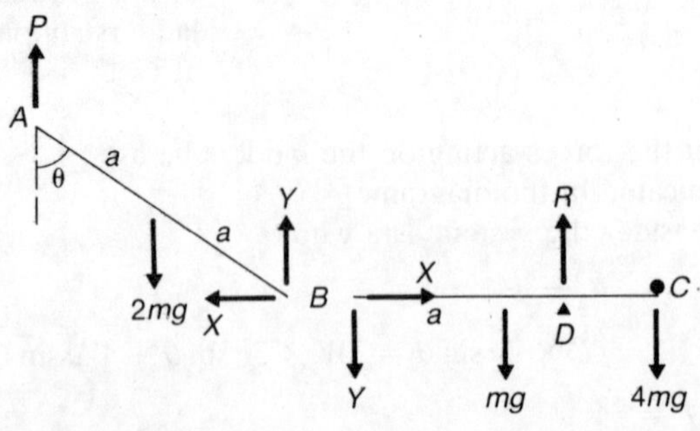

$B\circlearrowleft$: $P \times 2a\sin\theta = 2mga\sin\theta$. $\therefore P = mg$. ■

$\rightarrow$: $X = 0$.

$\uparrow$: $P + Y = 2mg$. $\therefore Y = mg$.

Hence the force exerted on AB at B has magnitude mg and is directed vertically upwards. ■

[A] We include equal and opposite reactions on the rods at B. Also, as the peg at D is smooth, the reaction R at D will be perpendicular to BC.

Consider the rod BC.
Let $BD = d$.

$D\circlearrowleft$: $\quad Yd + mg(d - a) = 4mg(2a - d)$. [C]

$\therefore d + (d - a) = 8a - 4d$.

$\therefore d = \dfrac{3a}{2}$.

Hence $BD = \dfrac{3a}{2}$. ■

[B] This angle is not given in the data and so we do not expect it to occur in the answers. However, we include it in order to show the examiner that we have taken moments correctly.
[C] We choose to take moments about D in order to eliminate the reaction R, which we have not been asked to find. If necessary, we could find R by resolving vertically for rod BC.

So $\quad R = Y + mg + 4mg = 6mg$.
Then

$B\circlearrowleft$: $\quad R(BD) = mga + 4mg(2a)$.

$\therefore 6BD = a + 8a. \qquad \therefore BD = \dfrac{3a}{2}$.

13.

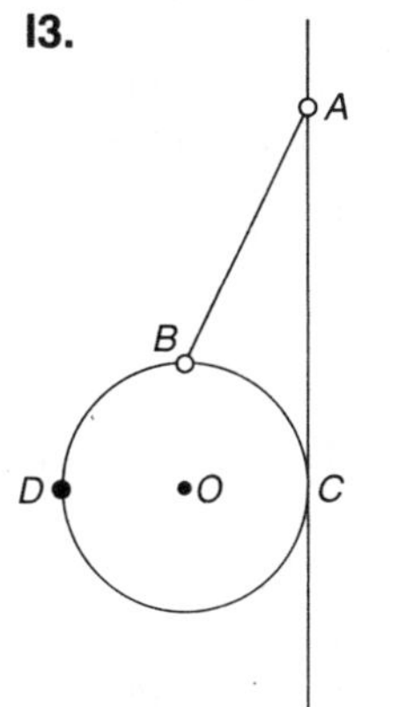

The diagram shows a uniform sphere, of centre O, radius a and mass m, in contact with a rough vertical wall at C. A particle of mass $\frac{1}{2}m$ is attached to the sphere at D, where CD is a diameter of the sphere. The sphere is held in equilibrium with CD horizontal by means of a rod AB of mass m which is smoothly hinged to the sphere at B, which is vertically above O, and is smoothly hinged to a point A on the wall which is a distance $3a$ above C. Find
(i) the smallest value of the coefficient of friction between the sphere and the wall which is consistent with equilibrium,
(ii) the horizontal and vertical components of the forces acting on rod at A and B. □

Consider the system as a whole

$A\circlearrowleft$: $\quad R \times 3a = mg \times \frac{1}{2}a + mga + \frac{1}{2}mg \times 2a. \qquad \therefore R = \frac{5}{6}mg$.

Consider the sphere

$B\circlearrowleft$: $\quad Fa + \frac{1}{2}mga = Ra. \qquad \therefore F = \frac{1}{3}mg$.

For equilibrium at C: $\quad F \leqslant \mu R, \qquad \therefore \mu \geqslant \dfrac{2}{5}$. ■

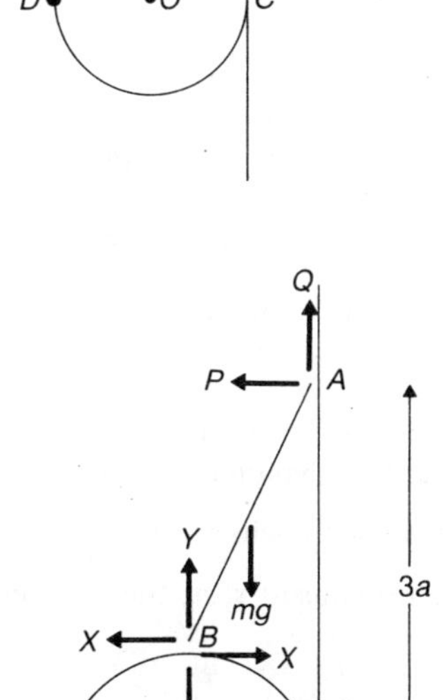

Hence the smallest value of the coefficient of friction, between the sphere and the wall, which is consistent with equilibrium is $\dfrac{2}{5}$.

$\uparrow$: $\quad Y + mg + \frac{1}{2}mg = F. \qquad \therefore Y = -\dfrac{7}{6}mg$.

$\rightarrow$: $\quad X = R. \qquad \therefore X = \dfrac{5}{6}mg$.

$\therefore$ at B the forces on the rod are $\dfrac{5}{6}mg$ horizontally and away from the wall and $\dfrac{7}{6}mg$ vertically downwards. ■

Consider the rod AB

$\uparrow$: $Q + Y = mg$. $\therefore Q = \frac{13}{6}mg$.

$\rightarrow$: $X + P = 0$. $\therefore P = -\frac{5}{6}mg$.

$\therefore$ the forces on the rod at A are a horizontal force of $\frac{5}{6}mg$ towards the wall and a vertically upward force of $\frac{13}{6}mg$. ■

---

14. A wedge of mass m is placed with one face on a rough horizontal table and a particle of mass m is held at rest on a second rough face of the wedge which is inclined at an angle arc tan $\frac{5}{12}$ to the horizontal. The particle is released.

Given that μ_1 is the coefficient of static friction and the coefficient of dynamic friction between the particle and the sloping face of the wedge and that μ_2 is the coefficient of the static friction between the wedge and the table, show that when $\mu_1 = \frac{1}{2}$ the system will remain in equilibrium. Given that the wedge remains at rest when $\mu_1 = \frac{1}{3}$, show that $\mu_2 \geqslant \frac{4}{111}$. □

Let the forces acting on the particle and the wedge be as shown in the diagram. A

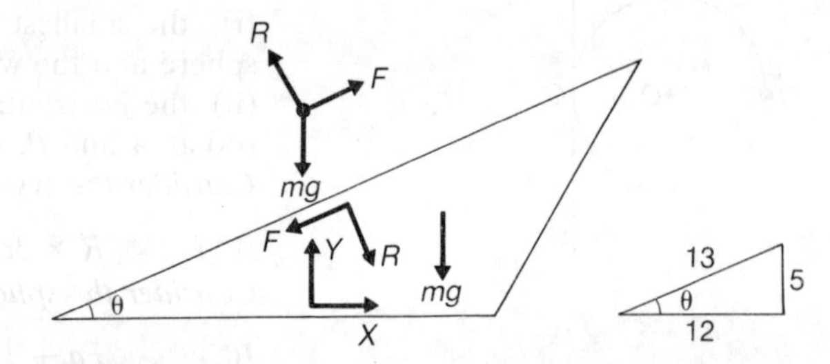

(i) Consider the particle.

Resolving perpendicular to the plane.

$R = mg\cos\theta$. $\therefore R = \frac{12}{13}mg$.

Resolving parallel to the plane.

$F = mg\sin\theta$. $\therefore F = \frac{5}{13}mg$.

For equilibrium; $F \leqslant \mu_1 R$. $\therefore \mu_1 \geqslant \frac{5}{12}$.

As $\frac{1}{2} > \frac{5}{12}$, the particle will not move relative to the wedge. ■

(ii) As $\frac{1}{3} < \frac{5}{12}$ the particle will slide down the face of the wedge.

$\therefore F = \frac{1}{3}R = \frac{4}{13}mg$. B

A It is advisable to draw an 'exploded' diagram, with the wedge and particle separated so that it is clear how the forces R and F act on the bodies.

B When the particle moves the friction $F = \mu R$, where μ is the coefficient of dynamic friction. Also, the value of R remains unchanged at $\frac{12}{13}mg$ because, as the wedge remains at rest, the motion is perpendicular to R.

Consider the wedge:

$$\uparrow: \quad Y = F\sin\theta + R\cos\theta + mg. \quad \therefore\ Y = \frac{333}{169}mg.$$

$$\rightarrow: \quad X + R\sin\theta = F\cos\theta. \quad \therefore\ X = -\frac{12}{169}mg. \quad \boxed{C}$$

For equilibrium of the wedge,

$|X| \leqslant \mu_2 Y. \qquad \therefore\ \mu_2 \geqslant \frac{4}{111}.$ ■

$\boxed{C}$ A negative value of X simply means that it acts in the opposite direction. However we must now remember to use F in the form $|X| \leqslant \mu_2 Y$.

14.4 EXAMINATION QUESTIONS AND SOLUTIONS

Q1.

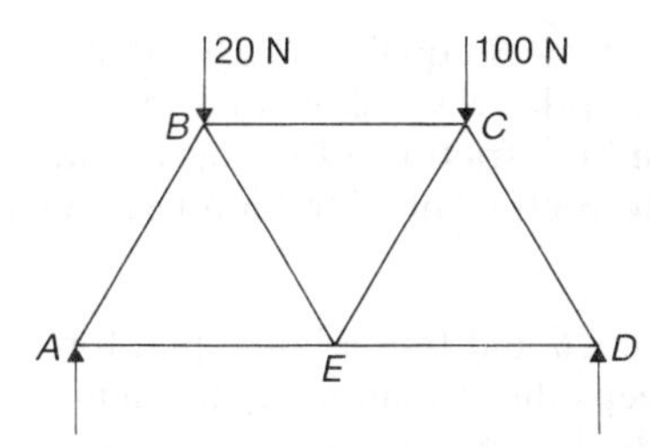

The diagram shows a framework consisting of seven equal smoothly jointed light rods AB, BC, DC, DE, AE, EB and EC. The framework is in a vertical plane, with AE, ED and BC horizontal and is simply supported at A and D. It carries vertical loads of 20 N and 100 N at B and C respectively.

(a) Calculate the reactions at A and D.
(b) Find the forces in AB, AE and BC.

Show also that, for all possible values of the vertical loads at B and C, the force in BC is proportional to the sum of those loads.

(AEB 1985)

Q2.

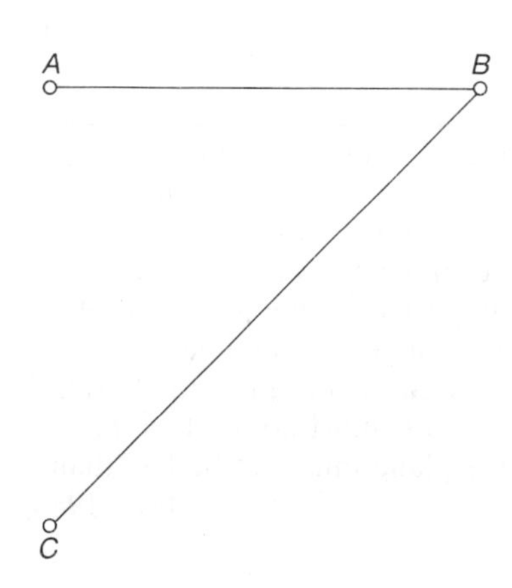

The diagram shows two uniform rods AB and BC of lengths $2l$ and $2l\sqrt{2}$, respectively, and each of weight w per unit length. The rods are smoothly hinged to each other at B and to fixed points at A and C where C is at a distance $2l$ vertically below A. Find, in terms of w and l, the horizontal and vertical components of the force exerted on the rod BC at B.

When, in addition, an anticlockwise couple of moment G is applied to the rod BC, the direction of the force exerted on this rod at B is then along the rod. Find, in terms of w and l, the value of G and the horizontal component of the force exerted on the rod BC at C.

(JMB 1986)

Q3.

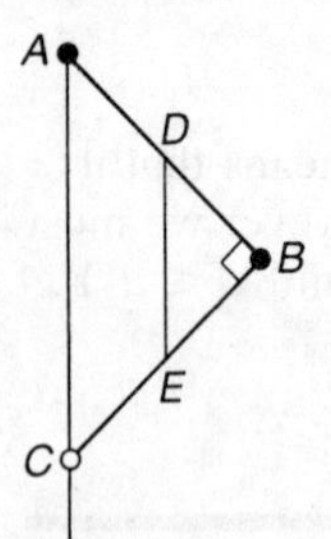

Two uniform rods AB and BC, each of weight W and length $2a$, are smoothly jointed together at B. The rod AB is freely hinged at the fixed point A to a thin smooth vertical wire. A small light ring threaded on the wire is attached to the rod BC at C. The rods are held at right angles to each other, as shown in the diagram, by means of a light inextensible string whose ends are attached to the midpoints D and E of AB and BC respectively. Show that the tension in the string is of magnitude $2W$. Find the horizontal and vertical components of the force exerted at B by AB on BC.

(JMB 1986)

Q4.

AC is a fixed rough straight horizontal wire of length greater than $4a$. A straight uniform rod AP of mass $4M$ and length $2a$ is smoothly jointed to the wire at A. The end P of the rod is attached by means of a light inelastic thread PB of length $2a$ to a bead B of mass M which can slide on the wire.

(a) The system is in equilibrium with $\widehat{PAB}$ equal to θ. By taking moments about A for the rod AP, or otherwise, show that the tension in the string is $Mg/\sin\theta$. Find the frictional force at B and show that $\mu \geqslant \frac{1}{2}\cot\theta$, where μ is the coefficient of friction between the bead and the wire.

(b) With $\widehat{PAB} = \theta$, suppose $\mu = \frac{1}{4}\cot\theta$ and that a horizontal force S applied at P in the plane APB keeps the system in equilibrium. If B is on the point of slipping towards A, show that the tension in the string is $Mg/(3\sin\theta)$ and find the value of S.

(c) When $\widehat{PAB} = \theta$, $\mu = \frac{1}{4}\cot\theta$ and no horizontal force is applied at P, what is the minimum value of the mass of B required to ensure equilibrium?

(WJEC 1986)

Q5.

Two identical uniform circular cylinders, A and B, rest on a rough horizontal plane with their axes parallel and horizontal. They are positioned so that they just fail to make contact with each other along a generator. A third identical cylinder C, whose axis is parallel to the axes of A and B, is supported symmetrically by A and B. The mass centres of the three cylinders lie in a vertical plane perpendicular to their generators. Show that, for equilibrium, the coefficient of friction between A and C must not be less than $(2 - \sqrt{3})$ and that between A and the plane must not be less than $\frac{1}{3}(2 - \sqrt{3})$.

(JMB 1986)

SOLUTIONS

S1.

Let the loads at B and C be X N and Y N respectively. [A]
Let the tensions in AB, BE, BC be T_1 N, T_2 N, T_3 N and the reactions at A and D be P N and Q N as indicated. [B]
Let each rod be of length $2a$.

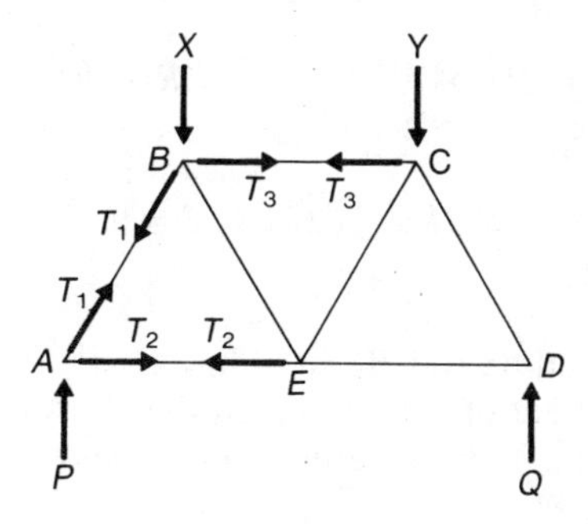

Consider the whole system

$D\circlearrowleft$: $\quad P \times 4a = X \times 3a + Ya.$

$\therefore P = \frac{1}{4}(3X + Y).$

$A\circlearrowleft$: $\quad Q \times 4a = Xa + Y \times 3a.$

$\therefore Q = \frac{1}{4}(X + 3Y).$

When $X = 20$ and $Y = 100$, $P = 40$ and $Q = 80$.
$\therefore$ the reactions at A and D are 40 N and 80 N. ■

Consider the joint A [C]

$\uparrow: T_1 \cos 30° + P = 0. \therefore T_1 = -\frac{1}{2\sqrt{3}}(3X + Y).$

$\rightarrow: T_2 + T_1 \cos 60° = 0. \quad \therefore T_2 = \frac{1}{4\sqrt{3}}(3X + Y).$

Consider the joint B
Resolving perpendicular to BE, [D]

$T_3 \cos 30° = T_1 \cos 30° + X \cos 60°.$

$$\therefore T_3 = -\frac{1}{2\sqrt{3}}(X + Y). \qquad (1)$$

When $X = 20$ and $Y = 100$,

$T_1 = -\frac{80}{\sqrt{3}},$

$T_2 = \frac{40}{\sqrt{3}}$ and $T_3 = -20\sqrt{3}.$

So AB has a thrust of $\frac{80}{\sqrt{3}}$ N,

AE has a tension of $\frac{40}{\sqrt{3}}$ N,

BC has a thrust of $20\sqrt{3}$ N. [E]

From equation (1) we see that, for all possible loads at B and C, the thrust in BC is proportional to the sum of those loads. *25m* ■

[A] Although the question starts with loads of 20 N and 100 N at B and C the final part of the question is concerned with arbitrary loads at these points. Thus, in order to prevent the same work being done twice, it is advisable to work with arbitrary loads throughout and to insert specific values when required.

[B] *A tip*. You have been asked for forces only in AB, AE and BC. To avoid forgetting this, mark the three rods in the diagram in some way, so as not to waste time finding forces in other rods. One possible way of doing this is to mark the rods with a coloured pen.

[C] There are only two unknown forces acting at A.

[D] We resolve perpendicular to BE in order to eliminate the need to introduce a tension in BE. Alternatively: For the triangle of rods ABE,

$E\circlearrowleft$: $\quad T_3 \times a\sqrt{3} = Xa - P \times 2a$

$= Xa - \frac{3X + Y}{2}a.$

[E] Remembering that a negative tension is interpreted as a thrust. The question says 'find the forces'. So you *must* state which are tensions and which are thrusts, or you may lose marks for omitting to do so.

S2.
Let the forces on the rods be as indicated in the diagram.

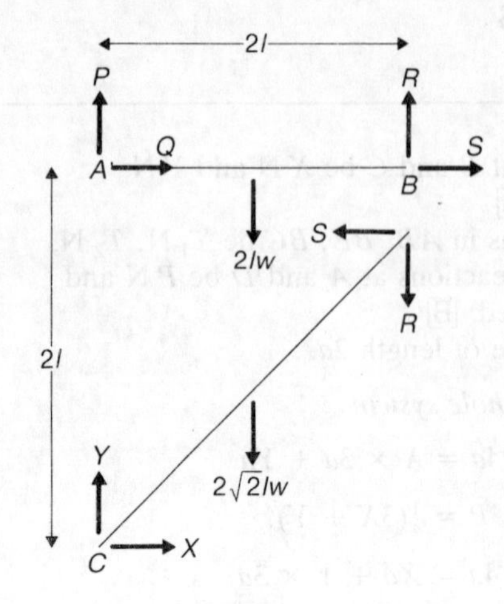

Consider the rod AB

A↺: $R \times 2l = 2lw \times l$. ∴ $R = lw$.

Consider the rod BC

C↺: $R \times 2l - S \times 2l + 2\sqrt{2}\,lw \times l = 0$.

∴ $S = wl(1 + \sqrt{2})$.

So the force exerted on rod BC at B has a horizontal component of $wl(1 + \sqrt{2})$ in the direction from B to A and a vertical component of wl downwards. ■

When the couple G acts on BC, let the forces on the rods be as indicated in the diagram. [A]

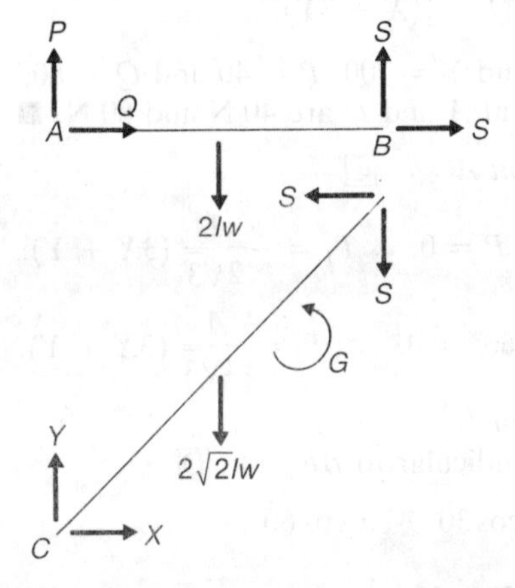

Consider the rod BC

C↺: $G = 2\sqrt{2}lw \times l$. [B]

∴ $G = 2\sqrt{2}l^2w$. ■

→: $X = S$.

Consider the rod AB

A↺: $S \times 2l = 2lw \times l$. ∴ $S = lw$.

∴ $X = lw$.

Hence the horizontal component of the force exerted on BC at C is lw in the direction from A to B. 11m ■

[A] Although we have used the same letters for the forces as were used in the first part of the question their values may, of course, be different. When inserting these forces it is advisable to start at B, for we are told that the force on BC at B is along the rod. Hence, as BC is inclined at 45° to the horizontal, we insert equal horizontal and vertical components. This then ensures that the force at B is along BC.

[B] The force at B is along BC; that is, it passes through C, so it has zero moment about C.

S3.

Let the tension in the string be T.
Then the forces on the two rods are as shown in the diagram. [A]

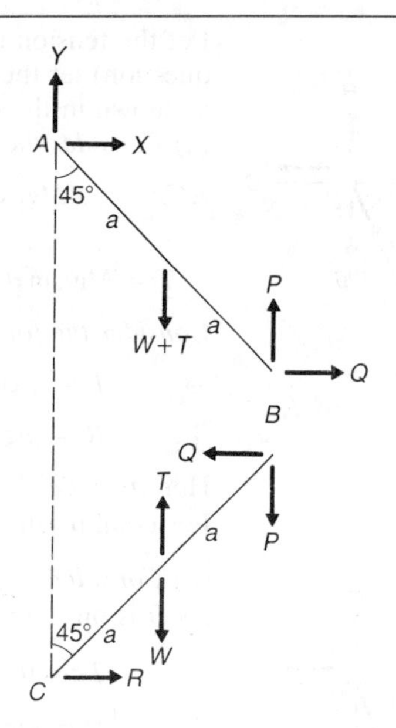

Consider the whole system

$A\circlearrowright$: $R \times 4a\cos 45° = (W + T)a\cos 45° + (W - T)a\cos 45°$.

$\therefore R = \frac{1}{2}W$.

Consider the rod BC

$B\circlearrowright$: $R \times 2a\cos 45° = (T - W)a\cos 45°$.

$\therefore T = 2R + W = 2W$.

$\therefore$ the tension in the string has magnitude $2W$. ■

$\rightarrow$: $Q = R = \frac{1}{2}W$.

$\uparrow$: $P + W = T$. $\therefore P = W$.

So the force exerted on BC at B has a horizontal component $\frac{1}{2}W$ towards the wire and a vertical component W downwards. *13m* ■

[A] As the wire is smooth the reaction on the rod BC at C will be perpendicular to the wire. That is, it will be horizontal.
Note however that the rod AB is *hinged* to the wire at A and so the force on AB at A is not necessarily normal to the wire.

S4.

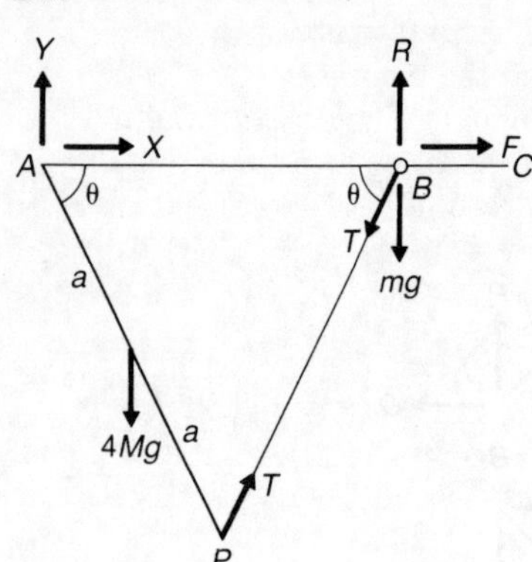

Let the tension in the string be T and (as its value varies in the question) let the bead B have mass m. The other forces acting are as shown in the diagram.

(a) *Consider the rod AB*

$$A\circlearrowleft: \quad 4Mga\cos\theta = T \times 2a\sin(\pi - 2\theta) = T \times 2a\sin 2\theta = 4Ta\sin\theta\cos\theta.$$

$\therefore T = Mg/\sin\theta.$ ■

Consider the forces acting at B

$$\rightarrow: \quad F = T\cos\theta = Mg\cot\theta. \qquad (1)$$

$$\uparrow: \quad R = mg + T\sin\theta = (m + M)g. \qquad (2).$$

Here $m = M$. $\quad \therefore R = 2Mg$.

For equilibrium, $F \leqslant \mu R$. $\quad \therefore \mu \geqslant \frac{1}{2}\cot\theta$. ■

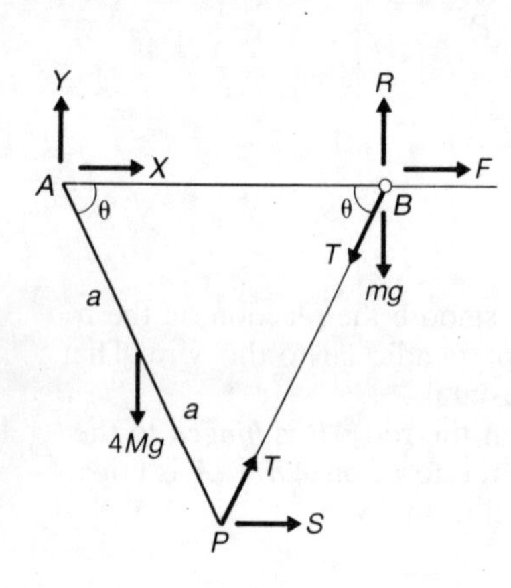

(b) *Consider the forces acting at B*

As B is on the point of slipping, $F = \mu R = \frac{1}{4}R\cot\theta$.

$$\rightarrow: \quad T\cos\theta = F = \tfrac{1}{4}R\cot\theta. \qquad \therefore R = 4T\sin\theta.$$

$$\uparrow: \quad R = mg + T\sin\theta. \qquad \therefore 3T\sin\theta = mg.$$

Here $m = M$. $\quad \therefore$ the tension in the string is $Mg/(3\sin\theta)$.

Consider the rod AP

$$A\circlearrowleft: \quad T \times 2a\sin(\pi - 2\theta) + S \times 2a\sin\theta = 4Mga\cos\theta.$$

$$\therefore S = \frac{4}{3}Mg\cot\theta.$$

(c) For equilibrium at B, $F \leqslant \mu R = \frac{1}{4}R\cot\theta$.

Therefore, from equations (1) and (2),

$$Mg\cot\theta \leqslant \frac{(m + M)}{4}g\cot\theta.$$

$$\therefore m \geqslant 3M.$$

Hence equilibrium is ensured with a minimum mass of B equal to $3M$. *27m* ■

S5.

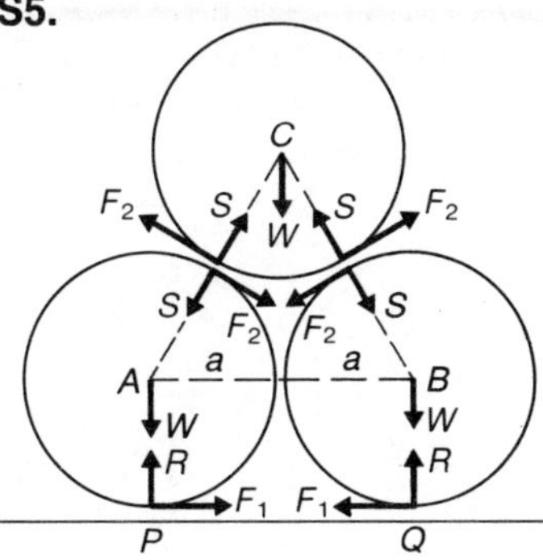

Let P and Q be the points of contact between spheres A and B and the horizontal plane. Using the symmetry of the system, we let the forces on each sphere be as shown in the diagram.
Note: The question says that the cylinders A and B just fail to make contact. We interpret this as meaning that we may take the distance AB to be $2a$ and, at the same time, assume that there is no reaction between the cylinders A and B.

Consider the sphere A

$$P\circlearrowleft: \qquad S \times a\sin 30° = F_2(a + a\cos 30°).$$

$$\therefore\ S = F_2(2 + \sqrt{3}). \qquad (1)$$

For equilibrium between A and C; $F_2 \leqslant \mu_2 S$, where μ_2 is the coefficient of friction between A and C.

$$\therefore\ F_2 \leqslant \mu_2 F_2(2 + \sqrt{3}).$$

$$\therefore\ \mu_2 \geqslant \frac{1}{2 + \sqrt{3}} = 2 - \sqrt{3}.\ ■$$

Consider the whole system

$$\uparrow: \qquad W + W + W = R + R. \qquad \therefore\ W = \frac{2R}{3}.$$

Consider the sphere A

$$A\circlearrowleft: \qquad F_1 a = F_2 a. \qquad \therefore\ F_1 = F_2. \qquad (2)$$

$$\uparrow: \qquad R = W + S\cos 30° + F_2\cos 60°$$

$$= \frac{2}{3}R + F_2\,(2 + \sqrt{3}).\frac{\sqrt{3}}{2} + F_2.\frac{1}{2}, \qquad \text{using (1).}$$

$$\therefore\ \frac{R}{3} = F_1\left(\sqrt{3} + \frac{3}{2} + \frac{1}{2}\right), \qquad \text{using (2),}$$

$$= F_1\,(2 + \sqrt{3}).$$

$$\therefore\ \frac{F_1}{R} = \frac{1}{3(2 + \sqrt{3})} = \frac{2 - \sqrt{3}}{3}.$$

For equilibrium at P, $F_1 \leqslant \mu_1 R$, where μ_1 is the coefficient of friction between A and the plane.

$\therefore\ \mu_1 \geqslant \frac{1}{3}\,(2 - \sqrt{3}).$ *18 m* ■

Chapter 15 Probability

15.1 GETTING STARTED

The probability of an event is the numerical value given to the likelihood that the event will actually happen. All probabilities lie between 0 and 1. An event which never occurs is assigned probability 0, and an event which is certain to occur has probability 1. Probabilities of events which may, or may not, occur are often assigned by considering what is likely to happen. For instance, if we toss a coin it is equally likely to come down with 'head uppermost' or 'tail uppermost'. So, discounting the possibility of its landing and balancing on its edge, we assign a probability of $\frac{1}{2}$ to each of the events 'head uppermost' and 'tail uppermost'. This means that if we toss the coin 1000 times we expect that on about 500 tosses the event 'head uppermost' will occur. Similarly if we draw one card from a standard pack of playing cards we assign the probability of $\frac{1}{4}$ to the event 'that the card will be a spade', $\frac{4}{52}$ to the event 'that the card will be a King' and $\frac{1}{52}$ to the event 'that the card will be the King of Spades'.

In this chapter we are concerned with estimating the result of a series of experiments or trials and of finding the likelihood, that is, the probability, that a particular result will occur. To do this we need to known how to combine probabilities and also how to decide whether the occurrence of a result affects the probability of another result being found.

So we must start by learning the basic addition and multiplication laws of probabilities.

15.2 BASIC DEFINITIONS

ESSENTIAL FACTS

F1. Possibility space

The set S of all possible outcomes of an experiment, or trial, is called a *possibility space* or *sample space*.

F2.

An **event** is a subset A of S, which contains all the outcomes in which a particular result (also called the 'event A') occurs.
When S contains a finite number $n(S)$ of equally likely outcomes, the probability, $P(A)$, that the event A will occur is given by

$$P(A) = \frac{n(A)}{n(S)} = \frac{\text{no. of outcomes in which } A \text{ occurs}}{\text{total no. of possible outcomes}}.$$

When an event A is certain to occur, $A = S$ and $P(A) = 1$. When an event A can never occur, $A = \varnothing$ (the empty set) and $P(A) = 0$.

F3.

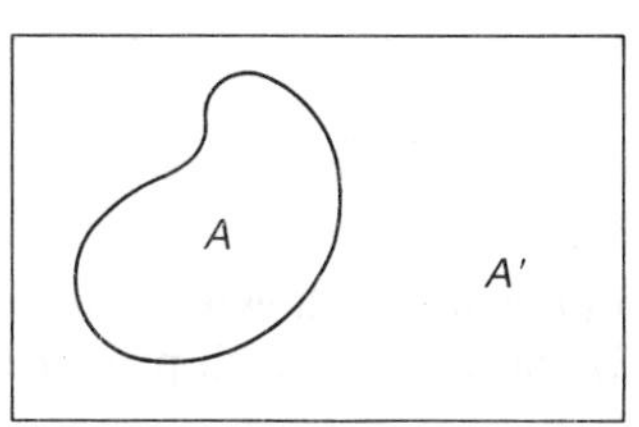

The set $A' = S - A$, that is, the complement of A with respect to S, represents the event 'not A'.
Then if $P(A) = p$, $P(A') = 1 - p$.

F4. Probability space

By assigning to each outcome s_i in the sample space S a probability p_i, where $0 \leqslant p_i \leqslant 1$ and $\sum_{i=1}^{n(S)} p_i = 1$, the space S is converted into a probability space. The probability $P(A)$ of any event A is the sum of the probabilities of all the outcomes contained in the subset A of S.

F5.

Let A and B be two subsets of S.
The set $A \cup B$ contains all the outcomes in which A or B (or both) occur and the set $A \cap B$ contains all the outcomes in which both A and B occur.
Thus $P(A \cup B)$ is the probability of event A or event B (or both) occurring and $P(A \cap B)$ is the probability that both event A and event B occur.

ILLUSTRATIONS

I1. If two dice are thrown, find the probability of the following events.
A: scoring a total of 3 *B*: the same score on each die
C: a 4 or more on each die *D*: scoring a total of 10 or more. □

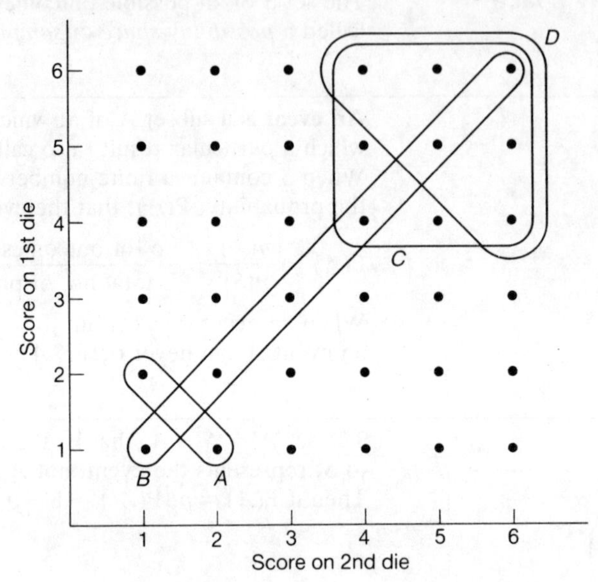

The possibility space consists of the 36 outcomes as shown on the diagram. There are two outcomes, (1, 2) and (2, 1), which produce a total score of 3.

$\therefore$ P(A) = 2/36 = 1/18. ■

There are six outcomes in which the score is the same on each die.

$\therefore$ P(B) = 6/36 = 1/6. ■

There are nine outcomes in which each die has a 4 or more.

$\therefore$ P(C) = 9/36 = 1/4. ■

There are six outcomes scoring a total of 10 or more.

$\therefore$ P(D) = 6/36 = 1/6. ■

I2. Six cards are to be selected from a standard pack of 52 playing cards. Find the probability that at least four of them will be hearts. □
The number of ways of selecting six hearts from 13 is ${}^{13}C_6$.
The number of ways of selecting five hearts from 13 is ${}^{13}C_5$.

The number of ways of selecting 1 card from the 39 which are not hearts is 39.

$\therefore$ the number of ways of selecting six cards, five of which are hearts, is $39 \times {}^{13}C_5$.

Similarly, the number of ways of selecting six cards, four of which are hearts, is ${}^{39}C_2 \times {}^{13}C_4$.

So it is possible to select six cards, of which at least four are hearts, in ${}^{13}C_6 + 39 \times {}^{13}C_5 + {}^{39}C_2 \times {}^{13}C_4$ ways.

The number of ways of selecting any six cards from 52 is ${}^{52}C_6$.

Hence the probability that at least four of the six cards are hearts is

$$\frac{{}^{13}C_6 + 39 \times {}^{13}C_5 + {}^{39}C_2 \times {}^{13}C_4}{{}^{52}C_6}$$

$$= \left(\frac{13!}{6!\,7!} + 39 \times \frac{13!}{5!\,8!} + \frac{39!}{2!\,37!} \times \frac{13!}{4!\,9!}\right) \times \frac{6!\,46!}{52!} = 0{\cdot}0286. \ ■$$

---

13. A coin is tossed three times. Given the events

A: three heads, B: at least two heads, C: at least one head

find (a) $P(A)$, $P(B)$, $P(C)$

(b) $P(A \cup B)$, $P(A \cap B)$

(c) $P(B \cup C)$, $P(B \cap C)$. □

The elements of the possibility space may be represented by the paths of a *tree diagram* which shows how all the outcomes may occur.

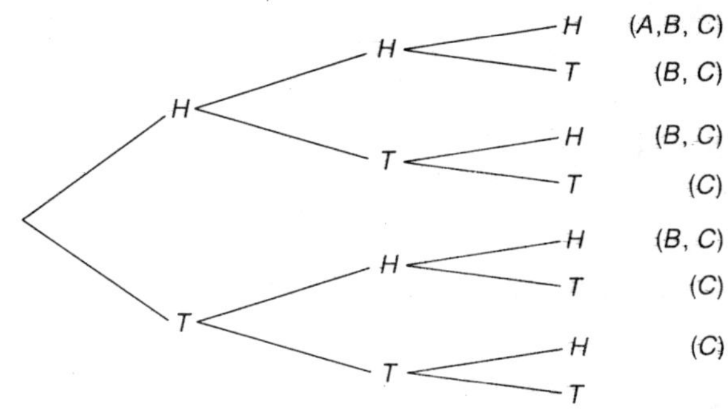

There are 8 equally likely outcomes.

A only occurs once. $\therefore P(A) = \frac{1}{8}$.

B occurs 4 times. $\therefore P(B) = \frac{4}{8} = \frac{1}{2}$.

C occurs 7 times. $\therefore P(C) = \frac{7}{8}$. ■

A or B (or both) occur 4 times. $\therefore P(A \cup B) = \frac{4}{8} = \frac{1}{2}$.

A and B both occur only once. $\therefore P(A \cap B) = \frac{1}{8}$. ■

B or C (or both) occur 7 times. $\therefore P(B \cup C) = \frac{7}{8}$.

B and C both occur 4 times. $\therefore P(B \cap C) = \frac{4}{8} = \frac{1}{2}$. ■

15.3 THE COMBINATION LAWS

ESSENTIAL FACTS

In the following facts A and B are two different events which may occur in a possibility space.

F1. The addition law

$\mathrm{P}(A \cup B) = \mathrm{P}(A) + \mathrm{P}(B) - \mathrm{P}(A \cap B)$.

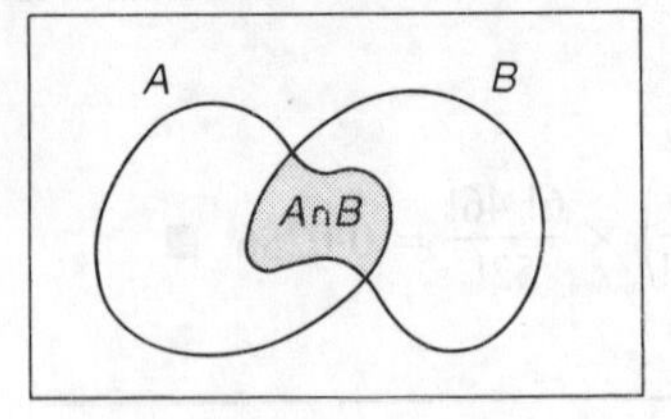

F2. Mutally exclusive events

The events A and B are *mutally exclusive* if $\mathrm{P}(A \cap B) = 0$. This means that the two events cannot *both* occur; if one occurs, the other does not.

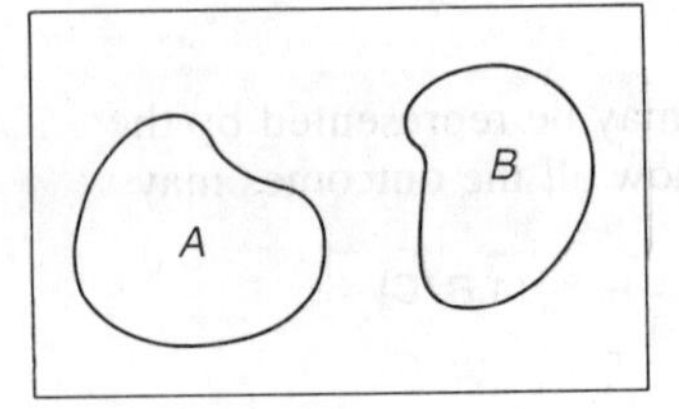

F3. Exhaustive events

The events A and B are said to be *exhaustive* if $\mathrm{P}(A \cup B) = 1$. This means that at least one of the events must occur in any trial.

F4. Independent events

The events A and B are independent if and only if $\mathrm{P}(A \cap B) = \mathrm{P}(A).\ \mathrm{P}(B)$.

ILLUSTRATIONS

I1.

The events A, B and C are such that

$$\mathrm{P}(A) = \frac{1}{10},\ \mathrm{P}(B) = \frac{1}{5},\ \mathrm{P}(A \cup C) = \frac{7}{20} \text{ and } \mathrm{P}(B \cup C) = \frac{7}{15}.$$

Given that A and B are mutually exclusive and that B and C are independent, find $\mathrm{P}(A \cup B)$, $\mathrm{P}(A \cap B)$, $\mathrm{P}(A \cap C)$ and $\mathrm{P}(B \cap C)$. Determine whether or not A and C are independent. □

A and B are mutually exclusive. $\therefore\ \mathrm{P}(A \cap B) = 0$. ■

Thus $\mathrm{P}(A \cup B) = \mathrm{P}(A) + \mathrm{P}(B) - \mathrm{P}(A \cap B)$

$$= \frac{1}{10} + \frac{1}{5} - 0 = \frac{3}{10}.\ ■$$

B and C are independent. $\therefore$ $P(B \cap C) = P(B).\ P(C) = \frac{1}{5}P(C)$.

However, $P(B \cap C) = P(B) + P(C) - P(B \cup C)$

$$= \frac{1}{5} + P(C) - \frac{7}{15}.$$

$$\therefore \frac{1}{5}P(C) = P(C) + \frac{1}{5} - \frac{7}{15}.$$

$$\therefore P(C) = \frac{1}{3}.$$

So $P(B \cap C) = \frac{1}{5} \times \frac{1}{3} = \frac{1}{15}$. ■

$$P(A \cap C) = P(A) + P(C) - P(A \cup C)$$

$$= \frac{1}{10} + \frac{1}{3} - \frac{7}{20} = \frac{1}{12}. \ ■$$

Finally; $P(A).\ P(C) = \frac{1}{10}\frac{1}{3} = \frac{1}{30} \neq P(A \cap C)$.

Hence A and C are not independent. ■

12.

A bag contains just 12 balls, of which 8 are blue and 4 are white. One ball is drawn at random from the bag and replaced; a second ball is drawn at random and replaced and then a third ball is drawn at random. Find the probability of drawing 2 blue balls and 1 white ball.
Event C is that the 3 balls drawn include at least 2 blue balls.
Event D is that the 3 balls drawn include at least 2 white balls.
Event E is that the 3 balls drawn include at least 1 blue ball and at least 1 white ball.
Find $P(C)$, $P(D)$ and $P(E)$. Determine whether any two of the events C, D and E are either independent or mutually exclusive. □

On each occasion a ball is drawn from a bag containing 8 blue balls and 4 white balls. Thus the ball drawn at any instant is not influenced by the previous draws. That is, the events of drawing a ball are independent. Thus the probability of drawing two blue balls in two draws is $\left(\frac{8}{12}\right) \times \left(\frac{8}{12}\right) = \left(\frac{2}{3}\right)^2$.

The probability of drawing a white ball is $\frac{4}{12} = \frac{1}{3}$.

✗ $\therefore$ the probability of drawing 2 blue balls and one white ball is $\left(\frac{2}{3}\right)^2 \times \left(\frac{1}{3}\right) = \left(\frac{4}{27}\right)$. ✗

What we have calculated is the probability of drawing a blue ball first, a blue ball second and a white ball third. However, this is only one way in which 2 blue balls and one white ball may be drawn. In fact this particular combination may be drawn in 3 different ways. To see this we note either that 2 blue balls and one white ball may be arranged in $\frac{3!}{2!\,1!} = 3$ ways, or that the white ball may be drawn first, second or third.

✓ We thus see that 2 blue balls and 1 white ball

have a probability of $\left(\frac{2}{3}\right)^2 \times \left(\frac{1}{3}\right) \times 3 = \frac{4}{9}$ of being drawn. ✓

Notation: Let P(XB, YW) be the probability of drawing X blue balls and Y white balls.

Thus $P(2B, 1W) = \frac{4}{9}$.

Event C occurs with either 2 blue balls and 1 white ball or with 3 blue balls.

Thus $P(C) = P(2B, 1W) + P(3B, 0W)$

$$= \frac{4}{9} + \left(\frac{8}{12}\right) \times \left(\frac{8}{12}\right) \times \left(\frac{8}{12}\right)$$

$$= \frac{4}{9} + \frac{8}{27} = \frac{20}{27}. \ \blacksquare$$

Similarly,

$P(D) = P(1B, 2W) + P(0B, 3W)$

$$\times = \left(\frac{8}{12}\right) \times \left(\frac{4}{12}\right) \times \left(\frac{4}{12}\right) + \left(\frac{4}{12}\right) \times \left(\frac{4}{12}\right) \times \left(\frac{4}{12}\right). \times$$

We have forgotten that the event $(1B, 2W)$ can occur in 3 ways.

$$P(D) = \left(\frac{8}{12}\right)\left(\frac{4}{12}\right)\left(\frac{4}{12}\right)3 + \left(\frac{4}{12}\right)\left(\frac{4}{12}\right)\left(\frac{4}{12}\right) = \frac{7}{27}. \quad \blacksquare$$

Also $P(E) = P(2B, 1W) + P(1B, 2W)$

$$= \frac{4}{9} + \left(\frac{8}{12}\right) \times \left(\frac{4}{12}\right) \times \left(\frac{4}{12}\right) \times 3 = \frac{2}{3}. \ \blacksquare$$

For event C *and* event D to occur, we must have 2 blue balls *and* 2 white balls.

This is not possible.

$\therefore P(C \cap D) = 0$ and so C and D are mutually exclusive.

However P(C). P(D) $= \frac{20}{27} \cdot \frac{7}{27} \neq$ P($C \cap D$).

$\therefore$ C and D are not independent. ■

Event C *and* event E occur when we have 2 blue balls and 1 white ball.

$$\therefore \text{P}(C \cap E) = \text{P}(2B, 1W) = \frac{4}{9} \neq 0.$$

$\therefore$ C and E are not mutually exclusive.

Also P(C). P(E) $= \frac{20}{27} \times \frac{2}{3} = \frac{40}{81} \neq$ P($C \cap E$).

$\therefore$ C and E are not independent. ■

Event D *and* event E occur when we have 1 blue ball and 2 white balls.

$$\therefore \text{P}(D \cap E) = \left(\frac{2}{3}\right) \times \left(\frac{1}{3}\right) \times \left(\frac{1}{3}\right) \times 3 = \frac{2}{9} \neq 0.$$

$\therefore$ D and E are not mutually exclusive.

Also P(D). P(E) $= \frac{7}{27} \frac{2}{3} = \frac{14}{81} \neq$ P($D \cap E$).

$\therefore$ D and E are not independent. ■

Alternative solution

The elements of the possibility space may be represented by the paths along a tree diagram, as in 15.2I3.

However, in this case, we also insert the relevant probability along the side of each path. Then the probability of any particular element in the space is determined by multiplying the probabilities appearing on its path.

Thus the probability of the event 'blue ball, white ball, blue ball' is given by $\frac{8}{12} \times \frac{4}{12} \times \frac{8}{12} = \frac{4}{27}$.

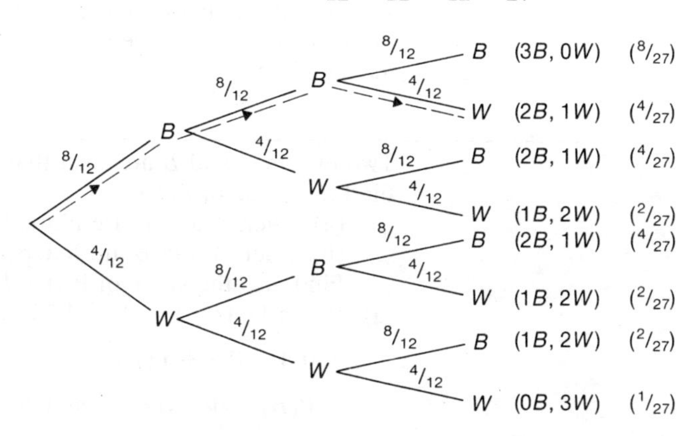

There are 8 elements in the possibility space. The composition of each element and its associated probability are given in columns at the side of the tree diagram.

By inspecting the columns, we see that

$$P(2B, 1W) = 3 \times \left(\frac{4}{27}\right) = \frac{4}{9}.$$

$$P(C) = \frac{8}{27} + 3 \times \left(\frac{4}{27}\right) = \frac{20}{27}.$$

$$P(D) = 3 \times \left(\frac{2}{27}\right) + \frac{1}{27} = \frac{7}{27}.$$

$$P(E) = 3 \times \left(\frac{4}{27}\right) + 3 \times \left(\frac{2}{27}\right) = \frac{2}{3}.$$

Similarly,

$$P(C \cap D) = 0.$$

$$P(C \cap E) = 3 \times \left(\frac{4}{27}\right) = \frac{4}{9}.$$

$$P(D \cap E) = 3 \times \left(\frac{2}{27}\right) = \frac{2}{9}.$$

The results regarding independence and mutual exclusiveness of the events then follow as before.

Note: It is instructive to repeat this question in the case when the bag contains exactly 6 blue balls and 6 white balls.

In this case you should obtain the following results.

$$P(2B, 1W) = \tfrac{3}{8}.$$

$$P(C) = \tfrac{1}{2}, \qquad P(D) = \tfrac{1}{2}, \qquad P(E) = \tfrac{3}{4}.$$

$$P(C \cap D) = 0, \qquad P(D \cap E) = \tfrac{3}{8}, \qquad P(E \cap C) = \tfrac{3}{8}.$$

It then follows that only C and D are mutually exclusive; but now events C and E are independent and events D and E are independent.

---

13. Two events A and B are such that $P(A) = 0{\cdot}5$ and $P(A \cup B) = 0{\cdot}8$. Find the value of $P(B)$

(a) when A and B are mutually exclusive,

(b) when A and B are independent.

Find also the value of $P(A' \cap B)$. □

(a) When $P(A \cap B) = 0$, $\quad P(A \cup B) = P(A) + P(B)$.

$$\therefore 0{\cdot}8 = 0{\cdot}5 + P(B).$$

$\therefore P(B) = 0{\cdot}3$ when A and B are mutually exclusive. ■

(b) When $P(A \cap B) = P(A).\ P(B)$,

$P(A \cup B) = P(A) + P(B) - P(A).\ P(B)$.

$\therefore\ 0{\cdot}8 = 0{\cdot}5 + P(B) - 0{\cdot}5P(B)$.

$\therefore\ P(B) = 0{\cdot}6$ when A and B are independent. ■

Let A and B be events in a possibility space S.
Then $A' + A = S$ and so $A' \cap B + A \cap B = S \cap B = B$.

It follows that $P(A' \cap B) + P(A \cap B) = P(B)$.

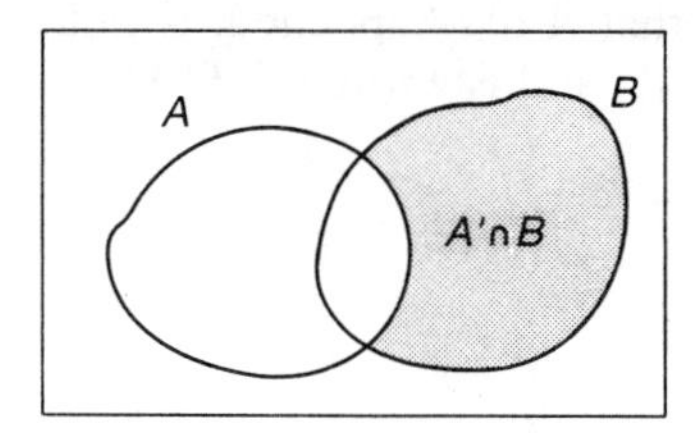

$$\begin{aligned} \therefore\ P(A' \cap B) &= P(B) - P(A \cap B) \\ &= P(B) - P(A) - P(B) + P(A \cup B) \\ &= P(A \cup B) - P(A) \\ &= 0{\cdot}8 - 0{\cdot}5 = 0{\cdot}3.\ ■ \end{aligned}$$

---

14. A card is selected at random from an ordinary pack of 52 playing cards. Denoting by A the event that the card is an ace, B the event that the card is black and C that the card is a club, determine whether or not each pair of A, B and C are independent. □
There are 52 possible cards which may be drawn.

An ace may be drawn in 4 ways. $\therefore\ P(A) = \frac{4}{52} = \frac{1}{13}$.

A black card may be drawn in 26 ways. $\therefore\ P(B) = \frac{26}{52} = \frac{1}{2}$.

A club may be drawn in 13 ways. $\therefore\ P(C) = \frac{13}{52} = \frac{1}{4}$.

There are two black aces. $\therefore\ P(A \cap B) = \frac{2}{52} = \frac{1}{26}$.

There are 13 black clubs. $\therefore\ P(B \cap C) = \frac{13}{52} = \frac{1}{4}$.

There is only one ace of clubs. $\therefore\ P(C \cap A) = \frac{1}{52}$.

Now $P(A).P(B) = \frac{1}{13}\frac{1}{2} = \frac{1}{26} = P(A \cap B)$.

$\therefore$ Events A and B are independent.

$P(B).\ P(C) = \frac{1}{2}\frac{1}{4} = \frac{1}{8} \neq P(B \cap C)$.

$\therefore$ Events B and C are not independent.

$P(C).P(A) = \frac{1}{4}\frac{1}{13} = \frac{1}{52} = P(C \cap A)$.

$\therefore$ Events C and A are independent. ■

15.4 CONDITIONAL PROBABILITY

ESSENTIAL FACTS

F1. The multiplication law

Let A be an event in a possibility space S with $P(A) > 0$.
The probability of an event B, given that A has happened, is called the **conditional probability** of B upon A, and is written $P(B|A)$.

$$P(B|A) = \frac{P(A \cap B)}{P(A)}.$$

The identity

$$P(A \cap B) = P(B|A)\, P(A)$$

is called the **multiplication law**.

$$\frac{P(B|A)}{P(A|B)} = \frac{P(B)}{P(A)}.$$

So $P(A|B) \neq P(B|A)$ [unless $P(A) = P(B)$].

F2.

Using 15.3 F4, it follows from F1 that

A and B are independent if and only if $P(B|A) = P(B)$.

This relation may be written:

The probability of B is unaffected by the occurrence of A.

F3.

Let $A_1, A_2, \ldots, A_n$ be mutually exclusive and exhaustive events in the sample space S.
Let B be an event of S. Then

$$P(B) = P(B|A_1)\, P(A_1) + P(B|A_2)\, P(A_2) + \cdots + P(B|A_n)\, P(A_n).$$

ILLUSTRATIONS

I1.

A factory A supplies glass decanters of which 10% are faulty. A second factory B also supplies glass decanters of which only 5% are faulty. A retailer stocks a large number of decanters of which 40% are from factory A and 60% are from factory B. Whenever a decanter is sold it is selected at random from stock. Find the probability that

(a) when a decanter is sold it will be faulty,
(b) a decanter, found faulty when sold, has been made in factory A.

A woman buys 4 decanters from the retailer. Find the probability that at least one decanter will be faulty. □
Define events A, B and F as follows.

A: a decanter is made in factory A,
B: a decanter is made in factory B,
F: a decanter is faulty.

For the stock held by the retailer, the data tells us that

$$P(A) = 0{\cdot}4,\ P(B) = 0{\cdot}6,\ P(F|A) = 0{\cdot}1,\ P(F|B) = 0{\cdot}05.$$

So when a decanter is sold, we have (using F3)

(a) $P(F) = P(F|A).\ P(A) + P(F|B).\ P(B)$
$= 0{\cdot}1 \times 0{\cdot}4 + 0{\cdot}05 \times 0{\cdot}6 = 0{\cdot}07.$ ■

(b) $P(A|F).\ P(F) = P(A \cap F) = P(F|A)\ P(A).$

$\therefore\ P(A|F) \times 0{\cdot}07 = 0{\cdot}1 \times 0{\cdot}4.$

$\therefore\ P(A|F) = \frac{4}{7}.$ ■

When 4 decanters are bought;

P(at least one faulty) = 1 − P(none faulty).

Now, probability of a single decanter not being faulty is $P(F')$ and $P(F') = 1 - P(F) = 0{\cdot}93$.
The decanters are selected at random, so the probabilities at each choice are independent.

$\therefore$ P(none faulty out of 4 purchased) $= (0{\cdot}93)^4$.

$\therefore$ P(at least one faulty) $= 1 - (0{\cdot}93)^4 = 0{\cdot}25$ to 2dp. ■

---

12. A die is thrown twice. Find the probability that the sum of the faces which turn up is greater than 10, given that,
(a) the first throw is a six, (b) exactly one of the throws is a six. □
Let S be the score obtained.
(a) We only need to find the probability of throwing a 5 or 6 on the second throw; this is $\frac{2}{6} = \frac{1}{3}$. ■
(b) Let A be the event 'exactly one throw is a 6'.
Let B be the event '$S > 10$'.

$$P(B|A) = \frac{P(B \cap A)}{P(A)}$$

$$= \frac{\text{no. of ways of } S > 10 \textit{ and } \text{exactly 1 six}}{\text{no. of ways of getting exactly 1 six}}.$$

$$= \frac{1 + 1}{1 \times 5 + 5 \times 1} = \frac{1}{5}.$$ ■

15.5 EXAMINATION QUESTIONS AND SOLUTIONS

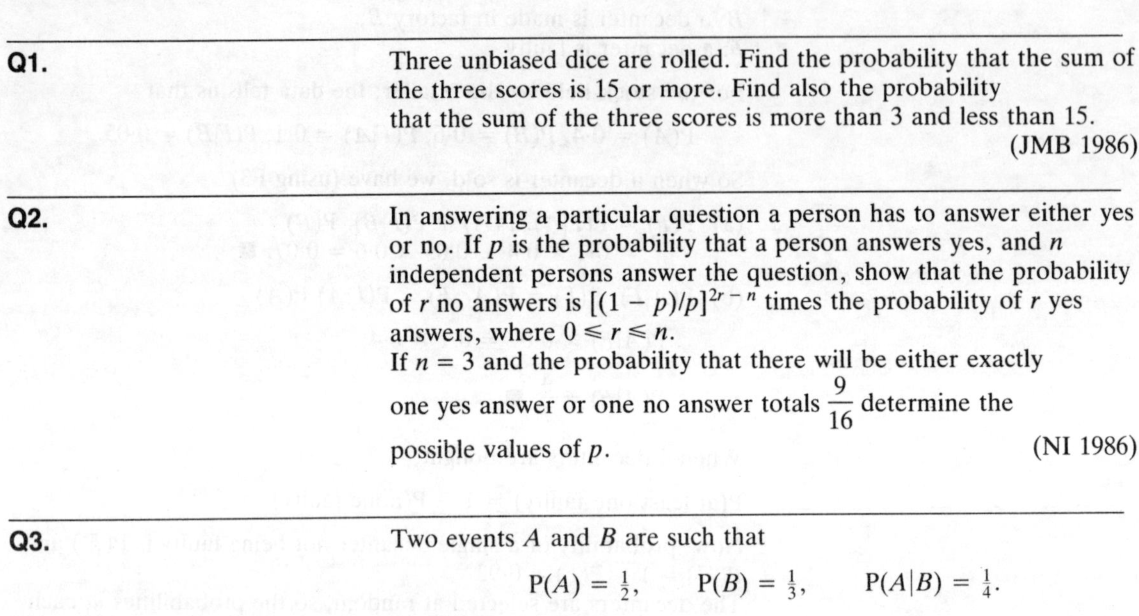

Q1. Three unbiased dice are rolled. Find the probability that the sum of the three scores is 15 or more. Find also the probability that the sum of the three scores is more than 3 and less than 15.
(JMB 1986)

Q2. In answering a particular question a person has to answer either yes or no. If p is the probability that a person answers yes, and n independent persons answer the question, show that the probability of r no answers is $[(1-p)/p]^{2r-n}$ times the probability of r yes answers, where $0 \leqslant r \leqslant n$.
If $n = 3$ and the probability that there will be either exactly one yes answer or one no answer totals $\frac{9}{16}$ determine the possible values of p.
(NI 1986)

Q3. Two events A and B are such that

$$P(A) = \tfrac{1}{2}, \qquad P(B) = \tfrac{1}{3}, \qquad P(A|B) = \tfrac{1}{4}.$$

Evaluate

(i) $P(A \cap B)$,
(ii) $P(A \cup B)$,
(iii) $P(A' \cap B')$.

Another event C is such that A and C are independent, and

$$P(A \cap C) = \frac{1}{12}, \qquad P(B \cup C) = \frac{1}{2}.$$

Show that B and C are mutually exclusive.
(JMB 1984)

Q4. (i) Random events X and Y are mutually exclusive and events X and Z are independent. Given that

$$P(X) = \frac{1}{4}, \qquad P(Y) = \frac{1}{5}, \qquad P(X \cup Z) = \frac{1}{2}, \qquad P(Y \cap Z) = \frac{1}{15},$$

find

(a) $P(X \cup Y)$, (b) $P(X \cap Y')$, (c) $P(X \cap Z)$, (d) $P(Y \cap Z')$.

(ii) A bag contains 12 balls, of which 4 are red, 5 are blue and 3 are white. Three balls are to be drawn at random and without replacement. Find the probability that

(a) all 3 balls will be of the same colour,
(b) all 3 balls will be of different colours. (LON 1986)

Q5.

An unbiased cubical die has four faces numbered 1 and two faces numbered 2. Two boxes are numbered 1 and 2; the box numbered 1 contains three red discs and two blue discs, and the box numbered 2 contains two red discs and three blue discs. Given that the die is rolled and three discs are then drawn at random without replacement from the box with the same number as that uppermost on the die, calculate the probabilities that

(i) two red discs and one blue disc are drawn from the box numbered 1,
(ii) two red discs and one blue disc are drawn,
(iii) the discs came from the box numbered 1, given that two of the drawn discs were red and one disc was blue.

(JMB 1984)

Q6.

(a) Events A and B have probabilities 0.4 and 0.5 respectively and the probability of either A or B is 0.7. Determine the conditional probability of A given B. Are A and B

(i) mutually exclusive,
(ii) independent?

(b) Let C and D be events with probabilities $P(C) \neq 0$ and $P(D) \neq 0$. Show that

$$P(D|C) = \frac{P(C|D)\,P(D)}{P(C)}.$$

For the purpose of house insurance an insurance company divides a city into three districts D_1, D_2, D_3. The probability that a house owner in district D_1 makes a claim in any one year is 0.1. The corresponding figures for districts D_2 and D_3 are 0.05 and 0.15 respectively. If 50% of the insured houses are in D_1, 30% in D_2 and 20% in D_3, determine the probability that a randomly chosen house owner makes a claim in a given year.
On a certain day the company receives a claim from one of its house owners. By using the result for $P(D|C)$ given above, or otherwise, determine the probability that the house owner making this claim lives in district D_3. (NI 1986)

SOLUTIONS

S1. As each die has six sides there are $6 \times 6 \times 6 = 216$ ways in which a score may be produced.

The number of ways in which a score of 15 or more may be produced is best found by systematically listing all possible ways.

Die 1	Die 2	Die 3
6	6	3, 4, 5, 6
	5	4, 5, 6
	4	5, 6
	3	6
5	6	4, 5, 6
	5	5, 6
	4	6
4	6	5, 6
	5	6
3	6	6

Hence there are 20 ways of producing a score of 15 or more.

$\therefore$ probability of scoring 15 or more is $\dfrac{20}{216} = \dfrac{5}{54}$. ■

There are $216 - 20 = 196$ ways of producing a score less than 15. Of these, only one, when each die scores 1, gives a score of 3 or less. Hence there are 195 ways of producing a score less than 15 and greater than 3.

$\therefore$ probability of a sum greater than 3 and less than 15 is $\dfrac{195}{216} = \dfrac{65}{72}$.

13m ■

S2. The probability that a person answers yes is p.

$\therefore$ the probability that a person answers no is $(1 - p)$.

r people can be selected from n people in ${}^nC_r$ ways.

$\therefore$ the probability of r yes answers (and $n - r$ no answers) is ${}^nC_r p^r(1 - p)^{n-r}$ and the probability of r no answers (and $n - r$ yes answers) is ${}^nC_r(1 - p)^r p^{n-r}$.

$$\text{So } \frac{\text{P}(r \text{ no answers})}{\text{P}(r \text{ yes answers})} = \frac{(1 - p)^r p^{n-r}}{p^r(1 - p)^{n-r}} = \left(\frac{1 - p}{p}\right)^{2r-n}.$$

$$\therefore \text{P}(r \text{ no answers}) = \left(\frac{1 - p}{p}\right)^{2r-n} \times \text{P}(r \text{ yes answers}). \blacksquare$$

When $n = 3$;

P(1 yes and 2 no answers) $= 3p(1 - p)^2$

and P(1 no and 2 yes answers) $= 3(1 - p)p^2$.

$\therefore$ P(exactly 1 yes or 1 no answer) $= 3p(1-p)^2 + 3(1-p)p^2$
$= 3p(1-p)$.

$\therefore 3p(1-p) = \dfrac{9}{16}$.

$\therefore 16p^2 - 16p + 3 = 0$. $\qquad \therefore (4p-1)(4p-3) = 0$.

So the possible values of p are $\frac{1}{4}$ and $\frac{3}{4}$. *11m* ■

S3.

(i) $P(A \cap B) = P(A|B)\,P(B) = \dfrac{1}{4}\dfrac{1}{3} = \dfrac{1}{12}$. ■

(ii) $P(A \cup B) = P(A) + P(B) - P(A \cap B) = \dfrac{1}{2} + \dfrac{1}{3} - \dfrac{1}{12} = \dfrac{3}{4}$. ■

(iii) $P(A' \cap B')$ = Probability of *not A* **and** *not B*.
= 1 − Probability of A or B or both
$= 1 - P(A \cup B) = \frac{1}{4}$. ■

A and C are independent.

$\therefore P(A \cap C) = P(A).\,P(C)$.

$\therefore \frac{1}{12} = \frac{1}{2}P(C)$.

$\therefore P(C) = \frac{1}{6}$.

$P(B \cap C) = P(B) + P(C) - P(B \cup C) = \frac{1}{3} + \frac{1}{6} - \frac{1}{2} = 0$.

$\therefore$ B and C are mutually exclusive. *11m* ■

S4.

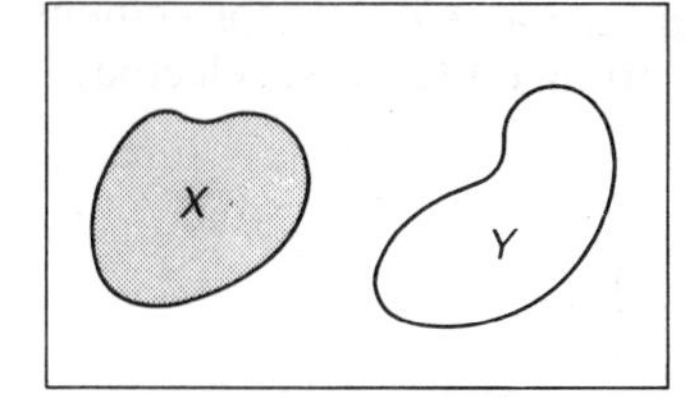

(i) $P(X \cap Y) = 0$ and $P(X \cap Z) = P(X).\,P(Z)$.

(a) $P(X \cup Y) = P(X) + P(Y) - P(X \cap Y) = \dfrac{1}{4} + \dfrac{1}{5} - 0 = \dfrac{9}{20}$.

(b) A Venn diagram with two non-intersecting subsets helps us to see that when X and Y are mutually exclusive
$P(X \cap Y') = P(X) = \frac{1}{4}$.

(c) $P(X \cap Z) = P(X).\,P(Z) = P(X) + P(Z) - P(X \cup Z)$.

$\therefore \frac{1}{4}P(Z) = \frac{1}{4} + P(Z) - \frac{1}{2}$.

$\therefore P(Z) = \frac{1}{3}$.

Hence $P(X \cap Z) = \dfrac{1}{4}\dfrac{1}{3} = \dfrac{1}{12}$. ■

(d) A Venn diagram with two intersecting subsets helps us to see that

$P(Y \cap Z') = P(Y) - P(Y \cap Z) = \dfrac{1}{5} - \dfrac{1}{15} = \dfrac{2}{15}$. ■

(ii) (a)

P(3 balls of the same colour) $=$ P(3 red) $+$ P(3 blue) $+$ P(3 white).

P(3 red) $=$ P(1st red) $\times$ P(2nd red) $\times$ P(3rd red) $= \frac{4}{12} \times \frac{3}{11} \times \frac{2}{10}$.

P(3 blue) $=$ P(1st blue) $\times$ P(2nd blue) $\times$ P(3rd blue)

$$= \frac{5}{12} \times \frac{4}{11} \times \frac{3}{10}.$$

P(3 white) $=$ P(1st white) $\times$ P(2nd white) $\times$ P(3rd white)

$$= \frac{3}{12} \times \frac{2}{11} \times \frac{1}{10}.$$

$\therefore$ P(3 balls of the same colour)

$$= \frac{4 \times 3 \times 2 + 5 \times 4 \times 3 + 3 \times 2 \times 1}{12 \times 11 \times 10} = \frac{3}{44}. \blacksquare$$

(b) 3 differently coloured balls can be arranged in $3! = 6$ ways.

$\therefore$ P(3 different colours)

$= 6 \times$ P(1st red) $\times$ P(2nd blue) $\times$ P(3rd white)

$$= 6 \times \frac{4}{12} \times \frac{5}{11} \times \frac{3}{10} = \frac{3}{11}.$$ *25m* ■

S5. The die has 4 faces numbered 1 and 2 faces numbered 2.

$\therefore$ P(throwing a 1) $= \frac{4}{6} = \frac{2}{3}$ and P(throwing a 2) $= \frac{2}{6} = \frac{1}{3}$.

Let A be the event of selecting 2 red discs and 1 blue disc.

(i) P(2 red and 1 blue from box 1) $=$ P($A \cap$box 1)

$=$ P($A|$box 1). P(box 1)

Now ✗ P($A|$box 1) $=$ P(2 red from 3 red and 1 blue from 2 blue)

$$= \left(\frac{3}{5} \times \frac{2}{5}\right) \times \left(\frac{2}{5}\right).$$ ✗

There are two errors: The selections are made *without* replacement and 2 red discs and 1 blue disc can be arranged (that is, selected) in $\frac{3!}{2!} = 3$ ways.

✓ $\therefore$ P($A|$box 1) $= 3 \times \left(\frac{3}{5} \times \frac{2}{4} \times \frac{2}{3}\right) = \frac{3}{5}$. ✓

$\therefore$ P(2 red and 1 blue from box 1) $= \frac{3}{5} \times \frac{2}{3} = \frac{2}{5}$. ■

(ii) $P(A) = P(A|\text{box 1}).\ P(\text{box 1}) + P(A|\text{box 2}).\ P(\text{box 2})$

$$= \frac{2}{5} + 3 \times \left(\frac{2}{5} \times \frac{1}{4} \times \frac{3}{3}\right) \times \frac{1}{3} = \frac{2}{5} + \frac{1}{10} = \frac{1}{2}. \ ■$$

(iii) $P(\text{box 1}\ |A) = \dfrac{P(A \cap \text{box 1})}{P(A)} = \dfrac{\frac{2}{5}}{\frac{1}{2}} = \frac{4}{5}.$ *16m* ■

S6.

(a) $P(A) = 0{\cdot}4$, $P(B) = 0{\cdot}5$, $P(A \cup B) = 0{\cdot}7$.

$\therefore P(A \cap B) = P(A) + P(B) - P(A \cup B) = 0{\cdot}4 + 0{\cdot}5 - 0{\cdot}7 = 0{\cdot}2.$

$P(A|B) = \dfrac{P(A \cap B)}{P(B)} = \dfrac{0{\cdot}2}{0{\cdot}5} = 0{\cdot}4.$ ■

(i) $P(A \cap B) \neq 0$. $\therefore$ A and B are not mutually exclusive. ■

(ii) $P(A|B) = 0{\cdot}4 = P(A)$. $\therefore$ A and B are independent. ■

Alternatively: $P(A).\ P(B) = (0{\cdot}4) \times (0{\cdot}5) = 0{\cdot}2 = P(A \cap B)$.

$\therefore$ independent.

(b) $P(D|C) = \dfrac{P(D \cap C)}{P(C)}$

$$= \frac{P(C|D).\ P(D)}{P(C)}. \ ■$$

Let C and D_i be the events 'a claim is made' and 'a house in D_i is chosen'.

$P(C|D_1) = 0{\cdot}1$, $P(C|D_2) = 0{\cdot}05$, $P(C|D_3) = 0{\cdot}15$

$P(D_1) = 0{\cdot}5$, $P(D_2) = 0{\cdot}3$, $P(D_3) = 0{\cdot}2$.

$\therefore P(C) = P(C|D_1).P(D_1) + P(C|D_2).P(D_2) + P(C|D_3).P(D_3)$

$= (0{\cdot}1) \times (0{\cdot}5) + (0{\cdot}05) \times (0{\cdot}3) + (0{\cdot}15) \times (0{\cdot}2) = 0{\cdot}095.$

$\therefore$ P(district D_3 given a claim is made) $= P(D_3|C)$

$$= \frac{P(C|D_3).P(D_3)}{P(C)}$$

$$= \frac{(0{\cdot}15) \times (0{\cdot}2)}{0{\cdot}095}$$

$$= \frac{6}{19}.$$ *22m* ■

FORMULAE

$\sec^2\theta = 1 + \tan^2\theta$

$\cos(\alpha + \beta) = \cos\alpha\cos\beta - \sin\alpha\sin\beta$

$\sin(\alpha + \beta) = \sin\alpha\cos\beta + \cos\alpha\sin\beta$

$\cos 2\theta = \cos^2\theta - \sin^2\theta$

$\quad = 2\cos^2\theta - 1 = 1 - 2\sin^2\theta$

$\sin 2\theta = 2\sin\theta\cos\theta$

$\text{cosec}^2\theta = 1 + \cot^2\theta$

$\tan(\alpha + \beta) = \dfrac{\tan\alpha + \tan\beta}{1 - \tan\alpha\tan\beta}$

$\tan 2\theta = \dfrac{2\tan\theta}{1 - \tan^2\theta}$

$p\cos\theta + q\sin\theta = \sqrt{p^2 + q^2}\cos(\theta - \alpha)$, where $\cos\alpha = \dfrac{p}{\sqrt{p^2 + q^2}}$, $\sin\alpha = \dfrac{q}{\sqrt{p^2 + q^2}}$

Function	Derivative
x^n	nx^{n-1}
$\cos x$	$-\sin x$
$\sec x$	$\sec x \tan x$
$\arccos x$	$-\dfrac{1}{\sqrt{1 - x^2}}$

Function	Derivative
e^x	e^x
$\sin x$	$\cos x$
$\text{cosec}\, x$	$-\text{cosec}\, x \cot x$
$\arcsin x$	$\dfrac{1}{\sqrt{1 - x^2}}$

Function	Derivative
$\ln x$	$\dfrac{1}{x}$
$\tan x$	$\sec^2 x$
$\cot x$	$-\text{cosec}^2 x$
$\arctan x$	$\dfrac{1}{1 + x^2}$

SYMBOLS USED IN THE TEST AND ON DIAGRAMS

⊕→	positive sense of direction along a given line
→	velocity
↠	acceleration
⌒ (small arc arrow)	angular velocity
⌒ (large arc arrow)	angular acceleration
➞	force
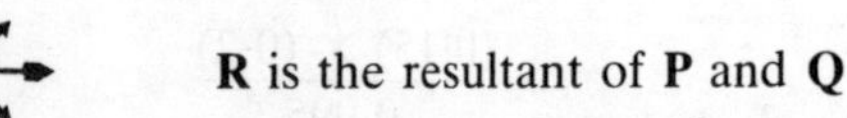	**R** is the resultant of **P** and **Q**
↻	couple
N1 (N2, N3)	Newton's First (Second, Third) Law
↑ : ↗	Resolving forces in the indicated directions
A↻:	Taking moments about A
'P↗ and ←R act'	Forces P and R act in the indicated directions
'G↻ acts'	A couple G acts in the indicated sense

UNITS AND DIMENSIONS

Quantity	Dimensions	Unit	Symbol
Length	L	metre	m
Mass	M	kilogram	kg
		1 tonne =	10^3 kg
Time	T	second	s
Speed	LT^{-1}		$m\,s^{-1}$
Acceleration	LT^{-2}		$m\,s^{-2}$
Angular speed	T^{-1}		$rad\,s^{-1}$
Angular acceleration	T^{-2}		$rad\,s^{-2}$
Momentum	MLT^{-1}		$kg\,m\,s^{-1}$
Force	MLT^{-2}	newton	$N = kg\,m\,s^{-2}$
Impulse	MLT^{-1}		$N\,s = kg\,m\,s^{-1}$
Energy	ML^2T^{-2}	joule	$J = kg\,m^2\,s^{-2}$
Power	ML^2T^{-3}	watt	$J\,s^{-1} = kg\,m^2\,s^{-3}$
Elastic modulus	MLT^{-2}	newton	$N = kg\,m\,s^{-2}$
Stiffness (spring constant)	MT^{-2}		$N\,m^{-1} = kg\,s^{-2}$

Symbol	Meaning	Symbol	Meaning
□	End of part (or all) of a question	✓	Correction of '✗' statement
■	Answer to part (or all) of a question	?	This method is not recommended
✗	This statement may lose marks	*22m* ■	Target time for answering, 22 minutes

Index

References to page numbers are in roman type, followed where appropriate by sub-section indications in *italic* type. Important definitions or principal mentions of an item are indicated in **bold**.